The Art of Genes

Enrico Coen was born in Liverpool in 1957 and studied genetics at the University of Cambridge. In 1984 he joined the Genetics Department at the John Innes Centre, Norwich, where he is today, working on the genetic control of flower development in *Antirrhinum* (snapdragon). His awards include the Science for Art Prize (1996), the EMBO Medal (1996), and the Linnean Gold Medal (1997). In 1997 he was made honorary Professor in Biology at the University of East Anglia. Professor Coen was e'

The Art of Genes

How Organisms Make Themselves

ENRICO COEN

John Innes Centre
Norwich

OXFORD
UNIVERSITY PRESS

OXFORD
UNIVERSITY PRESS

Great Clarendon Street, Oxford OX2 6DP

Oxford University Press is a department of the University of Oxford.
It furthers the University's objective of excellence in research, scholarship,
and education by publishing worldwide in

Oxford New York

Athens Auckland Bangkok Bogotá Buenos Aires Calcutta
Cape Town Chennai Dar es Salaam Delhi Florence Hong Kong Istanbul
Karachi Kuala Lumpur Madrid Melbourne Mexico City Mumbai
Nairobi Paris São Paulo Singapore Taipei Tokyo Toronto Warsaw

with associated companies in Berlin Ibadan

Oxford is a registered trade mark of Oxford University Press
in the UK and in certain other countries

Published in the United States
by Oxford University Press Inc., New York

British Library Cataloguing in Publication Data
Data Available

Library of Congress Cataloging in Publication Data
Data Available

ISBN–13: 978–0–19–286208–2
ISBN–10: 0–19–286208–1

4

Typeset by
Footnote Graphics, Warminster, Wilts

Printed in Great Britain by
Clays Ltd, St Ives plc

To Lucinda, Doris and Ernesto,
and to the memory of my friend Dan Branch

Preface

Over the past twenty years there has been a revolution in biology: for the first time we have begun to understand how organisms make themselves. The mechanisms by which a fertilised egg develops into an adult can now be grasped in a way that was unimaginable a few decades ago. Yet this revolution has been a curiously silent one. Our new picture of how organisms develop has been inaccessible to all but a small community of biologists. This is largely because the jargon and technical complexities have prevented many of the new and exciting findings from being communicated to a wider audience. Moreover, as scientists have concentrated on unravelling the details of the story, many of the broader implications of our new found knowledge have remained unvoiced. In my view this is particularly unfortunate because the study of development provides one of the most fertile meeting grounds for science, art and philosophy.

This book is an attempt to redress this situation. I have tried to give a broadly accessible picture of our current knowledge of how organisms develop, and the implications of these findings for how we view ourselves. The book is aimed at a wide audience, from the general reader with a curiosity about science, to the experienced biologist who may not have had time to follow many of the latest results or to consider their various ramifications.

In trying to accomplish this task, I have used a few key metaphors to convey the gist of what is going on as an organism develops, while at the same time providing detailed explanations of the basic mechanisms involved. At first sight, it may seem that little is to be gained by using these metaphors, but I would ask the reader to be patient as their true merit will start to become apparent later in the book (Chapter 7 onwards). They will then allow many of the latest and most complex ideas in development to be explained in an economical and accessible way, allowing the fundamental issues to be met head-on.

Inevitably, in trying to reach the general reader I have had to cover some well-established biological principles early on in the book (particularly in Chapters 2 and 5). I have done my best to make these explanations as clear and self-sufficient as possible; and have provided a glossary for quick reference at the end of the book. For those encountering these ideas for the very first time, it may seem like a lot to take on board, but I would encourage them to persist as they will reap rich rewards.

I am greatly indebted to the following people for their helpful comments on all or parts of this book: Michael Akam, Jorge Almeida, David Baulcombe, Rosa Beddington, Desmond Bradley, Keith Chater, Doris Coen, Ernesto Coen, Dick Flavell, Roger Freedman, Brian Harrison, Nick Hopwood, Jonathan Howard, Jonathan Jones, Hagai Karchi, Jane Langdale, Maria Leptin, Maddy Orme, Neil Shillito, Daniel St Johnston and Harold Woolhouse. I would also like to thank Antonio García-Bellido, Robb Krumlauf, Ed Lewis, Christiane Nüsslein-Volhard and Mathew Scott for giving me helpful interviews about their work.

I am grateful to the John Innes Centre for allowing me to take sabbatical leave to write the book and the Gatsby Charitable Foundation for its generous support. Many thanks also to Theresa Warr for keeping me organised, John Grandidge and Claire Walker at OUP, Melissa Spielman for her enthusiasm during copy-editing, and Nigel Orme for his patience and hard work in producing the illustrations that complement the text so well. As a first-time author, I am particularly indebted to my Editor, Michael Rodgers, for providing the perfect combination of support, encouragement and guidance. Finally, I would like to thank my family for putting up with me during the long period of gestation needed for the birth of this book.

Norwich   E.C.
October 1998

Contents

CHAPTER 1

Painting a picture

There are many different ways of making things, from the highly mechanised and automatic, like the manufacture of a car, to the more open-ended and creative, as when a work of art is produced. All of these processes are designed or carried out by humans; they all reflect the way the human mind works and organises things. Yet there is another form of making that underlies all these others: the making of an adult from an egg. As biological organisms, our ability to make or create anything depends on our body and brain having first developed from a microscopic fertilised egg cell. This is a very curious type of making, one that occurs without human guidance: eggs turn themselves into adults without anyone having to direct the process. The same is true for all the other organisms we see around us: acorns can grow into oak trees and chickens hatch from eggs with no extra help. Organisms, from daisies to humans, are naturally endowed with a remarkable property, an ability to *make themselves*.

Now as soon as you try thinking about how something might make itself, you encounter a fundamental paradox. The act of making assumes that the maker precedes and is distinct from whatever is being made. A builder has to be there before a house is built and is clearly not an integral part of the house. Saying that something makes itself implies it is both the maker and, at the same time, the object being made. It suggests that something can be its own cause, an incomprehensible concept normally reserved for the Almighty. The paradox is neatly illustrated in a picture by M. C. Escher showing two hands apparently drawing themselves (Fig.1.1). For a hand to draw anything it has to be there in the first place. But in Escher's picture, each hand depends, for its own existence, on what it is drawing, the other hand. We end up with a vicious circle.

There have been many attempts to resolve the paradox of how organisms make themselves, how an egg turns itself into an adult. Some have tried to deny that the process is truly one of making; the adult is in some sense already in the egg to begin with and therefore doesn't have to be made. Another view is that it is not really a self-generating process; instead, there is a separate guiding force that controls all the making. Yet another view, perhaps the most prevalent today, accepts that organisms make themselves but they do this by somehow following a program or set of instructions in the egg. In my view, none of these solutions is satisfactory.

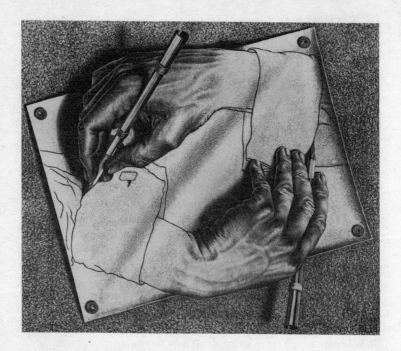

Fig. 1.1 *Drawing Hands* (1948), M. C. Escher.

There is, however, a different way of looking at the problem that has emerged from recent scientific research. In this book I want to describe this perspective by explaining some of the newly found principles that lie behind the formation of organisms. In unravelling this story we shall need to take a fresh look not only at how organisms develop, but also at how this is related to other types of making, from the manufacture of a car to the creation of a masterpiece. Far from being paradoxical, we will see that the development of organisms is the most basic form of making known to us, and, moreover, one that can help to illuminate all others. Before going any further, though, it will help to take a closer look at some of the solutions to the problem of self-making that have been offered in the past.

New or old formation

A commonly held scientific view during the seventeenth and eighteenth centuries was that organisms did not make themselves at all. Instead, they were thought to be already *preformed* in miniature within the fertilised egg. There was no new formation of structures when an egg grew into an adult, only the growth and unfolding of microscopic parts that were already there from the beginning. If, however, you were preformed in your mother, you must have been present in an even more minute form within her ovary when she was preformed in her mother. Tracing our lineage back in time we have to become smaller and smaller and enclosed within an increasing number of nested miniatures. According to this theory of *preformation*, in the beginning there was an individual of each species of animal or plant that contained within it all the other individuals of that species that would ever live. The age of the earth was thought to be fixed by the Bible at five to six thousand years, so it seemed possible to calculate how many members of each species had already been unpacked from the original founder. In the case of humans, Albrecht von Haller, a strong advocate of preformation, worked out that on the sixth day, God must have created at least two hundred billion human beings within Eve's ovaries (he assumed an average world population of one billion humans with a generation time of thirty years).

The original version of preformation theory assumed that the nested miniatures were contained within the mother's egg. Another possibility was raised by the discovery of spermatozoa in the late seventeenth century. Some scientists proposed that these tiny mobile organisms, swimming about in the seminal fluid, contained the encased miniature beings. After penetrating the egg, one of them could be nourished and eventually grow into an adult. Thus there were two opposed schools: the ovists, who believed that Eve's eggs were the repository of ourselves and our ancestors; and the spermists, who thought that we originally resided in Adam's sperm. Nevertheless, both schools were united in the belief that organisms were preformed.

It may seem surprising that the scientific community could have been satisfied with such a bizarre view: by attributing the original creation of encased beings to God, it appears to remove most of the problem from legitimate scientific enquiry. The relationship between science and religion was not, however, the same in the seventeenth and eighteenth centuries as it is today. Preformationists saw themselves as working firmly within the framework of Newtonian science. Isaac Newton was himself a devoutly religious man with a deeply held belief in the Creation. By studying nature, he thought scientists could come closer to appreciating the true wisdom of God's design. He believed that God had created an orderly universe obeying simple laws, like the law of gravity. Following the initial creation, the mechanical laws and forces, put there by God, looked after the behaviour of the universe, with perhaps a bit of divine intervention from time to time to keep things on track. Preformationists thought their view followed naturally from this. Although the initial creation of organisms as encased miniatures was a highly complex business, this was not too much of a problem because God, with his infinite creative powers, was directly involved at this stage. The important point was that once the miniature organisms had been created, they then developed according to simple laws. The development of adults from eggs was a simple mechanical process following the laws of geometry, the enlargement of a pre-existing structure. No special forces or complicated laws had to be invoked because all of the making had been carried out at the initial stages of creation. Once created, the process followed simply and inexorably, just as the planets revolved around the sun.

An alternative view to preformation became more widely accepted through the later eighteenth and early nineteenth centuries. It held that organisms were not already there in the fertilised egg but were formed by a process of true making. Organisms started from relatively simple beginnings. Complexity was then gradually built up through a process called *epigenesis* (Greek for 'origin upon'), until the final form emerged. For each individual there was a fresh formation of parts which slowly emerged as the egg grew into the adult: a process of genuine making rather than just one of enlargement. However, as the preformationists were keen to point out, this theory had the fundamental drawback that no simple physical mechanism could account for it. Whereas preformation was as simple as unpacking boxes, epigenesis seemed to need a special 'making force', a *vital force*, to do all the complicated business of making the organism. God would have had to create a force quite unlike any other, a force that was able to organise and make things. Alternatively God would have to interfere continually with the process of development, guiding it along himself every time an organism formed. A belief in true making therefore brought with it the notion of a rather extraordinary vital force. In its most extreme form, this idea led to the egg being thought of as almost a blank sheet, a *tabula rasa*, with all the information about the structure of an organism coming from the vital force that worked upon it.

Once you accept such a vital force, you can also imagine it assembling organisms in other ways, perhaps even spontaneously generating life from completely unorganised matter. Why limit the vital force to the development of eggs: why not also use it to explain the apparently spontaneous appearance of maggots on rotting meat or of microscopic organisms in broth that has been left for a while? The theory of epigenesis therefore became aligned with another theory prevalent in the seventeenth century: the theory of *spontaneous generation*. Eventually the idea of spontaneous generation started to be challenged through experiments such as those of Lazzaro Spallanzani in 1767, who showed that microscopic organisms only grew in flasks of boiled broth if they were left open to the air, not if they were kept sealed after boiling. This implied that these organisms were not being generated spontaneously within the broth by a vital force but were entering it from the surrounding air. Because they argued against a vital force, these experiments were also taken by many to be strong evidence against epigenesis. (The theory of spontaneous generation was only put finally to rest in the latter half of nineteenth century, through the work of Louis Pasteur.)

The theories of epigenesis and preformation can both be seen as attributing the creation of organisms to God, but they differed in their explanation of how this had come about. According to preformation, all the difficult aspects of making occurred at the initial creation, through the production of encased beings. After this, organisms formed by mechanical forces, operating in accordance with simple laws initially put in place by God. According to epigenesis, the story was different. The complexity of creation was not to be found in miniatures within the egg but in a special vital force, also devised by God, that was responsible for making organisms from eggs and perhaps from other things as well. It was a process of genuine making but one that ultimately depended on the creation of a special force. I have presented these views in their most extreme forms to make the basic assumptions clear. In practice, many scientists lay somewhere in between these extremes, borrowing some elements from each viewpoint.

You might think that the resolution of these two views would have depended on microscopic observation of what actually happened during the transformation of an egg into an adult. Are tiny miniatures really seen in the egg or sperm, or do the embryonic structures appear progressively? Some early preformationists did indeed claim to see a tiny man, called a homunculus, complete with arms, head and legs, tightly packed within every sperm. This was later discredited by detailed studies on the developing embryo, which showed a gradual appearance of organs and limbs rather than enlargement of preformed parts, apparently giving strong support for epigenesis. The preformationists countered, however, that the parts were so small or transparent that they could not easily be recognised early on. Preformationists did not necessarily believe that the encased miniatures were visible in the sperm or egg: they could be transparent and only gradually appear at later stages of growth. The argument between preformation and epigenesis therefore went back and forth, and mere observation of development was not enough to resolve the issue. It was other arguments, based on studies of heredity and evolution, that finally sorted out the controversy.

Heredity and evolution

A pioneer of these hereditary and evolutionary arguments was Pierre-Louis Moreau de Maupertuis, a French scientist of the mid-eighteenth century. Unfortunately, the outstanding insights of Maupertuis became neglected for a long time because he fell out with the French philosopher and writer, Voltaire, who subjected him to public ridicule and humiliation during his lifetime. Maupertuis's reputation never quite recovered from Voltaire's onslaught and his contributions have only come to be appreciated more recently.

Maupertuis made a detailed study of the inheritance of polydactyly, a rare condition in which people are born with extra digits on their hands and feet. By collecting information on the families of affected individuals he observed that a woman with this condition had passed it on to four of her eight children. One of her affected sons then passed it on to two of his five offspring, showing that this trait could be passed on either by men or women. Now according to preformation theory, encased miniature organisms had to be located in either the mother or the father but could not possibly be present in both parents at once. There was therefore no easy way to explain how mothers and fathers were equally able to pass a trait on to their offspring. Maupertuis concluded that preformation must be incorrect and proposed that both parents contributed hereditary particles which determined the characteristics of the offspring. The act of fertilisation allowed the particles from each parent to mix and unite with each other in various combinations, and so produce offspring that could bear traits found in either of the parents. For example, a child might have the hair and eye colour of its mother but a nose shaped like its father's. It is difficult to explain how such combinations could arise if the child was preformed in only one of the parents. Although Maupertuis tried to test many of his ideas further with breeding experiments using various animals, such as Iceland dogs, the precise behaviour of the hereditary particles was only elucidated much later, by Gregor Mendel in 1865, through his studies on plants.

Plants are much more prolific than dogs or other animals that were commonly chosen as subjects of breeding experiments. Plants are also easy to grow, self-fertilise and cross with each other. Shortly after Maupertuis died, Joseph Koelreuter refuted preformation using similar arguments to Maupertuis, by showing that in hybrids between different species of tobacco plants, both parents contributed equally to the character of their offspring. It did not seem to matter which species donated the pollen (i.e. acted as the male) or which received the pollen (acting as female); either way round the hybrid progeny looked the same. About a hundred years later, Mendel's careful breeding experiments with peas showed that this is because each parent plant contributes a set of hereditary factors, which we now call *genes*. Every parent, male or female, carries a set of genes that are shuffled and portioned out to its offspring. The characteristics of every individual depend on the *combination* of genes it inherits from its parents. Individuals cannot already have been preformed in either their mother or father because their characters are derived anew from the combined input of their parents.

Although the rules of heredity were taken as strong evidence against preformation, they also curbed some of the more extreme forms of epigenesis. Remember that epigenesis seemed to require a vital force that could make the adult from the egg. In the most extreme version, the egg could be thought of as a blank sheet, with all the information about the structure of the developing organism coming from the vital force. But if the fertilised egg starts off with genes donated by each parent, it is clearly not a blank sheet; it carries information from two individuals. If there was a vital force, its behaviour had to be highly circumscribed by heredity. Spontaneous generation would also be ruled out because organisms cannot develop from scratch, as they depend on genes being passed to them by parents.

Nevertheless, although its role might be constrained by heredity, a vital force still seemed to be needed to account for the formation of organisms. How could hereditary factors alone, blindly obeying the simple laws of mechanics, explain the orderly arrangement of organisms: the exquisite detail and harmony of a butterfly or an orchid? It seemed that either the hereditary factors would themselves have to have been endowed with some special organising force, or they would have to be guided by a separate force. Either way, it was difficult to escape from the idea that there is some sort of underlying vital force. The only way to get round this would be to demonstrate a source of organisation in the living world that was not ultimately dependent on a vital force. This could not be discovered by looking at heredity alone. It came from considering heredity in relation to a broader problem: evolution.

In 1751, more than a century before Charles Darwin published his theory of evolution, Maupertuis considered how variation in hereditary particles might account for the origin of species:

> [Species] could have owed their first origination only to certain fortuitous productions, in which the elementary particles failed to retain the order they possessed in the father and mother animals; each degree of error would have produced a new species; and by reason of repeated deviations would have arrived at the infinite diversity of animals that we see today; which will perhaps still increase with time, but to which perhaps the passage of centuries will bring only imperceptible increases.

Species could have arisen through an accumulation of errors in the transmission of hereditary particles, gradually modifying the features of organisms over time. Maupertuis realised that if species had evolved in this way and were not fixed for all time, it would be the final nail in the coffin for the idea of preformed encased miniatures. Preformation assumed that individuals only contained miniatures of their own kind so there was little room for variation, let alone the origin of new species. Species would have to be fixed according to their original creation rather than gradually evolving and changing. The idea that species had evolved through a gradual change in their hereditary make-up therefore undermined preformation. Eventually, however, the study of evolution was also to challenge certain forms of epigenesis by dispensing with the need for a vital force. Charles Darwin (and Alfred Russel Wallace) came up with an alternative mechanism to account for organisation in the living world: the theory of *natural selection*.

The theory of natural selection was based on three basic premises. (1) Individual members of a species *vary* to some extent from one to another. A population is made up of many different individuals, something that is most obvious in humans but also true of other organisms. (2) Much of the variation between individuals is *hereditary*, passed from one generation to the next. We have already seen that this depends on the transmission of hereditary factors—genes—although Darwin was not familiar with the details of Mendel's results. (3) Organisms have an *excessive rate of reproduction*, tending to produce more offspring than can possibly be sustained by their environment, with the inevitable result that many of them will die. If these three premises are true, the process of natural selection will occur in the following way. In every generation only a selection of individuals in a population will live to survive and reproduce. This selection will not be completely random but will favour individuals with certain characteristics, such as individuals with a greater ability to find food, or those that are better able to avoid being eaten. Now because individual variation is to some extent hereditary, individuals that finally make it to reproduce will pass some of their characteristics on to the next generation. This means that the characters that favoured an individual's chance of survival and reproduction will also be the ones that tend to be passed on. Repeating this process over many generations, with heritable variation arising and being selected every time, organisms will tend to evolve features that favour their survival and reproduction in the environment: in other words, adaptations.

The aspect of natural selection that most concerns us here is its implication for the way organisms develop. To make this clear, I need to distinguish between two sorts of process. On the one hand, there is *development*: the process whereby an egg grows into an individual adult. This occurs over the timescale of one generation. On the other hand, there is *evolution*: a process in which a population of individuals may change over many generations. Darwin's theory of natural selection was primarily a mechanism for explaining evolution; it showed how the adaptations we see today could have arisen through countless generations of natural selection acting on populations. But the process of development was also incorporated in this evolutionary picture. This is because the way an egg grows into an adult can itself be seen as an adaptation: individuals that develop in an orderly way are more likely to survive and reproduce than those that develop in a defective manner. Over many millions of generations natural selection could therefore have led to the evolution of the coherent patterns of development that we see today. The organised nature of development evolved through natural selection, acting within the bounds of physical and chemical laws. There need be no recourse to special vital forces to account for orderly development.

By the mid-twentieth century, biologists had therefore arrived at a position that might be called mechanistic epigenesis. Adults are not preformed within eggs as miniatures, they form gradually during the process of development. The fertilised egg, however, is not a blank sheet: it contains genes contributed by each parent, and these affect the characteristics of the final organism. The whole process has arisen as a consequence of natural selection acting over many millions of generations, rather than being the manifestation of a special vital force.

There is still, however, a major problem with this view: the mechanism by which the hereditary factors in the fertilised egg, the genes, lead to the forma-tion of adult features is left entirely unresolved. It is as if you have been presented with a magic trick, like a rabbit being pulled out of a hat. You know that it does not involve any real magic—no supernatural forces are involved—but you can't see how it was done. We witness this trick every time a child is born or when a seed grows into a plant. It is the trick that lies behind your very existence, and your ability to contemplate this or any other problem. Perhaps it is the greatest appearing trick of all time, and it is all done with no hands. Darwin's theory of natural selection suggests that no real magic need be involved; it is not necessary to invoke a vital force. But the mechanism of development—the way the egg transforms itself into an adult—still remains as obscure as ever. The nature of the problem can perhaps best be illustrated by looking at some of the more recent metaphors that have been used to try and account for development.

Modern metaphors

One of the most common metaphors for development is that the egg contains a set of instructions or a plan which is executed as the organism grows. Perhaps the instructions would say things like 'make a leg here' or 'make a nose there' or 'make flowers now'. It would be as if there is a tiny instruction manual in the egg, corresponding to the genes, and this is meticulously followed until the adult is eventually produced. The organism develops much as a car could be manufactured by someone following the right set of instructions.

It may seem that once a detailed set of instructions for how to make a car has been given, the structure of the car is completely specified. However, this makes the important assumption that someone is able to interpret and carry out the instructions. To make a car, it is not enough just to have a manual; someone has to be able to understand it and then put the right bits and pieces together. Following instructions is no small task. Understanding how a string of letters on a page relates to even a simple action, like taking two particular pieces of metal and connecting them with a bolt and nut, is far from trivial. We spend years as children learning a language and how to read books. Following a manual assumes all this prior knowledge and familiarity with language. Give any sort of manual to a monkey and it will not get very far.

The key point here is that our ability to interpret a manual is acquired independently of the manual itself. You cannot learn language or reading by looking at any manual: language has to be learned beforehand, by the complex process we experience as children. I am not talking here of learning a new language, like learning French once you know English—this clearly can be achieved by following a manual, a Teach Yourself French book. I am referring to the ability to understand and read any language at all, the first language you learn as a child. You cannot just give a series of elementary manuals to a newborn baby, leave it on its own for ten years, and expect the child to work out itself how to use language and start reading. Even if you gave the baby books containing lots of diagrams with arrows pointing here and there, it would still not get very far. How would it know what all the lines on the diagrams refer to? What does an arrow signify? In which order should the pictures be looked at? These are all things we take for granted when we know how to interpret pictures and words, but they would not be obvious to an uneducated child. Learning any sort of language is a complex process that involves a child interacting with its environment, including the other people around it. It cannot be derived alone from any sort of manual, no matter how beautifully written or illustrated.

If we were to accept the idea that an egg contains a set of instructions, we would therefore also need an independent agent that is able to interpret and carry them out. But if this agent is truly independent of the instructions, as the person is who follows a manual, where does it come from? We are postulating a highly complex agent, with the ability to interpret and carry out instructions, that exists independently from the instructions themselves. From an evolutionary point of view, either this complex agent had to be there from the beginning, in which case we are coming dangerously close to postulating a vital force, or it evolved by natural selection. If it arose by natural selection, though, variation in the agent would have to be passed on from one generation to the next, as this is one of the key requirements for natural selection to work. In other words, the agent would have to be transmitted by hereditary factors, genes. But the genes correspond to the instructions, so it turns out that the agent does have to depend on the instructions after all. We have ended up with a vicious circle. The ability to interpret instructions depends on the instructions! The problem here is that the instruction metaphor breaks down as soon as you try to understand how the instructions are followed. Development is simply not equivalent to someone following a manual because, unlike the case in the process of manufacture, there is no way of defining the interpretation and execution of instructions independently of the instructions themselves.

Perhaps the problem with the instruction metaphor is that it comes too near to human activity. We might be better off using a metaphor that avoids human involvement altogether. A favourite choice is the computer. The fertilised egg could contain a program, much like a computer program, that is executed as the organism gradually develops. The adult is the output of a carefully orchestrated program that has evolved over millions of years. There is no human involvement here: the process seems to be self-contained, and runs automatically like a machine. In pursuing this metaphor, however, a problem appears as soon as you think about the relationship between the program and the computer that is running it. To use computer jargon, we can distinguish between *hardware*, the actual bits and pieces of the computer (printed circuits, disks, wiring, etc.) and *software*, the various programs that can be run on the machine. Now the key point is that for computers, the hardware is independent of the software. The machinery of a computer has to be there before you can run a program; it is not itself a product of the program.

Compare this to what happens in the development of an organism. Here the output of the program, the final result, is the organism itself with its complex arrangement of organs and tissues. This means that the software, the program, is responsible for organising hardware, the organism. Yet throughout the process, it is the organism in its various stages of development that has to run the program. In other words, the hardware runs the software, whilst at the same time the software is generating the hardware. We are back to a circular argument because software and hardware are no longer independent of each other.

The problem here is that unlike organisms, computers do not make themselves. The components of a computer do not just organise themselves into the appropriate circuits. All computers have to be manufactured by an external agency: the human hand together with machines and tools that were themselves made by the human hand. By contrast, organisms develop without the guidance of an external agency, so there is no independence between software and hardware, between program and execution.

One way of trying to get round this problem is to continue with the computer analogy but to imagine a computer that really can make itself, where its hardware and software are interdependent. We could start by thinking of computers with mechanical arms, wielding tools so that they can start to modify themselves. This is certainly one approach, but I think it would eventually lead to either abandoning the distinctive notions of software and hardware, or modifying them so much that they cease to bear much relationship to their original meaning. Whatever the case, I do not believe that stretching the computer analogy in this way is very helpful for understanding development.

It may seem that we have run out of useful comparisons. Perhaps the development of organisms is just so different from anything else that comparisons with other processes, like following instruction manuals or running computer programs, are always doomed. As we have seen, each time we try to make some distinctions, like the separation between instruction and execution or hardware and software, we are confronted by the same old paradoxes. Maybe there is simply nothing we are familiar with that remotely resembles the process of development. In my view there is another way of looking at the problem. To appreciate this, we will need to go back to examine how humans make things.

A change in perspective

When someone makes something, we naturally separate the maker from the made, the subject that is doing the making as distinct from the object being made. A builder builds a house. A painter paints a picture. What could be more elementary? If we look over the shoulder of an artist in action, we readily distinguish between the materials such as canvas, paint and brushes, and the painter who sits in front of the canvas busily painting away. The artist seems to have a vision in mind and is simply using the materials as tools to transfer his or her ideas onto the canvas. The role of the artist and the materials are quite distinct. The artist is the creator and the materials are the slaves at the artist's beck and call. We couldn't have a clearer example of a separation of the maker and the made.

Now look at the same process from the *artist's point of view*. The artist is continually looking and being influenced by what he or she sees. As soon as some paint is mixed and put on the canvas, the artist sees a new splash of colour that wasn't there before. This is bound to produce a reaction in the artist who will interpret the effect in a particular way. Perhaps the colour is just right, or a bit too strong, or put in slightly the wrong place, or has a surprising effect by having been placed near another colour. The next action of the artist will be influenced by what is seen and may involve a modification of the colour, or maybe leaving it, or moving to a different part of the canvas. The artist is continually looking at what is happening, responding to the changing images on the canvas that enter his or her visual field, correcting or leaving what is there but never ignoring it. Artists cannot paint pictures with their eyes closed. If you ever watch someone lost in the act of painting, they are always looking with great intensity, reacting to what is before them. Each action produces a reaction which is in turn followed by another action. The same process is repeated again and again. As more marks are made, the effects are compounded, accumulating so that a whole history of brush strokes starts to influence the next one. Artists need not be consciously aware of this at all; as far as they are concerned, it is all part of one continuous activity. A deeply involved artist gets completely absorbed in the act of painting; the activity takes over as a self-generating process. The materials, the tools, the canvas just become an extension of the artist and the painting gradually develops from a highly interactive colour dance, rather than being a simple one-way transfer of a mental image from the artist onto a separate canvas. The distinction between the maker and the made that the onlooker sees so clearly is far less obvious from the artist's point of view.

When seen from this perspective, the act of painting provides a very good example of a process which does not involve a clear separation between plan and execution. The artist need have no clear plan of all the colours and brush strokes to be executed. I have taken painting a picture as my example, for reasons that will become clear later on in this book, but a similar thing could be said of other types of human creativity. The philosopher R. G. Collingwood used the example of composing poetry to make the same point in his book *The Principles of Art*:

> suppose a poet were making up verses as he walked; suddenly finding a line in his head, and then another, and then dissatisfied with them and altering them until he had got them to his liking: what is the plan which he is executing? He may have had a vague idea that if he went for a walk he would be able to compose poetry; but what were, so to speak, the measurements and specifications of the poem he planned to compose? He may, no doubt, have been hoping to compose a sonnet on a particular subject specified by the editor of a review; but the point is he may not, and that he is none the less a poet for composing without having any definite plan in his head.

When someone is being creative there need be no separation between plan and execution. We can have an intuitive notion of someone painting a picture or composing a poem without following a defined plan. Yet the outcomes of such creative processes—the painting or the poem—are not random but highly structured. In this respect, I want to suggest that human creativity comes much nearer to the process of development than the notion of manufacture according to a set of instructions, or the running of a computer program.

Now as soon as a word like creativity is used, a few alarm bells might start to ring. Isn't this just bringing in vitalism again? We have already gone through various arguments against mysterious vital forces, yet it may seem as if I am ushering them in again through the back door. This would of course be a legitimate concern if I was suggesting that human creativity was itself imbued with some sort of supernatural spiritual force. This, however, is not what I am saying. Our ability to create anything depends on the activity of a remarkable biological structure, the human brain, and the way it interacts with its environment. The brain is itself a product of a developmental process that has evolved over countless generations, as Darwin himself pointed out. Our brain, including its creative potential, is a product of evolution. I am not suggesting that human creativity is a purely biological process with no cultural input. Clearly, what we create depends on how the brain develops and interacts with its environment. But our ability to create anything at all does depend on the way our brain works, on an underlying biological system that has evolved. In comparing the development of organisms to human creativity, I am not injecting a fresh dose of vitalism, I am simply drawing a comparison between two related processes.

You might wonder why I should wish to draw any sort of comparisons at all. Why not concentrate on development alone and forget trying to compare it with another type of process? After all, it is not as if we are remotely near to understanding what goes on in a human brain when something is being created, so why use something we don't understand as a point of comparison for development? My reasons are twofold.

First of all, although we do not understand the details, we can get some useful general intuitions from thinking about the way we create things. In my view these can be very helpful in gaining an overall sense of what is being achieved during the process of development, and they may also prevent us from being misled into making other less appropriate comparisons. My aim is to use comparisons with creativity not as an explanation of development, but as a viewpoint to help guide us through some of the latest scientific ideas and results on how organisms develop.

The second reason is that the comparison can also be illuminating the other way round: by understanding the basic principles of development we can begin to look at all other forms of making, including human creativity, in a new light. We shall be able to see creativity from a new perspective; not as an isolated feature of human activity, but as something that is itself grounded in the way we develop.

To pursue this approach, we will need to get beneath the surface of developing organisms and start to look at processes from within. This may seem like an almost impossible task. I can interview artists and ask how they set about painting a picture, but how can I possibly get inside the alien world of an organism that is gradually developing? I could of course watch from the outside but this would be no more revealing than looking over the shoulder of someone painting. Somehow a dialogue with the organism has to be opened up that allows us to access its inner secrets. Remarkably enough, there is a way of doing this. One of the great biological success stories of the last two decades has come from the interrogation of organisms about how they carry out development. I do not mean that plants and animals have been rounded up for verbal questioning. The interrogation has been carried out in a different language: the language of genes. It is this story and its fundamental implications for all forms of making, from biological to human, that I want to tell.

CHAPTER 2

Copying and creating

In comparing the development of organisms to a creative process, such as painting a picture, it may appear that I have overlooked a very important distinction: creativity involves originality and inventiveness that seem without parallel in biological development. An artist does not continually paint the same picture again and again: each creation is different from the previous one. As Leonardo said, 'The greatest defect in a painter is to repeat the same attitudes and the same expressions'. Organisms, though, seem to develop with much greater consistency. To be sure, every individual is slightly different from the next; even identical twins do not look exactly the same. But the extent of variation seems much less than that between different original paintings. The development of a mouse or an oak tree appears to be more highly circumscribed and defined from the outset than the process of creating a picture.

Perhaps the development of organisms is more like copying the same picture again and again rather than a creative process. After all, we use the word *reproduction* in both art and science with this type of comparison in mind. In art, it refers to the process of making copies from an original; in biology it is the production of new individuals every generation. The outcome of both processes is comparable: you end up with lots of things that look quite similar to each other—many copies of the *Mona Lisa* or many rabbits.

In this chapter I want to explore the extent to which this comparison between reproduction in art and biology is valid. We shall see that there is an element of similarity between the two types of reproduction, but there is also a fundamental difference that will eventually bring us back to the issue of how development compares with creative processes.

Reproduction in art

To reproduce a work of art, you need to be able to copy it in some way. There are various ways of doing this. Look at Fig. 2.1, which shows a class of children being taught how to copy a leaf, taken from a teachers' handbook on drawing of 1903 (I first came across this picture in Ernst Gombrich's classic book *Art and Illusion*). Every child is copying the leaf on the blackboard with remarkable consistency, conjuring up our worst images of Victorian discipline. Although this example might be on the extreme side, it shows how the notion of copying

Fig. 2.1 Copying in a Victorian classroom.

is generally associated with discipline and slavish imitation rather than imagination. Making a good copy of a picture seems to require excellent technique and rigour but not the creativity that might go into producing an original.

There are other ways of reproducing a picture, apart from copying by hand. Most modern reproductions of paintings are made by photographing the original. Here copying has become almost entirely a technical exercise: a question of mastering the camera and printing process. The picture is automatically transformed into a negative image, from which as many positive prints can be manufactured as needed, so long as you have the appropriate equipment and expertise. An equivalent type of copying is used to reproduce sculptures, by first taking a mould, a negative, and then making a cast to get back to something that looks like the original sculpture.

In all these cases of art reproduction, whether it is done by hand alone or with the aid of various devices, there is always a process of *copying* from an original or template. The final goal is already there before you start—the leaf on the blackboard is there for all to see. The aim is simply to make something that resembles it as closely as possible. In the most successful case, you end up with a replica that might almost be substituted for the original. In *creating*, however, the aim is not to produce a replica of what is already there. You might of course be inspired by a beautiful woman sitting before you, but the final picture is not simply a copy of the woman; it is a two-dimensional image on a canvas that the artist has created for the first time. We would have no difficulty in distinguishing the painting of Mona Lisa from the person in the flesh. The aim of the artist is not to produce something that could ideally replace the subject of the painting, but to create a special type of image. Copying and creating are different sorts of process. Which of these does biological reproduction come closer to? Before we can deal with this question, I shall first need to describe some of the basic principles underlying biological reproduction.

Reproduction by division

A good place to begin looking at biological reproduction is with microbes. Any glass of pond water or clump of soil is teeming with microscopic organisms, many of which consist of single *cells*. These *unicellular* organisms are of the order of one-thousandth to one-hundredth of a millimetre long. Although tiny, they are far from simple. They are able to grow by taking up and metabolising various nutrients from their environment and can even exhibit what might be called 'behaviour', such as swimming towards or away from a particular chemical. These organisms come in many shapes and sizes and are remarkably diverse in their range of lifestyles, occupying habitats from hot springs to the deepest reaches of the oceans. Nevertheless, they are all constructed on similar principles. Each has a membrane going all the way round it that keeps the fluid contents of the cell from spewing out. The membrane provides a boundary, allowing the cell to maintain its individuality, separating it from the outside. This means that anything that enters or leaves the cell has to pass through the membrane. The cells are far from being just bags full of simple fluid, though. They contain a range of complex structures and molecules—some of which we will encounter later on—that are essential for their various activities.

The activity of these unicellular organisms that most concerns us here is their ability to reproduce by a process called *cell division*. When an individual cell grows to a certain size, it starts to narrow in the middle. The narrowing continues until eventually the cell becomes divided into two separate cells, like a drop of water splitting in two. Under ideal conditions, a single-celled bacterium, such as *Escherichia coli*, which lives in your gut (normally without harmful effects), can grow and divide once every twenty minutes, allowing it to multiply to more than one million individuals in a mere seven hours. The problem of reproduction for many unicellular organisms therefore boils down to the question of how a cell can grow and divide into two.

I should mention that reproduction for unicellular organisms is not quite as monotonous as this: most of them also undergo some sort of sex once in a while. This usually involves two individual cells coming together—in some cases completely fusing—to produce a hybrid cell. It can be compared to cell division in reverse, two cells becoming united as one, rather than one cell dividing in two. Why unicellular organisms, or any organisms for that matter, have sex is a complicated question that has been argued about extensively in scientific literature. We need not go into it here, except to say that one undoubted consequence of sex is that it allows an exchange of genetic information between individuals. This means that even for unicellular microbes, the process of reproduction involves more than just cell division.*

Multicellular organisms

Fascinating though the reproduction of microbes might be, it seems to be a far cry from the way you or I reproduce. We cannot go forth and multiply by splitting ourselves in half. Our style of reproduction is more complex than that of a unicellular microbe, but nevertheless there is a common element that links the two. Both microbes and ourselves are based on the same type of building blocks, cells, but whereas individual microbes may comprise a single cell, we are made of many billions of cells; we are *multicellular* organisms.

The realisation that plants, animals and microbes are based on the same fundamental cellular units is one of the greatest unifying discoveries to have been made in biology. It has a curious history that went in fits and starts. We can begin back in 1665, when Robert Hooke described the microscopic structure of cork as a network of tiny chambers, or *cellulae*. The Latin word *cella* means 'a small room' and we still use it in this way when talking of a hermit's cell or prisoner's cell. The term seemed appropriate at the time because Hooke was not looking at living cells but at their relics: the cells in cork are dead, so he was seeing the skeletal outline of the previously living cells of a plant. The cell was therefore originally thought of as being a passive container rather than a living entity. This concept gradually shifted as the contents of cells were looked at more carefully, so that the cell eventually became identified as a living unit of life rather than just a vessel. A key advance made in the early nineteenth century was the discovery that most living cells contain a small body, called the *nucleus*, suspended in the fluid of the cell. By comparing the nuclei of plant and animal cells, the botanist Matthias J. Schleiden and animal physiologist Theodor Schwann, realised in the 1830s how both types of cell might be formed on very similar principles. Schwann recalls that at the time he had been working

---

*In some unicellular organisms, such as *Acetabularia*, reproduction is further complicated by their undergoing significant changes in shape during their life cycle.

on the nerves of tadpoles and frogs and had noticed nuclei in the cells of a particular structure called the notochord:

> One day, when I was dining with Mr Schleiden, this illustrious botanist pointed out to me the important role that the nucleus plays in the development of plant cells. I at once recalled having seen a similar organ in the cells of the notochord, and in the same instant I grasped the extreme importance that my discovery would have if I succeeded in showing that this nucleus plays the same role in the cells of the notochord as does the nucleus of plants in the development of plant cells.

Plant and animal cells contained a similar looking nucleus, suggesting to Schleiden and Schwann that both types of cell might have been generated by a common process. This led them to propose that both plants and animals were constructed from the same elementary units, cells, formed by a single universal mechanism. In other words, trees, frogs, worms and people were all made of cells that had arisen in the same way. Unfortunately, although they did a great service in unifying plant and animal biology, the mechanism that Schleiden and Schwann believed to be responsible for cell formation turned out to be wrong. They thought that cells formed anew by growing within pre-existing cells, and their strength of conviction (particularly Schleiden's, for whom modesty was not a strong point) misled biologists for more than twenty years. It eventually became clear that all cells multiply by division rather than forming anew: *every cell in a plant or animal has arisen by division of previous cells.*

In parallel with the work on multicellular plants and animals, the cellular nature of microscopic organisms was also becoming clear in the mid-nineteenth century. These tiny creatures, each comprising a single cell, shared many features with the individual cells of multicellular organisms. The cell therefore became the fundamental unit of all life.

This unity can now be viewed as reflecting a common evolutionary past. The first cells—the common ancestors of all life on earth—are thought to have arisen as unicellular organisms about three and a half billion years ago. For the next three billion years or so, life continued to be dominated by unicellular creatures. The distinction between plants and animals is thought to have occurred whilst life was still in this unicellular phase. Unicellular plants sustained themselves using energy trapped from sunlight, through a process called photosynthesis. Unicellular animals survived by feeding off others. These different lifestyles led to specialisations in cell construction that are still evident in the cells of plants and animals today. The self-sufficient lifestyle of plants was compatible with having a hard protective casing or cell wall. Animal cells, however, had to retain mobility and flexibility to catch and engulf their prey and could not afford to be surrounded by a cumbersome rigid wall. When complex multicellular plants and animals evolved, about half a billion years

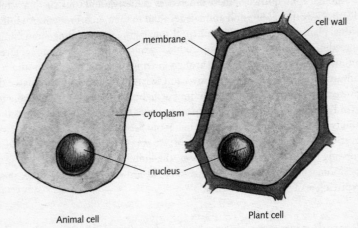

Fig. 2.2 Generalised animal and plant cell.

ago, these basic differences in cell construction were retained. Look at Fig. 2.2, which compares a generalised plant and animal cell. Both types of cell have an outer membrane surrounding the cell fluid, called *cytoplasm*, and a nucleus within. The nucleus is surrounded by its own membrane, containing pores that allow molecules to pass between the nucleus and cytoplasm. In addition, plant cells produce a hard outer coating, the cell wall (cell walls provide the major ingredient of paper and wood). It was these walls—the protective covering—that Robert Hooke saw when he described cells for the first time by looking at cork down a microscope.

The differences between the cells of plants and animals have had many repercussions on the way these organisms are constructed. Multicellular plants are supported by a mesh-work or lattice of cell walls that extends throughout their body. They can continue to grow and develop extra parts throughout their life, like ambitious neighbours who are forever adding extensions to their house, because each new addition carries its own internal lattice of supporting cell walls. In contrast, animals are supported by a framework of bones or a toughened outer covering, made by specialised cells in restricted locations of the body. Development tends to be concentrated in the early phase of an animal's life when all its parts are formed in the appropriate arrangement. A child is born with all the essential limbs and organs in place and this basic arrangement is maintained for the rest of its life. (Some animals do undergo a major reconstruction during their life by going through a second phase of development. For example, when caterpillars turn into butterflies they undergo redevelopment within the confines of a pupa by a process called metamorphosis.)

Another consequence of these differences in lifestyle and construction is the distinct way that plants and animals respond to their environment. An individual plant explores its environment through growth: extending, branching, spreading and invading the space around it whilst rooted to the same spot. An animal achieves similar results through moving its whole body. A moth flies towards light, whereas a plant grows and twists in its direction. Animals can protect themselves by running, hiding or fighting. A plant is far less able to avoid damage but can tolerate enormous losses to its body because it has the ability to keep growing, to the point that it can become a chore to mow the lawn every week. Some animals, such as crabs, are able to regenerate lost parts, like a missing limb. However, in these cases, the limb is regenerated at its original position, whereas plants generally respond by growing extra parts rather than replacing on a new-for-old basis.

Reproduction through development

We can now return to the problem of how multicellular organisms, such as humans, mice and oak trees, reproduce. Like unicellular organisms, their reproduction is based on cell division but instead of making many separate individuals, the cells stay together to gradually build up a multicellular individual. We can summarise the whole process with a *life cycle*. In the case of humans, for example, parents each contribute a cell: a sperm cell from the father and an egg cell from the mother. These two cells fuse together to form a fertilised egg. This slowly develops in the womb, first dividing to give two cells, then four, and so on. After many more rounds of division, the fertilised egg has grown into an embryo and eventually a new adult is formed, comprising many billions of cells. All of this depends on more and more cell divisions. Continuing with the life cycle: if the adult is female, some cells from her ovaries will divide in a special way to produce egg cells. If male, cell divisions in the testicles will give sperm cells. Finally, the sperm and egg cells are brought together and fuse to produce the fertilised egg for the next generation, closing the cycle. The life cycle involves alternation between a single-cell phase, the fertilised egg, and a complex multicellular adult phase. The two are linked by numerous cell divisions and an occasional sexual fusion.

A similar cycle underlies the reproduction of a flowering plant. When a grain of pollen from one flower lands on the female part of another flower, a sperm cell from the pollen fuses with an egg cell in the mother. The fertilised egg then divides repeatedly, doubling the number of cells each time until a tiny plant embryo forms. The process temporarily stops at this point and the plant releases the embryo with a protective hard outer coat, in the form of a seed. In the right conditions, as when the seed is planted in the ground, cell divisions resume in the embryo so that it eventually grows into a new plant, with flowers which produce more egg cells and pollen grains. As with humans, single-cell phases alternate with multicellular phases. One difference between flowering plants and humans is that many flowers are hermaphrodites, producing both male and female organs; this sometimes allows an individual to fertilise itself if pollen lands on female organs from the same plant. (There are also some plants, like willows and stinging nettles, that are more like us in having separate sexes.) Some multicellular organisms, such as aphids and dandelions, can side-step the requirement for sexual fusion during the life cycle and are able to develop from unfertilised eggs, although in these species there is still alternation of single-cell and multicellular phases.

To understand the mechanism by which multicellular organisms reproduce, it is not enough to know how a cell can grow and divide; we also need to know how a single cell, the fertilised egg, can give rise to a multitude of different types of cells in the complex arrangements that form the mature individual. The human body contains many organs and tissues, each made of various types of cells—nerve cells, blood cells, hair cells, etc.—each arranged in a precise way. These different cell types can be distinguished by characteristics such as size, shape, structure and behaviour. By behaviour I mean some of the more dynamic properties of cells like whether they move, grow, divide or even die. In a similar way, the various parts of a plant are made up of many different cell types with different properties, although unlike those of animals, plant cells do not usually move relative to each other because they are fixed in position by their cell walls.

The problem of development is to understand how the complex pattern and arrangement of different cell types that make up a mature organism can arise from a single cell in a consistent way each generation. This problem applies to multicellular organisms, and not to unicellular organisms that reproduce by simple cell division. Throughout this book I shall use the term *development* in this sense: to refer to the process whereby a single cell gives rise to a complex multicellular organism.

To see more clearly what the process of development involves, I will need to introduce three types of molecule that play a fundamental role in it: proteins, DNA and RNA.

Proteins as guiding shapes

The properties of every plant or animal cell depend on the types of *protein* it contains. We normally come across proteins as part of our diet. Proteins, though, play a much more pervasive role in our lives than this might lead us to believe. All the processes in the body, such as digestion, secretion, moving, sensing and thinking, depend on the activity of different types of protein molecule. Without proteins we would not be able to do anything.

The most important feature of protein molecules that allows them to encourage all these things to happen is their *shape*. The way in which a protein's shape can influence events is rather central to this book, so I need to be very clear about the principles involved.

If you put an empty bucket outside and let it fill with rain, you might say the bucket is holding the water. Obviously this does not mean it is actively doing anything about the water, trying desperately to keep it all together. The shape of the bucket leads to the water being held in a particular way as long as the rain pours down to fill it. The bucket facilitates or guides the way that the downpour of water is collected. Without the bucket being there, the water would never heap up on its own to form a bucket-shaped mound. The process is driven by an energy source that is outside the bucket: the rain pouring down from above, or more remotely, the sun's energy that evaporated water from the earth's surface and led to the formation of clouds. We could imagine more complex combinations of shapes, such as a mountainside covered with buckets, perhaps connected together by a network of other shapes, in the form of tubes or pipes that guide rain water from bucket to bucket and finally into a reservoir. Further pipe shapes could guide the water to drive a turbine and generate energy in a different form. Each shape facilitates one course of events rather than another but does not itself provide the required energy to drive the process along.

In a similar way, each cell contains many thousands of different types of proteins, each one with a different shape, according to the process it guides. These processes are at a sub-microscopic scale, the scale of molecular reactions. Molecules in a fluid are always on the move, continually jostling around very rapidly and bumping into each other. The higher the temperature, the faster molecules move around; and at the temperature needed to sustain life, there is quite a commotion in the cell's interior. For a molecular reaction to happen, say for molecules A and B to join together, the molecules need to come together in the right way. Normally, when A and B happen to bump into each other, nothing might happen because they do not meet in a suitable manner and they just career off again into the distance. But suppose A encounters a large molecule, a protein, that has a shape with a nice little pocket that A fits into very comfortably (Fig. 2.3). We could imagine, for example, that the pocket in the protein matches the shape of the A molecule, like a lock matching a key. The A molecule may stick to the protein and not career off. If the protein has another nearby pocket that matches molecule B, then when B is bumped into, it will also tend to stick. There is a reasonable chance that the protein will have both A and B stuck to it at the same time and, if they are held in the right way,

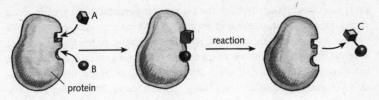

Fig. 2.3 Protein (enzyme) catalysing a chemical reaction.

they will react with each other, joining up to form a new molecule, C. In this way, the shape of the protein, the structure of its pockets and crevices, can facilitate a reaction: A and B coming together to make C. Once C has formed, it may leave the protein, freeing up the pockets to join another pair of A and B molecules. Because the protein is not consumed by the reaction but simply helps to guide it along, it is said to act as a *catalyst*. All of the molecular events catalysed by a protein happen extremely quickly: a protein may promote 1000 reactions like this every second. This is because the molecules move around and react with each other at such a mind-boggling rate.

Proteins that catalyse these sorts of reactions are called *enzymes*. In the example shown, C is the product of the reaction; but it is also possible for the reaction to go in the reverse direction, breaking C down into A and B. The direction which any reaction takes will largely depend on the energy involved in making or breaking chemical bonds between the molecules and on the amount (concentration) of A, B and C molecules in the cell.

I should mention that matching the shape of a molecule by a protein is not just a question of complementing its three-dimensional shape. Molecules also have small electrostatic charges distributed over them, so that some regions of a molecule tend to be more positively charged and others more negatively charged. In matching the shape of molecules, proteins also match their distribution of charges, so that a positive region of a molecule lies next to a negative region of the protein. These charges can be very important in slightly deforming the shape of molecules, by pulling or repelling parts of them, and thus facilitating particular types of reaction. To simplify matters, though, I shall use *shape* throughout this book as a general term to cover both the three-dimensional structure of a molecule and its particular charge distributions.

Proteins do not provide any energy to drive reactions; they are just catalysts by virtue of their particular shapes. Through its compatibility with various molecules, the shape of a protein can encourage or facilitate one reaction occurring rather than another. Each different type of reaction usually needs a protein with a different shape. The protein that helps A and B come together would not help D and E; they would need their own protein to help them along.

Every cell therefore needs many thousands of different *types* of protein, each with a distinct shape, to guide its numerous internal reactions. Proteins with various shapes will be mentioned throughout this book, so it is very important to remember that the role of a protein, the process it guides, is an automatic consequence of its shape. It is not actively 'doing' anything other than facilitating one thing or another happening, just as a bucket helps water collect in one place rather than another. Nevertheless, I may lapse into saying that a protein does this or that as a form of shorthand. It is so much easier to say a protein 'does X' than 'its shape facilitates X happening'. This should be taken in the same spirit as saying a bucket 'holds water' rather than 'its shape facilitates water assuming a bucket-like conformation'.

As with the bucket filling with water, the energy that drives all these proteins ultimately comes from the sun (with the exception of some bacteria that obtain their energy from inorganic compounds). Solar energy is captured within the cells of plants, where it is guided to produce sugars from carbon dioxide and water. The light energy is effectively converted into a different form of energy, stored in the chemical bonds of the sugar molecules. The chemical energy and components in the sugars are then channelled in all sorts of different directions, by other types of protein in the plant, to make molecules such as carbohydrates, fats and more proteins. These products are in turn essential for sustaining animal life: when an animal eats a plant, the energy and chemical components of the plant are channelled by the animal's proteins to make its own carbohydrates, fats and proteins. In other words, the energy that drives the internal reactions of animals comes from the food they eat, which ultimately depends on energy from the sun.

There is one more key feature of proteins that I need to mention: their shape is not completely rigid but can *change*, depending on which other molecules happen to be bound to them. When a molecule binds in a pocket, it can cause a change in the protein's overall conformation or shape. In most cases, when the molecule leaves, the protein will return to its original shape. These reversible changes in protein shape underlie much of our behaviour. The movement of every muscle in your body depends on countless muscle proteins changing their shape back and forth very quickly. Similarly, the transmission of electrical signals in your brain and nerve cells depends on rapid reversible changes in the shape of particular proteins. As with the other processes I have mentioned, the energy to drive all these events does not come from the proteins themselves, but ultimately comes from the sun.

Given that the combination of protein shapes in a cell is responsible for many of its properties, a major part of trying to understand development has to do with explaining how some cells of the body come to contain different proteins from others. Why is it that cells forming in the brain region have proteins appropriate to brain cells whereas those in the liver region have proteins relevant to liver cells? There are thousands of cell types in the body, all arranged in a very precise manner, so the problem of how each comes to have its own particular spectrum of proteins becomes rather daunting. I will try to address this question in later chapters. Here, I want to consider how proteins themselves are made and how they get their shape.

Making a biological copy

To understand how the structure of proteins is determined, I need to introduce another key molecule, *DNA*. DNA is a very long molecule made up of two strands; it is double-stranded, the way that string is often made from two inter-twining strands (Fig. 2.4). Each strand is itself made of a sequence of molecular subunits, called *bases*, of which there are four different types, symbolised with the letters A, C, G and T (the letters actually stand for the chemical names of the bases, adenine, cytosine, guanine and thymine). The bases are strung together in a particular order, just as letters are ordered in every word; although unlike our alphabet, which has 26 letters, there are only 4 types of letters in DNA. Nevertheless, a DNA molecule can carry a lot of information with its four-letter alphabet because it is enormously long.

Most cells have several DNA molecules of various lengths, each individual DNA molecule being called a *chromosome*. In many organisms the chromosomes are in pairs, with members of the same pair having the same length and a comparable DNA sequence. The numbers and lengths of chromosomes are often different between species. For example, humans have 23 pairs of chromosomes in the nucleus of each cell, giving a total of 46. Fruit flies have 4 pairs of chromosomes per cell, and *Antirrhinum* (snapdragon) plants have 8 pairs. A typical human chromosome has about one hundred million bases in it, making it several centimetres long. If we were to magnify this to the width of string, with the bases spaced at 1 mm intervals, it would stretch for about a

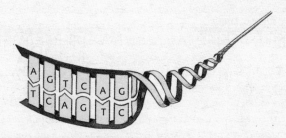

Fig. 2.4 Structure of DNA showing pairing between bases on opposite strands.

hundred kilometres. Although very long, each DNA molecule is very tightly packed, allowing all 46 chromosomes to be stored within the nucleus of a single cell.

The most important feature of DNA, which gives it such a unique place in life, is that it can be *replicated* or copied. This happens in a manner that corresponds quite closely to the process of reproduction in art. One way to make replicas of a small statue would be to pour a rubber mixture over the statue to obtain a mould. After the mixture has set, the rubber mould could be carefully peeled off the statue and then filled with a suitable casting material, like plaster, that will eventually harden. The cast could then be revealed by peeling off the mould. Once we have a cast we can make a new mould from it and so continue to replicate mould or cast as we wish. All that is needed to make replicas is two complementary materials that can match each other: the mould and the cast. It doesn't really matter whether we start from a mould or a cast, we can make as many copies as we want.

DNA replicates on a similar principle, except that in this case the mould and the cast are made of the same material. Like a cast and a mould, the two strands of DNA are *complementary* to each other. The bases that make up each strand fit like jigsaw pieces into their counterparts on the other strand according to the following rules: A fits with T, G fits with C, as shown schematically in the enlarged part of Fig. 2.4. This means that if A is present at a particular position on one strand we can be sure that T will be located at the corresponding position on the opposite strand. Similarly, C on one strand will always be opposite G on the other. Knowing the sequence of bases on one strand, we can therefore predict the complementary sequence of the other strand.

To replicate DNA, the two strands need to be separated, just as we need to take the mould off the cast if we want to make new copies of a statue. Once separated, each strand can then be used as a mould or cast to produce a new complementary strand. The separate strands act as templates, ensuring that only matching bases are lined up along their length to form a new matching strand, as shown in Fig. 2.5. We eventually end up with two identical

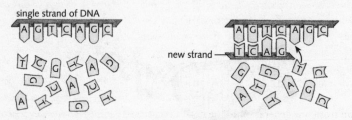

Fig. 2.5 Copying a single strand of DNA by incorporating matching bases to form a double-stranded molecule.

double-stranded DNA molecules, having started off with only one, so there is now an additional pair of casts and moulds.

This mechanism of replication follows quite naturally from the structure of DNA, the complementarity between its two strands. This is why the discovery of the structure of DNA, by James Watson and Francis Crick, was such a watershed: it pointed the way to how biological copying might occur. As Watson and Crick wrote in a famous passage towards the end of their paper in 1953: 'It has not escaped our notice that the specific pairing we have postulated immediately suggests a possible copying mechanism for the genetic material.'

The replication of DNA does not occur spontaneously: it requires an input of energy and some guidance (just as statues don't proliferate without someone separating and pouring the casts and moulds). There are specific proteins that guide the energy and components needed for DNA copying, ensuring that the individual bases that make up the new DNA strands are incorporated efficiently and effectively (the bases are themselves made by chemical reactions catalysed by yet other proteins). It is important to emphasise, though, that although these proteins help to ensure that DNA is copied, they do not themselves provide the information that is stored in the DNA sequence. The person who pours the moulds and casts when copying a statue does not provide the information in the statue, its shape and form. He or she is merely allowing an existing structure to be copied. In the same way, the proteins that catalyse DNA copying do not determine the sequence of the DNA—that is already provided—but simply help the sequence to get replicated.

The process of DNA replication normally occurs every time a cell divides so that each daughter cell inherits one set of DNA molecules. A dividing human cell containing 23 pairs of chromosomes would therefore contribute a replica of all 46 chromosomes to each cellular offspring. This sort of division, *mitosis*, is typical of most cells in your body. There is a less common type of division, *meiosis*, which is nevertheless very important because it is essential for sexual reproduction. In the case of humans, meiosis occurs in some of the cells of the testicles and ovaries. During the division of these cells, the two members of each chromosome pair become separated from each other, so that you end up with progeny cells, sperm and eggs, having only 23 rather than 46 chromosomes in their nuclei. Each parent therefore only puts half of its chromosomes into each of its sperm or egg cells. When the sperm fertilises the egg, the 23 chromosomes from each parent come together to give a total of 46 once again. Every child therefore inherits 23 chromosomes from its father and 23 from its mother.

Now although DNA might impress us with its ability to be copied and passed on from one generation to the next, this property imposes some severe constraints on its other potential roles. The materials used for making moulds or casts are carefully chosen to have certain features. The mould and cast should not stick to each other too much and the mould has to be flexible enough to be easily removed from the cast without breaking either cast or mould. Both moulding and casting materials also need to change from a fluid to a solid state so that they can set easily. These requirements are even more difficult to satisfy if the mould and the cast are to be made of the same material, as is the case for DNA. Now, although DNA meets all these requirements very well, it is at the expense of other properties, such as being able to fold up into complicated shapes. The features of DNA which make it so suitable for being copied render it much less useful for doing anything else in the cell. This means that for the information contained in DNA to have any significance for the behaviour and properties of cells, it needs to be converted into a different form: the currency of proteins.

From DNA to RNA to proteins

Proteins are made of molecular subunits, called *amino acids*, strung end to end. There are 20 different types of amino acids, 20 letters in the protein alphabet. The shape of the protein, and hence what it guides, depends on the particular *sequence* or order of its amino acids. A typical protein might contain several hundred amino acids strung together in a very specific order. Although they are joined together in a linear chain, the amino acids nudge, pull and push against each other, owing to their own individual shapes, adjusting their positions so that the chain folds on itself until the protein ends up with a stable arrangement or overall shape. Every type of protein has its own diagnostic sequence of amino acids and hence its own shape that allows it to catalyse particular events in the cell.

Proteins and DNA are similar to each other in both being built from sequences of molecular subunits—amino acids and bases respectively—but are very different in other ways. Whereas proteins are able to fold up into all sorts of wonderful shapes, allowing them to promote particular events in the cell, DNA is structurally much more limited, being more like a long piece of string that sits in the nucleus. On the other hand, this wearisome feature of DNA is precisely what allows it to do something that proteins cannot: DNA can get copied very easily. The information in DNA—the sequence of its bases—can get replicated and passed on every time a cell divides. By contrast, once a protein is made it cannot be used as a template to make more copies of itself. Considered alone, proteins and DNA each suffer from a major deficiency, but by working together they can make up for each other's weaknesses. I now want to describe how this happens.

We can start by notionally dividing the sequence of a long DNA molecule or chromosome into shorter stretches, say a few thousand bases long, each stretch being called a *gene*. A chromosome can be 100 million bases long, so we could theoretically divide it up into many thousands of genes, following each other along its length. Each gene can be thought of as a word, a few thousand letters long, written in the four-letter alphabet A, T, G and C. A chromosome of 100 million bases will therefore contain many thousands of such gene words along it. It has been estimated that humans have about 70 000 genes in each set of 23 chromosomes, while fruit flies have about 12 000 genes in their set of 4 chromosomes, and a plant such as *Arabidopsis* (a small weed commonly found in gardens) has about 25 000 genes in its set of 5 chromosomes.

Like words, genes are characterised by the particular sequence of letters they contain. The most important part of a gene, as far as proteins are concerned, is a stretch called the *coding region*. This region contains information, coded in the four-letter alphabet of DNA, that when properly translated can lead to a particular type of protein being made. There is a precise correspondence between the sequence of bases in the coding region and the sequence of amino acids in the resulting protein. Thus, each type of protein is encoded by a distinct gene. Once you know the nature of this correspondence, called the *genetic code*, you can convert any sequence of DNA into a protein sequence, just as once you know Morse code you can convert a series of dots and dashes into a word written in letters. The genetic code is organised in triplets, so that a set of three DNA bases, such as ATG, corresponds to a particular amino acid, called methionine in this case. Another triplet, GCA, corresponds to a different amino acid, called alanine. In this way, the DNA sequence in the coding region is translated as a series of consecutive triplets, to specify a sequence of amino acids in a protein. This means that a stretch of DNA has to have three times as many subunits along its length as the protein it codes for: a coding region 300 bases long will be needed to give a protein of 100 amino acids. There are also some triplets with roles similar to punctuation, indicating where translation of the DNA information should begin or end.

We need not be concerned here with all the details of how the information from a gene is translated into a protein, but there is one important further aspect that needs to be covered. The translation of information from DNA to protein does not happen directly but involves an intermediary: another type of molecule called *RNA*. Like DNA, the RNA molecule is made of a string of four types of bases, each type of RNA base matching the shape of one of the DNA bases. But whereas DNA is double stranded and very long, an RNA molecule is made up of a single strand and tends to be much shorter: a typical RNA molecule is only a few thousand bases long, compared to the hundreds of millions of bases that can go into DNA.

As shown in Fig. 2.6, for a gene to make a protein, the coding region of a gene is first copied or *transcribed* into RNA molecules. The process of transcribing

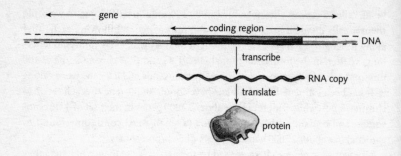

Fig. 2.6 From DNA to RNA to protein.

DNA into RNA is not too different from the way DNA is replicated. The two strands of DNA are separated, and one of them acts as a template for the assembly of a string of complementary RNA bases in the appropriate order. For any gene, only part of the DNA—the sequence covering the coding region—is transcribed into RNA in this way. The RNA copies are then free to leave the nucleus and enter the surrounding cell fluid, the cytoplasm. Here, the information in the RNA copies is *translated* into a sequence of amino acids, to make a protein. Depending on its shape, the protein may remain in the cytoplasm, or it may enter the nucleus, or it may be secreted out of the cell. To make a protein, information therefore has to be transferred from *DNA to RNA to protein*.

There is an interdependence between DNA and proteins. As far as the structure of each type of molecule is concerned, information flows in the direction from DNA through RNA to protein. This means that if the DNA sequence of a coding region is altered, the amino acid sequence of the encoded protein may consequently change. The reverse does not apply: a change in the sequence of a protein has no direct effect on the sequence of DNA. Structural information does not flow from proteins back to DNA. On the other hand, the information in proteins—their shape—is needed to guide the various events in a cell, including the replication of DNA, transcription from DNA to RNA and translation of RNA into proteins, as well as numerous other processes that occur in a cell. This mutual dependence between DNA and proteins is based on the very different properties of each type of molecule: they are specialised in complementary ways.

You might be wondering how such a system ever evolved in the first place, because neither proteins nor DNA can function alone. One view is that this system evolved as a specialisation from a more primitive form of life—existing billions of years ago—that was not based on DNA and proteins but on RNA. The properties of RNA are in many ways intermediate between DNA and proteins: like DNA it is able to be copied, but it is more versatile in its shape and can catalyse a limited number of chemical reactions in the cell. Perhaps, then, life was originally based on a single type of molecule, RNA, that was able both to replicate and to promote particular reactions. According to this view, DNA and proteins evolved later on as a specialisation that allowed these different aspects of life to be carried out more effectively.

Reproduction in biology and art

Let me summarise the story so far. I began this chapter with the question of how closely reproduction in art and biology resemble each other. In art, reproduction involves the process of copying. In exploring biological reproduction, we have also come across a copying process: the replication of DNA. There is an original template—a set of DNA molecules—that is dutifully copied and passed on from parents to offspring every generation.

Nevertheless, although we have revealed a copying process involved in biological reproduction, what is being copied is very remote from a complex organism. Forty-six balls of string, each marked with a sequence based on four colours, denoting A, G, C and T, are hardly the same as a human being. There is currently an international effort, the Human Genome Project, aimed at determining precisely the DNA sequence of all human chromosomes. The human sequence will not be uniquely defined, because all individual humans have a slightly different sequence, but it should be possible to get a sequence that is typical of human DNA. Even when we know this enormously long string of letters, however, it will not tell us how we are made. To understand our reproduction as complex organisms, we have to be able to relate this sequence in some way to the process by which a fertilised egg cell develops into a living human being.

The relationship of a DNA sequence to a fully developed organism is not captured by any comparison with artistic reproduction: humans are not related to DNA as copies are to an original. If development were like copying, we would have to imagine that there was some sort of miniature human in the fertilised egg that was being used as the original, a view that takes us back to the idea of preformation discussed in the previous chapter. We have seen, though, that the subject of biological copying is not a tiny human, but something with an entirely different structure: DNA. This raises the fundamental problem of how this sequence of DNA is related to the complex biological form that eventually develops from the fertilised egg.

Now although this has no parallel with reproduction in art, there is, in my view, an artistic activity that comes near to it in many ways: the process of *creating*. By creating, I mean the highly interactive process that goes on when, say, an artist paints an original picture. Unlike copying, the final aim is not there in front of the artist to begin with. The picture emerges through an interaction between the artist, the canvas and the environment. As I will show in this book, the process by which a DNA sequence eventually becomes manifest in a complex organism has many more parallels with the interactive process of creating an original than with making a replica or copy. Of course, there is still the issue of open-endedness: the extent to which the development of organisms is more highly circumscribed than painting a picture. I shall eventually return to this problem later on in the book when I have further clarified what is going on during development.

From DNA to organism

To unravel the problem of development, we will need to understand more about the DNA language: how the sequence of bases in the DNA molecule is related to the organism that eventually develops from a fertilised egg. Before proceeding any further, I want to outline a general method for deciphering the meaning of DNA that will be employed throughout this book.

Deciphering DNA is rather like learning a completely new type of language. As children, we learn language in several ways. One is by *association*—people pointing to objects and saying the relevant word like 'table', 'chair', 'flower'. This may help to convey the significance of a noun but it does not cover other aspects of language such as verbs and sentence structure. We learn these more easily through listening to and exploring how words are *used*. By continually hearing and using strings of words in various contexts we eventually start to get a more comprehensive knowledge of the meaning and structure of language.

To understand the meaning of DNA we have to somehow find a way to establish gene associations and uses. It turns out that there is a method for doing this but it is almost the reverse of the way we usually acquire language. To get an idea of how it works, I shall try to draw out a parallel procedure in terms of our own language.

Imagine a society that is identical to ours except that there are no tables. I shall refer to this as a *mutant* society, deviating from ours in one respect: the absence of tables in this case. No one would ever feel the need to refer to tables in this society, and a word for *table* would not be part of the language. If external observers were to compare this society with our normal one they would notice two things that distinguished it: (1) there are no tables and (2) the word *table* is missing from the language. By associating these two things the observers might conclude that the word *table* is used to refer to the objects that are present in the normal but missing in the mutant society. In this way the significance of the word *table* could be guessed at. The important point about this way of deciphering a language is that the tables and the word *table* go together. Remove one and the other automatically vanishes. It is a form of learning by association; an indirect form of pointing, like a 'spot the difference' game where you show two pictures that only differ in one detail, such as the apples being missing from a tree. As long as you can associate the missing objects with a word, in this case *apples*, the significance of the word can be deduced.

In our mutant society, we first removed the tables and then found the word *table* was not necessary. But what if it also worked the other way round? Suppose that when you remove the word *table* from a language, the tables disappeared. In this case the mutant society is the direct result of removing the word. Obviously this is not the way our language works: we cannot remove words and watch objects disappear. But although such a notion might sound peculiar to us, from the external observers' point of view, the situation is the same as before. They are presented with the same two societies and would still come to the conclusion that *table* refers to the objects missing from the mutant society.

This sort of reverse logic is a very important method used by biologists to learn the significance of particular DNA words. The missing words correspond to *mutations* in genes. Although DNA is normally replicated with remarkable fidelity, occasionally a mistake is made in the copying process, such as a wrong base being put in (say an A instead of a G) or a small stretch of DNA failing to get copied and therefore being deleted in the daughter molecules. When such a mutation occurs it can have lasting consequences because it will be faithfully replicated and passed on to all subsequent DNA copies—just as if we were to scratch a cast and then make a mould from it, all subsequent casts made from that mould would carry the scratch. Mutations are normally very rare: typically one error is made for every billion bases copied. In some conditions, as when DNA is exposed to ultraviolet radiation, the rate of mutation can increase due to the DNA having been damaged. That is why excessive sunbathing can be harmful: it results in DNA mutations in the skin cells that can sometimes lead to cancers. In this case, the mutation is restricted to the person in which it occurs (i.e. it is not transmitted to offspring) because it affects only cells that remain in his or her body. If, however, a DNA mutation arises in cells of the *germline*, which gives rise to sperm or eggs, it can get passed on to the next generation. When a sperm or egg cell carrying a mutation contributes its share of DNA to a fertilised egg, the individual that develops from the fertilised egg will carry the mutation in *all* cells of its body. Should the individual have offspring, it will transmit the mutation through its own sperm or egg cells to the next generation (actually the mutation will be passed on to only half of the offspring because each sperm or egg cell produced by the individual will carry only a half-share of chromosomes). In this way, the mutation may eventually accumulate in a population of individuals.

Some mutations in the germline will have little effect on organisms that later develop because they change a relatively unimportant bit of DNA, but other mutations may have significant consequences. For example, if the mutation is in the coding region of a gene, it may change the sequence of amino acids in the encoded protein and hence its shape. The altered protein may no longer be able to function properly in guiding a particular process in the cell. With this process missing or defective, the organism that develops may also be altered. In this way, a DNA mutation can lead to an observable difference in the organism, and because DNA is replicated, the difference will be heritable. The inherited differences in eye or hair colour, flower or leaf colour, height, shape and the numerous other characteristics that distinguish individuals are a consequence of variation in their DNA sequence that arose at some point from mutations. Essentially all heritable variation, the raw material for evolution, is a consequence of mutations in DNA.

Mutations are very important for getting at the meaning of DNA because they provide us with a way to link DNA with an organism's appearance. By associating the alteration in an organism's characteristics with a mutation in a particular piece of DNA, we get an indication of what that piece normally signifies. It's like learning a language backwards, removing words and seeing which objects disappear. If I breed a lot of normal red-flowered *Antirrhinum* (snapdragon) plants and find a mutant with *white* flowers, I can say that a gene signifying *red* has been altered. If I find a *bald* mutant mouse, I might conclude that a gene signifying *hairs* had become defective.

In all these cases, the effect of the mutation is the *opposite* of what the gene normally signifies because it reflects a defect in the gene. Mutations typically show what happens when the action of a gene is negated or removed rather than displaying its positive effects. This can cause some confusion because mutations and genes are usually named according to how they were first noticed rather than what they normally signify. For example, when a mutant fruit fly was noticed with white instead of the normal red eyes, it was said to have a mutation in the *white* gene, even though normally this gene is involved in making the eyes red. So in many cases, a gene's name indicates the opposite of what it normally signifies for the organism because of the topsy-turvy way that we discover its meaning.

Although the study of mutations in this way can be very revealing, it by no means gives us a complete picture of the DNA language. If all we did was point to objects and say their names, a child might learn nouns by association but very little else. The way words are used and combined in various ways to form meaningful sentences would not be appreciated. There are also further complications with our method of learning by association. In a mutant society without tables, there would be knock-on effects on other objects; for example there would be fewer chairs and no table-cloths. In trying to work out the meaning of *table* by association, we might get confused and think the word also refers to some missing chairs and table-cloths. If we only have simple associations to go by, it will sometimes be difficult to distinguish these knock-on effects from the primary defect in our mutant. We therefore need an additional method for working out the meaning of DNA: we need to determine how the genes are *used* during the life of the organism.

We have already taken some steps in this direction by considering how genes code for proteins, via an RNA intermediate. To get further, we need to follow the activity of those genes that are particularly relevant to an organism's development. One of the main problems with trying to do this is in identifying the small stretch of DNA, the gene, that relates to a particular feature of development. In the 1970s and 80s, methods were developed for *gene cloning*, which allowed specific genes to be isolated from organisms and multiplied in bacteria. Once isolated in this way, specific genes together with their corresponding RNA and protein molecules could start to be studied in great detail. This eventually led to a much better understanding of how genes are used during development. One of the main purposes of this book is to explain how this approach has given us much deeper insights into what the DNA language means.

In practice, the study of genes in this way has been carried out on relatively few organisms. The choice of which organism to study has often been a source of lively debate and rivalry among biologists. Ask a biologist why he or she has chosen a particular organism for study and you will hear of its many advantages. The arguments biologists use to justify their choice of organism have varied through history, as biological problems and techniques have changed. A favourite organism for early developmental studies was the chick. As Ernst Haeckel pointed out in 1874:

> Hens' eggs are easily to be had in any quantity, and the development of the chick may be followed step by step in artificial incubation. The development of the mammal is much more difficult to follow, because here the embryo is not detached and enclosed in a large egg, but the tiny ovum remains in the womb until the growth is completed.

Although the chick was a favourite for a long time, other organisms, such as frogs, newts and sea urchins, became very popular in the late nineteenth and early twentieth centuries. This is because scientists became interested in seeing what happened when bits of embryos were surgically removed or interfered with. Frogs and sea urchins are particularly amenable for this sort of approach because their embryos are readily accessible and can survive such assaults rather well.

More recently, the detailed study of genes and mutations imposed yet new criteria on the choice of which organism to investigate: organisms that can grow and breed in very large numbers and have a short generation time became desirable. The fruit fly, *Drosophila melanogaster*, often to be seen circling around dustbins or wine glasses, is a particular favourite because it is easy to maintain hundreds of flies in a bottle, and the time between generations is only two to three weeks. Important work on the fruit fly was started in about 1910 by Thomas Hunt Morgan but it was not until the 1980s, with the advent of gene cloning, that it became the subject of extensive molecular studies. Other animal species that have been intensively studied in this way include the mouse, nematode worm (*Caenorhabditis elegans*) and zebrafish.

Many plant species also offer advantages for genetic studies because they are easy to self-fertilise, cross and breed in large numbers. It was considerations like these that led Gregor Mendel to use peas for his initial studies on heredity in 1865. More recently, many plant biologists have chosen to work on a garden weed, *Arabidopsis thaliana*. Working on a weed is not as surprising as it might first sound because some of the desirable properties of an organism for genetic studies, such as rapid growth and short time between generations, are also those that make a good weed. In the late 1980s it became apparent that these advantages could be combined with molecular approaches, and research on *Arabidopsis* spread through plant laboratories like gold fever. However, I have to admit to having a soft spot for one particular plant, which also happens to be the one I work on, *Antirrhinum majus* (snapdragon), because it turns out to have several advantages for the study of genes involved in flower development.

To summarise, the reproduction of multicellular life involves the development of a complex organism from a fertilised egg. At the heart of this process are DNA molecules that get copied every time a cell divides, much as a work of art might be reproduced. But this raises the fundamental problem of how DNA, a linear sequence of bases, is related to the elaborate three-dimensional organism that finally develops from the fertilised egg. To address this, we need various methods for deciphering the DNA language, revealing its meaning and significance for the organism. In practice, these methods have been applied to a few well chosen plants and animals. It is some of the basic lessons that have been learned from these studies, and what they say about the relationship between development and other processes, such as human creativity, that I want to describe in the following chapters.

CHAPTER 3

A question of interpretation

Making patterns is a common feature of human creativity. Whether it is arranging flowers in a vase, decorating a cake, designing wallpaper, or playing a series of musical notes, we delight in making ordered and balanced compositions. A good example is the mosaic in Fig. 3.1, where differently coloured stones have been carefully placed to give a more or less symmetrical and harmonious image. The end result often appears simple and pleasing. But the process of creating a pattern is far from straightforward: it is difficult to say what goes on in the mind of the pattern maker as various elements, such as the coloured stones in a mosaic, are tried out and played with. Eventually one arrangement is considered more pleasing but it is seldom easy to explain precisely why this is so, what makes someone prefer one composition to another. Making patterns involves a complex interplay that defies simple dissection.

There is a parallel problem in the natural world. Plants, animals and fungi display a remarkable array of highly ordered and coherent patterns, from butterfly wings and tropical fish to orchid flowers and toadstools. Many of these patterns are themselves borrowed in human designs, as with the peacock illustrated in the mosaic (Fig. 3.1). But how is this wonderful diversity of natural patterns produced? How do single cells develop into organisms with such elaborate designs? We shall see that like human pattern making, this also involves a complex interplay, but in this case some of the basic elements are easier to dissect. In this chapter I want to begin exploring this problem with what might seem like a rather esoteric type of pattern, the arrangement of bristles on a fly's back.

Bristle patterns

In the 1950s Curt Stern was studying the pattern of bristles on fruit flies at the University of California. An adult fruit fly normally has 40 bristles on its back, each at a very specific location (Fig. 3.2, left). The bristle pattern on the left and

Fig. 3.1 Italian mosaic, sixth century, San Vitale, Ravenna.

right halves of the fly are mirror images of each other, so the overall pattern comprises 20 pairs of symmetrically disposed bristles. The question that intrigued Stern was how such a coherent and reproducible pattern might arise through the action of genes. As I mentioned in the previous chapter, one way to analyse how genes affect an organism is to look at mutations. In this case, Stern was interested in mutations that changed the pattern of bristles, effectively giving the fly a different hair style. Several mutants of this type had been found, in which bristles were missing at certain positions but not others (Fig. 3.2, middle and right). Although the overall number of bristles was altered, the mutant patterns were nevertheless symmetrical because the missing bristles were always lost in pairs. If we number each pair of bristles from 1 to 20, then

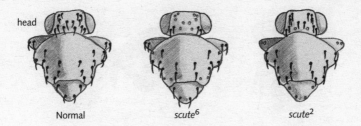

Fig. 3.2 Bristle patterns due to different versions of the gene *scute*. The effect of the normal version is shown on the left and compared to two mutant versions, *scute⁶* and *scute²*.

flies with one type of mutation might be missing pairs 1, 2, 3, 4, 5, 10, 17 and 18 whereas those with a different mutation would lack bristle pairs 5, 8 ,9, 19 and 20. The implication was that the bristles were being left out in a coordinated manner from each side of the insect's back.

These flies with different bristle patterns had mutations in a particular gene, called *scute* (scute means hardened scale or plate and refers to part of the insect's back affected by the mutation). It appeared that the *scute* gene could mutate in various ways, each mutation resulting in a different bristle pattern. Stern reasoned that maybe the *scute* gene was somehow involved in carefully planning and coordinating where bristles form. Perhaps each mutation produced a distinct version of *scute* that affected the overall planning differently, so that particular bristles were left out in a consistent and symmetrical way, just as someone planning a different version of a mosaic might alter particular stones on both sides, retaining the overall symmetry.

To test this idea, Stern wanted to see what would happen to the final bristle pattern when distinct versions of *scute* were imposed on different parts of the insect's back. If, say, all the cells in one half of an insect's back carried one version of the *scute* gene, and all those in the other half carried a different version, where would the bristles go when faced with two competing plans? Producing a fly that is part one thing, part another, sounds like the stuff of mythology, like the Chimera of Greek legends that was part lion, part goat, part dragon. There is, however, a remarkable case in which this sort of thing can occur quite naturally. In many species of insect, rare individuals, called *gynanders*, can be found which are part male and part female, the left half being one sex and the right half being the other. Gynanders are particularly striking when they occur in species with distinctly coloured sexes. For example, the left specimen in Fig. 3.3 is a gynandric wasp in which the left half is red and wingless, typical of the female of this species, whereas the right half is black and winged, resembling the male. The other specimen is a gynander of the Swallow-tail butterfly, *Papilio dardanus*: the male half on the right has a distinctive light

Fig. 3.3 The insect on the left is a gynander of the solitary wasp *Pseudomethora canadensis*, with a female left half (red and wingless) and a male right half (black and winged). The specimen on the right is a gynander of the butterfly *Papilio dardanus* with the left half female and the right half male.

yellow colour and swallow tail on the hind wing, whereas the left half has colouring characteristic of females. Such peculiar individuals were highly sought by butterfly collectors, even tempting some unscrupulous characters to fake them by cutting a male and female in half and sticking them together. Unlike these fakes, however, the boundary between male and female parts in naturally occurring gynanders does not go straight up the midline but wanders over to varying extents from one side to the other.

Stern wanted to exploit gynanders to help him breed a fly that had distinct versions of *scute* in different parts of its back. To explain how he did this I need to give some background on how sex is determined in flies. Recall that each cell of an organism typically has a nucleus containing several very long molecules of DNA—chromosomes—that carry all the genes. Humans have 23 pairs of chromosomes, one set coming from each parent, whereas fruit flies have 4 pairs of chromosomes. The difference between male and female flies has to do with one particular pair of chromosomes, called the *sex chromosomes*. In females, these two chromosomes are similar to each other and are referred to as X chromosomes. Males, however, are distinct in only having one of these X chromosomes (the other sex chromosome present in males is quite different and is called the Y chromosome). It is the *number of X chromosomes* that determines the sex of the fly: an individual with 2 X chromosomes will be female, whereas a fly with only 1 X chromosome will be male.* Fly gynanders arise during the divisions of a fertilised female egg with 2 X chromosomes in its nucleus. Occasionally, the nucleus divides abnormally, so that instead of copying both of the X chromosomes to its daughter nuclei, one X gets lost, giving one nucleus with 2 X chromosomes and the other with only 1 X. When the adult fly finally develops, approximately half of its cells will carry one type

---

*In humans, the situation is slightly different because it is the presence/absence of the Y chromosome that determines sex rather than the number of X chromosomes.

of nucleus, and half will carry the other. The half that has nuclei with 2 X chromosomes will be female, whereas the half that carries 1 X will be male. Normally such gynanders are very rare because X chromosomes do not usually get lost during division, but Stern used flies with a special type of ring-shaped X chromosome that goes missing more often, so that gynanders arose quite frequently. Now because the *scute* gene happens to be located on the X chromosome, Stern also managed to arrange things such that the female part of the fly would end up expressing one version of the *scute* gene whereas the male part would have a different version.

Stern now looked to see what would happen to the bristles in gynanders with part of their back having one version of *scute*, promoting one sort of overall bristle pattern, and the other part with a different version of *scute*, trying to impose an alternative overall pattern. He expected that things might get a bit confused as these two conflicting patterns tried to assert themselves, imagining that perhaps a new pattern or at least a compromise between the two might arise. To his surprise nothing of the sort happened. Each part of the fly behaved independently from the other part: whether or not a bristle formed in a region of the back depended only on which particular version of *scute* that region carried. The regions behaved autonomously and appeared to completely ignore which version of *scute* their neighbouring regions might have. Stern was dumbfounded:

> I remember well how I brooded over my mosaics [gynanders] that showed the paradoxical autonomy in patterning until I suddenly realised that the paradox was not in Nature but in my preconceived views. Instead of pretending 'The facts do not fit my theory—the worse for the facts,' or complaining, 'life has double-crossed the . . . writer,' I had to admit the wisdom of Goethe: 'Nature . . . is always true . . . she is always right . . . and the mistakes and errors are always those of man.'

He had to abandon the idea that the *scute* gene was affecting the overall planning of bristles and came up with an alternative explanation for his results. An analogy should help to explain why Stern was so surprised and the conclusion he eventually came to.

A pattern of hidden colours

You have been invited to a football stadium with lots of other people to send a birthday message to someone flying overhead in a helicopter. The stadium has already been carefully marked out with a grid of squares, each square being painted a distinct colour, making the stadium look like a colourful patchwork. You are given two pages of instructions and a large piece of card, black on one

Multi-coloured grid

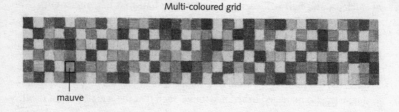

mauve

Fig. 3.4 Football stadium with a grid of coloured squares (*above*) being used according to the first page of instructions to make a black-and-white message (*below*). The mauve-coloured square you are standing on is highlighted.

side and white on the reverse, and told to go and stand on a square. After choosing a square, you look at the first page of instructions, which has a list of all the colours on the grid. The colour of the square you happen to be standing on is *mauve*, and next to mauve on this page is written the instruction 'hold the cardboard above your head with the white side uppermost' which you obey. Some of the other people also hold up cards with white on top, whereas others have black on top, depending on the colour of the square they are standing on. From the helicopter, the birthday girl looks down and sees the pattern of black and white squares making the word 'HAPPY', though from your position you have no idea what the final message looks like (Fig. 3.4).

Someone then shouts out 'all change to page two'. You turn to page two, where the instruction next to mauve says 'hold the cardboard with the black side uppermost'. Some of the other people also flip their cards over, whereas others are instructed not to change. From the helicopter the girl looks down again and sees the pattern of squares change to make 'BIRTHDAY', but again you are ignorant of what she sees (Fig. 3.5).

Now apart from being thrilled, the birthday girl might wonder how the pattern of black and white squares was produced by the people on the ground. We can assume that the coloured squares everyone is standing on are hidden from her view, so she has to base her explanation just on the pattern of black and white squares that she sees. She might come up with the plausible hypothesis that everyone down below knew what the final message was to be and that they carefully planned and discussed with each other which side of the card

Fig. 3.5 Black-and-white message produced by following the second page of instructions in the football stadium.

they would show each time. This was not of course how the messages were actually produced but she has no way of knowing this.

On her next birthday she takes the helicopter ride again but this time there has been a major bungle in the stadium. The people standing in one half have mistakenly been given page two of the instructions to read first instead of page one. Half the stadium is therefore holding up cards to make 'HAPPY' whilst the other half is sending 'BIRTHDAY'. She looks down and sees a message that seems to be in two parts (Fig. 3.6).

Remembering the words displayed in the previous year, she realises what has happened: there has been a mix up with the messages. Most importantly, she now knows that her earlier hypothesis has to be wrong. If the people on the ground were really discussing with each other and planning what to send, how could they possibly produce a nonsensical two-part message, with such a clear line of distinction between the two halves? Surely, through their discussions they would have realised that the message did not make sense and would have changed it accordingly. Even a bit of neighbourly consultation might have ensured that at least each of the letters was correctly formed, avoiding a letter that was part 'P' and part 'T'. She would now realise that each individual was behaving independently of his or her neighbours and was simply holding up a black or white card according to some prescribed formula. Perhaps she would work out that there was something like a grid on the ground with each square distinguished in some way, and that the people were simply behaving according to their square, irrespective of what their neighbours were up to.

This is essentially how Stern came to his conclusion about bristle patterns.

Fig. 3.6 Message in two parts, with people to the left reading page one and those on the right reading page two. A thick line is drawn to make the boundary between the two half-messages easier to see.

Given the obvious symmetry and consistency of each mutant bristle arrangement, he expected that each pattern was generated by a change in planning and communication between the cells of the fly's back. However, when he made his gynanders that had distinct versions of the *scute* gene in different regions, he saw a pattern made up of two separate parts, as if the two versions had been cleanly cut and pasted together. Cells in each region behaved according to their version of *scute*, rather than modifying their pattern in relation to their neighbours. The *scute* gene was therefore not involved in planning and communication but was affecting the way cells responded to an underlying grid, or *prepattern* in Stern's terminology. In terms of the analogy, the prepattern corresponds to the patchwork of coloured squares, each version of the *scute* gene corresponds to people responding in a particular way to the colour of their square, and the observed bristle pattern is equivalent to the final display of black and white squares.

We can think of the fly's back as therefore containing a pattern of *hidden colours* that the *scute* gene responds to. Precisely what these hidden colours are and how a gene can respond to them will become clearer in later chapters. The important point here is that although we cannot see the colours, we can infer their existence in an abstract sense, as a way of explaining the observed two-part bristle patterns. The hidden colours do not have to be arranged as regular squares, however. You could imagine deforming a grid in various ways so that each coloured region varies in shape and size, giving a highly irregular patchwork. What matters is that there are distinctive regions in the patchwork for the *scute* gene to respond to, whatever their particular shapes and sizes might be. I have chosen colours to denote the distinctions in the patchwork but you might well imagine using numbers or letters. There are, however, some advantages to using the colour metaphor over other symbols, as will become clearer in later chapters. One is that colours completely fill the regions they refer to, whereas numbers and letters can only be understood in relation to boundaries. That is why colours are so useful on maps as a way of defining territories and countries, rather than names alone, for which you always need to know which particular outlines they relate to.

The key feature of Stern's explanation is that the prepattern, the patchwork of hidden colours, is exactly the same in each fly irrespective of the final bristle pattern. Mutations in *scute* affect the observed bristle pattern by simply changing the *response* to the prepattern. I shall summarise this by saying that each version of *scute* makes a different *interpretation* of the prepattern. At first sight this may look like a strange use of words because we normally use interpretation to denote a purely mental process. The notion of a gene making an interpretation is rather central to this book, so before going any further I need to be very clear about the sense in which I am using the word.

Stains and ink blots

The notebooks of Leonardo da Vinci are full of advice to aspiring artists on how to develop their painting technique. In a famous passage, he recommends looking at stained walls for inspiration:

> Look at walls splashed with a number of stains, or stones of various mixed colours. If you have to invent some scene, you can see there resemblances to a number of landscapes, adorned with mountains, rivers, rocks, trees, great plains, valleys and hills, in various ways. Also you can see various battles, and lively postures of strange figures, expressions on faces, costumes and an infinite number of things, which you can reduce to good integrated form. This happens on such walls and varicoloured stones, (which act) like the sound of bells, in whose pealing you can find every name and word that you can imagine.
>
> Do not despise my opinion, when I remind you that it should not be hard for you to stop sometimes and look into the stains of walls, or ashes of a fire, or clouds, or mud or like places, in which, if you consider them well, you may find really marvellous ideas.

When Leonardo looks at a series of stains and sees a landscape with a battle in progress, he is clearly making a rather individual interpretation. Someone else looking at the same set of stains might interpret them in an entirely different way. But it is this very ambiguity in the stains, their ability to be interpreted in different ways, that can lead the observer to make new discoveries. If he was advising someone on how to paint a landscape, why not look at a real landscape rather than a stained wall? Unlike a real landscape, the greater ambiguities in the wall allow for a free ranging and creative mind to see novel forms that might not be arrived at by the more obvious routes. In a limited sense the forms are already there on the wall to begin with: once the shape of a face is pointed out, it is there for all to see. Nevertheless, the act of staring at the wall and coming up with particular shapes clearly adds something, because one response is selected out of an enormous possible number. Interpretation is therefore not a neutral event, but a highly selective process.

The same principle, put to a different purpose, underlies the use of ink blots in a psychological test, described by H. Rorschach in the 1920s. In this test, a person is shown a series of ink blots and asked to interpret them in whatever way he or she chooses. Not all ink blots will do for the test: they are purposely chosen to be suggestive but ambiguous, in order to stimulate various types of interpretation. Perhaps on being shown the blot in Fig. 3.7, someone might see a butterfly or a bat, or something else.

Now the responses to particular blots are clearly not random. If they were random—bearing no relation whatever to the material being shown—we would not call them interpretations but simply random utterances. The tricky

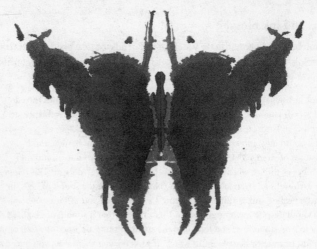

Fig. 3.7 Ink blot.

part is trying to explain what leads someone to select one particular interpretation. If a person is shown the ink blot in Fig. 3.7 and says it looks like a bat, we can be reasonably confident that he has previously seen a bat or at least a picture of a bat. But most people also know what butterflies look like, so why might someone say 'bat' instead of 'butterfly'? In some cases we might be able to come up with a reasonable answer, such as that he saw a vampire movie the previous night and bats were therefore on his mind. But in most cases it would be difficult to decide. (The Rorschach test avoids this problem because it does not depend on trying to account for particular responses. The test is based on a compilation of interpretations given by 'normal' subjects, and those given by people with various psychological problems. Diagnosis proceeds by seeing whether a subject's profile of responses falls within or outside the 'normal' range, as defined by previous compilations.)

The factors that lead to one interpretation over another can be so subtle and complex as to defy solution. The best general comment we could make is that a subject's response has to do with his or her state of mind. This in turn depends on how the subject's brain has developed and responded to previous experiences, going back to childhood and beyond. In other words, the response is *historically informed*: it depends in some way on the biological and environmental past of the individual. But this is so complicated, involving so many interactions, contingencies and chance events, that finding an explanation for one response as compared to another may effectively be impossible. This is true whether we are talking about ink blots, stains on a wall, clouds or any other highly ambiguous shape.

These examples serve to illustrate three key features of interpretations: (1) In each case we can identify a *common frame of reference* that might be shown to anyone, some stains on a wall or an ink blot. (2) Individuals respond to this common information in a *selective* way, giving one response amongst many. (3) The selections are not random but are *historically informed*, depending on the individual's biological and cultural past, usually in a very complex way.

We can now return to the question of how a gene can be said to interpret something. Like the other examples of interpretations, the various bristle patterns conferred by the *scute* gene depend on a common frame of reference, the pattern of hidden colours. This is the same in each fly, irrespective of what the final bristle pattern looks like. Based on this common underlying patchwork, each version of the *scute* gene then ensures that only one specific pattern of bristles is produced; it selects one outcome amongst many. The process therefore resembles interpretations in two key respects. I now need to explore possible parallels with the third feature of interpretations: the historical process that leads the *scute* gene to respond in one way rather than another.

The evolution of pattern

Why do organisms display particular patterns, such as a specific arrangement of bristles, rather than others? As with tracing the origin of mental interpretations, this is a complex historical problem. To illustrate this, it will be helpful to switch from bristles to a different type of pattern which is much better understood in terms of its evolution: the colour patterns of butterfly wings.

In 1848, the naturalists Henry Walter Bates and Alfred Russel Wallace left England on a joint expedition to explore the Amazon. Wallace returned after four years, but Bates stayed on for a further seven. During his long stay, Bates managed to accumulate an enormous number of biological specimens, including about 14000 species of insects. Bates was particularly struck by a curious feature of the butterflies: there was a surprisingly close resemblance in colour patterns between some distantly related species, suggesting to him that one group of butterfly species might be mimicking the other. Because some of the species were thought to be distasteful to predators, he proposed that other species might mimic them to fool predators into thinking that they were distasteful too. This is illustrated in Fig. 3.8, taken from his paper, where the bottom row shows two brightly coloured species that birds quickly learn to avoid eating as they contain noxious compounds and are therefore *unpalatable*. The butterflies in the top row look very similar superficially to those below them, but they are all *palatable* varieties of a distantly related species that birds

Fig. 3.8 Unpalatable species of butterfly (*bottom row*) and their palatable mimics (*top row*). The species shown are: *Leptalis amphione* (*top left*), *Leptalis orise* (*top right*), *Mechanitis polymnia* (*bottom left*) and *Methona psidii* (*bottom right*) (from Bates 1862).

are able to eat without ill effect. However, because they mimic the unpalatable species, these varieties will also be avoided by birds, based on their previous learning experiences. The palatable species is effectively hitching a ride on the learned response of birds to the other species (this only works effectively if the mimetic species is not too common relative to the unpalatable species: otherwise the birds would mostly encounter palatable mimics and would not learn avoidance). This phenomenon is called Batesian mimicry after its first proponent.

But why has this particular strategy evolved? Why does the palatable species go to all the trouble of mimicking the other, rather than just evolving an unpleasant taste of its own? To answer this question we have to look at why some butterflies taste unpleasant. In many cases the nasty taste comes from their having eaten noxious plant compounds during their earlier life as caterpillars. Based on this, we can come up with a plausible scenario for how unpalatability might have evolved. It starts with a particular species or group of plants that produce a noxious chemical to protect themselves from being eaten by animals. Being well protected, the plants thrive. Eventually, an insect species overcomes the plants' defence, by sequestering the poisonous chemicals out of harm's way, say in a particular gland. The insect is now free to gorge itself on a normally harmful group of plants without competition and it therefore spreads very successfully. As an added bonus, the insect is now harmful to any other animal that tries to eat it because of the stored chemical. This is one way, amongst other possibilities, in which some nasty-tasting butterflies may have evolved.

We are therefore faced with two possible evolutionary outcomes: in one case the butterfly acquires a nasty taste, in the other case it mimics the appearance of an unpalatable species. Which of these is more likely to occur?

The evolution of mimicry depends on several conditions. To begin with, there have to be heritable variations, mutations, leading to a chance resemblance between the palatable and unpalatable species that tends to fool a bird every so often. Although the initial resemblance might be very poor, natural selection acting on a series of such variations over many generations might eventually give a closer and closer match. Whether this series of events occurs depends on how different the species are to begin with and how mutations happen to affect wing patterns. There are also further issues that have to be borne in mind. For instance, in some butterfly species, male colour is thought to be used in signalling to other rival males. Thus if a mutation that changed colour pattern also reduced the ability of a male to compete, any advantage it might gain from mimicry could be outweighed by the disadvantage to mating. In other words, a change in pattern that is advantageous from one point of view will only be selected for if it does not jeopardise other important roles. Whether mimicry evolves is therefore contingent on a range of circumstances. There are many cases in which one or more of these will not be fulfilled, and for this reason it is perhaps not too surprising that mimicry has not evolved in many butterfly species, even though they are palatable.

What about the evolution of a nasty taste? In the scenario I gave, a caterpillar has to overcome a plant's defence by sequestering a noxious chemical. As with the evolution of mimicry, this would involve a series of mutations arising by chance and being acted upon by natural selection. Whether this is at all feasible in this case would depend on factors like the digestive system of the insect, the availability of a suitable site for sequestration, the insect's normal feeding pattern and how mutations happen to affect all of these processes. Only with the appropriate conditions might such a mechanism evolve.

Whether or not evolution will lead to a butterfly species being a mimic or acquiring a nasty taste (or perhaps neither of these, or even both), therefore depends on a constellation of factors concerning the species and its environment. The colour pattern we see displayed by a butterfly today is clearly not random, it is based on the past. But because evolution proceeds by modifications of what went before, the particular route it has taken will have been affected by numerous biological interactions, contingencies and chance events.

Many of the same principles apply to man-made patterns. A good illustration of this comes from Archibald Christie's book, *Pattern Design*. By the early

Middle Ages, the art of silk-weaving had achieved great perfection in many Islamic countries. It was common practice at these times for the Muslim silk-weavers to insert expressions of good-will in their fabrics, following an ancient tradition. The piece of silk shown in Fig. 3.9 includes an Arabic inscription of this type reading 'GLORY, VICTORY, AND PROSPERITY'.

As the art of silk-weaving spread westwards, other countries started to imitate the Muslim fabrics, including the various inscriptions. Being incomprehensible in medieval Europe, the written characters eventually degenerated through successive copying into meaningless scribbles (Fig. 3.10). Various stages in this degeneration can be followed in the silk fabrics themselves, as well as in representations of them in paintings or carvings. In the carving of a veil, shown in the right part of Fig. 3.10, the artist seems to have been dissatisfied with the

Fig. 3.9 Silk brocade with Arabic inscription, thirteenth or fourteenth century. Victoria and Albert Museum, London.

Fig. 3.10 Debased Arabic inscriptions. On the left is a woven silk fabric: Sicily, late thirteenth century. On the right is part of a border carved on a veil worn by a sculptured figure: France, fifteenth century.

confused pattern and tried to bring it nearer to convention by making it look more like western script. Some early writers on Gothic art spent fruitless hours trying to decipher the meaning of these strange characters, not realising that they were debased versions of Arabic.

In some cases, the instinctive desire for order and harmony led to further modifications of the letters into symmetrical patterns, known as 'mock Arabic' devices, as shown in the upper three silk fabrics in Fig. 3.11. Other developments resulted in a complete loss of any resemblance to script, and led to a distinctive type of pattern, as shown in the bottom part of Fig. 3.11.

Now if you were simply presented with some of the later versions of debased Arabic, without any knowledge of their origins, they might strike you as little more than attractive patterns. They may seem to have been devised purely on the basis of their immediate appeal. We only get a deeper understanding by tracing their idiosyncratic past, their history of migration from a culture that happened to use one form of script to a different culture that copied the patterns without comprehension of their original meaning. Through successive

Fig. 3.11 Debased Arabic inscriptions in various stages of formalisation. The top three patterns are from Italian silks of the fourteenth century. The bottom pattern comes from an initial letter in an illuminated manuscript Bible at Winchester Cathedral, England, late twelfth century.

copying, errors crept in and alterations were introduced, eventually leading to distinctive patterns. Although the later designs look quite different, they are still imbued with the influence of the original script.

Genes and interpretations

These examples of the evolution of butterfly patterns and human designs serve to show how patterns can change through a series of modifications, each step depending on what went before. The same can be said for fly bristles. The typical pattern of bristles displayed by a fruit fly depends on a complicated evolutionary trail, to which a whole range of different circumstances will have contributed. It is in this sense that we can say that the normal response of the *scute* gene to the hidden patchwork of colours is historically informed. The *scute* gene reacts in one way amongst many other possibilities because of a complex evolutionary past. We shall deal with precisely how this sort of response can evolve in a later chapter.

We can summarise by saying that the *scute* gene affects the bristle pattern by making a particular interpretation. Like other examples of interpretation, this involves making a selective and informed response to a common frame of reference, the pattern of hidden colours in the fly. As we shall see, the ability to interpret hidden colours is not just a feature of *scute*, it is a property of almost all the genes in an organism. Once appreciated, this will provide an essential key to understanding how the language of genes operates during development. But to get any further, we will first need to take a closer look at the underlying patterns of hidden colours.

CHAPTER 4

A case of mistaken identity

At three in the morning on the third of September, 1786, Johann Wolfgang Goethe jumped into a coach, assumed a false name, and set off for Italy. Goethe had just turned 37. In his youth, he had achieved great success with the publication of a tragic novel, *The Sorrows of Young Werther*. The book was so popular that a cult industry rapidly grew around it. There were Werther plays, operas and songs; even pieces of porcelain were made showing Werther scenes. In spite of his outstanding literary success, Goethe chose at the age of 26 to serve for a period in the court of Weimar, at the invitation of the Duke. At various times during the next eleven years he assumed responsibilities for the mines, the War Department, and the Finances of the Duchy. However, life in Weimar eventually proved too restrictive and by the time he was 37 Goethe felt impelled to escape incognito to a new environment.

Goethe travelled around Italy for about twenty months. During this time he developed various scientific theories concerning the weather, geology and botany. It may come as a surprise that so famous a poet should have concerned himself with science. Goethe, though, had far-ranging interests in nature. His scientific work was particularly important to him, and he dedicated much of his time to it. The aspect that most concerns us here is an important botanical idea he had during his Italian journey.

A unifying theme

To understand Goethe's idea and how he came to it we need to go back a few years to a discovery he made during his period at Weimar at the age of 34. Goethe had been struck by fundamental similarities in the structures of different organisms and became convinced that they were all formed in a common way. One of the most obvious illustrations of this was the similar arrangements of bones in the skeletons of many different animals. For instance, the human thigh bone, or femur, had an easily identified counterpart in a dog, bull, lion or any other mammal. However, although such a one-to-one

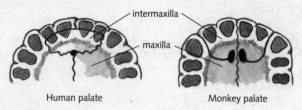

Fig. 4.1 Intermaxillary bone of human and monkey.

correspondence could be established for most bones in the body, there were some apparent exceptions. For example, monkeys had a bone in the middle of their face, called the *intermaxilla*, which appeared to be lacking in humans (this bone is also known as the premaxillary). This was often taken to be an important distinguishing mark that separated man from ape. But Goethe's belief in a fundamental unity between organisms encouraged him to look much more closely at the human skull. Eventually he discovered that the intermaxillary bone was also present in man but it had been overlooked because it was tucked away in the upper jaw and was closely joined with other bones (Fig. 4.1). His conviction in the commonality of forms had led him to discover something that others had missed. He was able to show that rather than being a distinguishing mark, the intermaxillary bone was actually a connecting link that unified man with other animals.

One piece of evidence that Goethe used to support his identification of the bone came from abnormalities. He noted that in individuals born with a cleft palate, the cleft almost always ran along the join between the proposed intermaxillary region and the surrounding bones, pointing to the intermaxillary bone as being a separate entity. He was using a rare congenital defect, the cleft palate, as a way of more clearly revealing what was normally going on. It was a type of argument he was to employ again in support of his botanical theories.

During his time at Weimar, Goethe also developed a profound interest in botany, helped by teachers from the nearby Academy at Jena. The local forests, gardens and estates provided an extensive flora on which he could practice and apply his botanical knowledge. But it was only when he went to Italy that a unifying idea about plants started to crystallise, as he explained in an autobiographical essay later in life:

> everything that has been round about us from youth, with which we are nevertheless only superficially acquainted, always seems ordinary and trivial to us, so familiar, so commonplace that we hardly give it a second thought. On the other hand, we find that new subjects, in their striking diversity, stimulate our intellects and make us realise that we are capable of pure enthusiasm; they point to something higher, something which we might be privileged to attain. This is

the real advantage of travel and each individual benefits in proportion to his nature and way of doing things. The well-known becomes new, and, linked with new phenomena, it stimulates attention, reflection and judgement.

Exposed to a new flora during his Italian journey, Goethe was stimulated to think more deeply about plants. As with his work on skulls, he was searching for a fundamental unity that lay behind the surface of things. He came to realise that there was a single underlying theme to plants, epitomised by the *leaf*. It seemed to him that the same theme occurred again and again throughout the life of every plant:

> While walking in the Public Gardens of Palermo, it came to me in a flash that in the organ of the plant which we are accustomed to call the *leaf* lies the true Proteus* who can hide or reveal himself in all vegetal forms. From first to last, the plant is nothing but leaf, which is so inseparable from the future germ that one cannot think of one without the other. Anyone who has had the experience of being confronted by an idea, pregnant with possibilities, whether he thought of it for himself or caught it from others, will know that it creates a tumult and enthusiasm in the mind, which makes one intuitively anticipate its further developments and the conclusions towards which it points.

> Knowing this, he will understand that my vision had become an obsessive passion with which I was to be occupied, if not exclusively perhaps, still for the rest of my life.

On returning to Germany, Goethe wrote up his idea in an essay, *The Metamorphosis of Plants*, published in 1790. He began by describing the typical life of a plant. After germination of the seed, a tiny shoot bearing one or two small leaves emerges from the ground. As the seedling grows, foliage leaves are successively produced, spaced out around the axis of the stem. At this stage all there is to the plant is stem and leaves (Goethe was not concerned with roots in his account). Eventually, however, the plant starts to form flowers. The question was how flowers might be related to the rest of the plant. Goethe proposed that the different parts of a flower were fundamentally *equivalent* to foliage leaves; it was just that instead of being spaced out along a stem, the parts of a flower were all clustered together.

Take a flower and look at its component parts. You will find that there are several types of organs, clustered around each other in concentric rings or *whorls* (Fig. 4.2). By whorl, I mean a region or zone of the flower that normally includes organs of one type (this is not quite the same as a botanist's definition but it will be more useful for our purposes). Many flowers have four whorls of organs. The outermost whorl comprises the *sepals*, usually small green leaf-like structures that protect the flower when it is in bud. Within these is a whorl of

*Proteus is a sea god of Greek and Roman mythology fabled to assume various shapes.

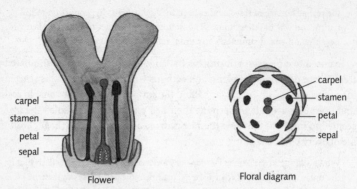

Flower

Floral diagram

Fig. 4.2 Section through a typical flower (*left*) showing organs arranged in concentric whorls of sepals, petals, stamens and carpels. The overall arrangement is also shown in diagrammatic form (*right*).

petals, usually the most obvious and attractive parts of a flower. Next come the *stamens*, the male sex organs that bear pollen. Finally, in the centre are the *carpels*, the female organs that when pollinated will grow to form fruits containing seeds. This concentric arrangement is shown diagrammatically in the right part of Fig. 4.2.

Goethe proposed that the floral organs, as well as all the foliage leaves, were simply different manifestations of a common underlying *theme*. This theme could be realised in different ways during plant growth, first as foliage leaves, then as the organs of a flower: sepals, petals, stamens and carpels. It seemed as though an underlying organ was simply passing through a series of different forms. He called this process of change *metamorphosis*, by analogy with the changes many insects experience (though unlike insect metamorphosis where the whole organism undergoes a change, Goethe's version is more abstract and refers only to parts of the organism expressing a change, the various leaf-like organs). Accordingly, above ground level, a plant was made solely of stems and a series of different types of organs based on a common theme.

In support of his claim, Goethe emphasised the many similarities between flower organs and foliage leaves. It is perhaps not too difficult to imagine that sepals are equivalent to leaves because they usually have a very leaf-like appearance. Petals are also not so different from leaves, give or take a bit of shape and colour. But what about the sex organs? The male organs (stamens) do not bear any obvious resemblance to leaves. In the case of the female organs (carpels) we sometimes get a faint leaf-like appearance when they have been fertilised and grow into fruits or pods containing seeds: a pea pod could be thought of as a leaf that has been folded lengthways and had the edges stuck together. But what about a tomato? Slice a tomato cross-wise and you will see two or more segments, each containing seeds. Is a tomato several leaf-like organs joined together? The tomato segments do not look like leaves, so it is not at all obvious that they are the same sort of thing. As with his studies on the human skull, Goethe turned to abnormalities to help resolve the issue.

Helpful monsters

Monstrous flowers are curiously attractive. For years gardeners have selected varieties with extra petals, sometimes called double-flowered forms. Roses, for example, have only five petals in the wild, yet many of the commonly cultivated garden varieties have many more than this. They have been selectively bred for their appeal to humans. In some cases, these abnormal flowers have extra petals at the expense of sex organs, so they can no longer reproduce properly by sexual means (many of them are propagated vegetatively, by taking cuttings).

Although considered attractive to gardeners, most botanists viewed these abnormalities with suspicion, as unruly freaks of nature that would not repay further study. The eighteenth-century philosopher Jean Jacques Rousseau, also a keen botanist, warned young ladies against the dangers of such flowers:

> Whenever you find them double, do not meddle with them, they are disfigured; or, if you please, dressed after our fashion: nature will no longer be found among them; she refuses to reproduce any thing from monsters thus mutilated: for if the more brilliant parts of the flower, namely the corolla [petals], be multiplied, it is at the expense of the more essential parts [sex organs], which disappear under this addition of brilliancy.

Rather than shunning these monstrosities, Goethe realised that they could provide important clues to understanding how flowers normally form. To Goethe, the monstrous flowers with extra petals in their centre suggested that the sex organs could somehow be transformed into petals. Surely this showed that the different organs of a flower were interconvertible and so fundamentally equivalent. If this conclusion was granted, then the obvious similarity between foliage leaves and at least some of the flower organs (sepals and petals) indicated that all of the organs of a plant should be lumped into the same equivalence group. The various parts of a flower were equivalent to each other and to other types of leaves; they were all variations on a common theme. As further confirmation of this idea, Goethe cited abnormal roses which, instead of sex organs, had an entire shoot emerging from their centre, bearing petals and leaves (Fig. 4.3). Here was a clear illustration of the equivalence between floral organs and leaves.

When Goethe wrote his essay on plant metamorphosis, he was not aware that

Fig. 4.3 Proliferating rose showing leafy shoot emerging from the centre.

some of the ideas had been arrived at twenty years before him, by Caspar Friedrich Wolff. Wolff was one of the founding fathers of the theory of epigenesis, the view that organisms develop by new formation rather than being preformed in the egg (Chapter 1). At the age of 26, Wolff had produced a doctoral dissertation at the University of Halle, *Theoria Generationis*, which was remarkable in its scope and insights for having such a young author. It included a range of original microscopic studies on the development of plants and animals. From his plant work, he had been struck by how various parts, such as leaves and floral organs, arise in a similar way at the growing tips of the plant (Wolff was the first to describe the plant growing tip). A few years later, in 1768, he considered this in the light of abnormal flowers:

> one observes that the stamens in the Linnaean Polyandria [species with many stamens in their flowers] are frequently transformed into petals, thereby creating double flowers, and conversely that the petals are transformed into stamens; from this fact it may be concluded that the stamens, too, are essentially leaves. In a word, mature reflection reveals that the plant, the various parts of which appear so extraordinarily different from one another at first glance, is composed exclusively of leaves and stem, inasmuch as the root is part of the stem.

Wolff had come to the same conclusion as Goethe: the various parts of a flower could be thought of as equivalent to leaves, and thus the whole plant above ground was made up of only stem and leaf-like organs. Later on, Goethe came across this work and acknowledged Wolff's precedence. Nevertheless, Goethe developed the idea of the equivalence of plant organs much more extensively than Wolff, and put it forward more coherently as a theory of plant development.

The reception of Goethe's theory was mixed. Some biologists regarded his ideas as of the utmost importance, and viewed him as a founding father of *morphology* (Goethe coined the term), the scientific study of shape and form. Others were less generous and saw Goethe's contribution as over-idealistic, trying to make nature conform to his poetic views, rather than being a serious scientific theory based on hard facts: they were the dabblings of an amateur rather than an important scientific effort. As I have mentioned, Goethe's own view was that his work on science was much more than a mere adjunct to poetry. He took his scientific studies very seriously and continued with them for the rest of his life, dedicating much of his later time to the study of optics.

One of the problems with assessing Goethe's botanical ideas has been that, until quite recently, his theory could not be followed up experimentally. He was much more concerned with giving a general intuition of how plants were formed than with laying the foundations of an experimental programme of investigation. It was only with the advent of new approaches to the study of flower development that many of his ideas have come to be appreciated again from a fresh perspective. I shall now outline some of this more recent work and eventually return to consider Goethe's contribution in the light of this.

Identity mutants

Many of the flower abnormalities of the type described by Goethe are caused by mutations in particular genes. Their significance became much clearer during the 1980s, when systematic collections of such mutants were obtained by screening many thousands of plants for exceptional individuals with abnormal flowers. The screens were mainly carried out in two species: *Arabidopsis thaliana* and the snapdragon, *Antirrhinum majus*. To show how these studies helped illuminate the nature of floral monstrosities, I need to describe three important classes of mutant that emerged from these screens, called *a*, *b* and *c*.

Remember that a flower normally has four concentric whorls of organs, which proceed from outside to inside in the order sepals, petals, stamens and carpels. In mutants of class *a*, the sepals and petals, which normally occupy the outer two whorls, are replaced by sex organs: carpels grow in place of sepals, and stamens in place of petals (Fig. 4.4). If we were to give a formula for the normal flower as *sepal, petal, stamen, carpel*, the class *a* mutant would be *carpel, stamen, stamen, carpel* (I have underlined the organs that are altered compared to normal). In other words, structures that are normally restricted to the inner regions of the flower, the stamens and carpels, have now taken over the outer

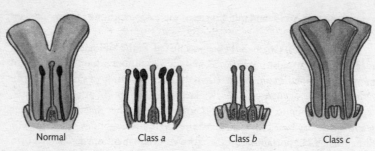

Fig. 4.4 Normal flower compared to three classes of mutant, *a*, *b* and *c*.

positions as well. I should emphasise that this does not involve any organs actually moving or changing position. Rather, the outer organs develop with an altered *identity*, as carpels and stamens rather than sepals and petals. Each organ grows and develops in the same location as in a normal flower, but the organs in the outer whorls assume the same identity as those that are normally found in the inner whorls.

The situation is somewhat reminiscent of a painting by René Magritte, showing a pair of shoes partially transformed into feet (Fig. 4.5). Shoes are worn on the outside of our feet, yet here they acquire some of the features of what is normally within them, some human skin and toes. A structure on the outside has started to assume the identity of something that is normally within. In this case, however, the transformation is purposely left incomplete, so that only part of each shoe resembles a foot. (Magritte was particularly fond of inverting the normal arrangement of things, so that many of his pictures provide good illustrations for the reverse logic of genetics, as we shall see again in later chapters.)

In mutants belonging to the next class, *b*, the identity of a different pair of organ types is affected: petals are replaced by sepals, and stamens are replaced by carpels, giving the formula *sepal, sepal, carpel, carpel* (Fig. 4.4). As with the previous class, two whorls are affected but in this case it is the pair lying between the outermost and innermost whorls, where the petals and stamens normally form.

In mutants of class *c*, the two inner whorls of the flower are affected: stamens are replaced by petals, and carpels are replaced by sepals, giving *sepal, petal, petal, sepal* (Fig.4.4). This is essentially the opposite of class *a* mutants: inner reproductive organs are now replaced by outer sterile organs. Some garden varieties with extra petals may belong to this class. In some cases, you can get numerous petals in this way because the normal flower contains many stamens, each of which is replaced by a petal. (I should mention that there are some additional complications with interpreting garden varieties. In some cases the

Fig. 4.5 *The Red Model*, René Magritte (1935). Musée nationale d'Art moderne, Paris.

transformations towards petals may not be complete, so you get only a pro-portion of the sex organs being replaced, sometimes imperfectly. This may be because the mutations have not fully inactivated the relevant gene. A further complication is that class *c* mutants can also have extra whorls within the flower, on top of the usual four, for reasons that are not yet fully understood.)

The ABC of hidden colours

What is remarkable about all these mutations is that they seem to result in almost *perfect* transformations in the type of organ made. We normally think of mutations as messing things up in some way, but here stamens, for example, appear to be replaced by perfectly formed petals. That is why roses with numerous petals in place of stamens can seem very attractive to us: their petals are still well-formed. How is it that a mutation, the inactivation of a gene, can lead to such a neat conversion?

We can get a helpful insight by considering a parallel situation in language. In many cases, if you remove a word from a sentence, the sentence will become grammatically incorrect and meaningless. As with many mutations, you end up with a mess. But there are some words that can always be removed without such ill-effects. Take the word *please*. Parents spend many hours indoctrinating children to say *please*. 'I want more juice' . . . 'What's the magic word, dear?' 'I want more juice *please*.' Both of these child's requests make perfect grammatical sense; it is just that one is considered rude and the other polite. The word *please* has a particular type of role: it provides a way of distinguishing between polite and rude sentences, rather than being essential for their grammatical structure. It is an arbitrary convention that requests are impolite unless they include the word *please*. We might say that without *please*, rudeness is assumed by *default*. The notion of a default allows us to see how the removal of a word can influence the significance of a sentence, whilst at the same time preserving its grammatical correctness. (It strikes me that our common convention is very inefficient: it would be better if the default state was polite and we should have to add extra words like *you numskull* to make a sentence rude, avoiding the needless waste of energy on teaching children to say *please* every time they ask for something.)

In a similar way, the genes affected in the mutant flowers have a special type of role that can be understood in terms of defaults. To see how this works, I need to describe a simple model that was designed to account for the three mutant classes, *a, b* and *c*. The basic elements of this model were arrived at independently by two research groups in the late 1980s: Elliot Meyerowitz, John Bowman and colleagues working on *Arabidopsis* at Caltech, California; and Rosemary Carpenter and me working on *Antirrhinum* at the John Innes Institute, Norwich. There are various ways of presenting this model, but here I want to describe it in terms of what I shall call *hidden colours*. It is important to bear in mind that these are abstract rather than real colours. Their only justification at this stage is to provide a convenient way of explaining the different types of floral mutant. We will get a more concrete understanding of what these hidden colours are in the next chapter.

According to the model, the flower can be symbolised as four concentric rings of hidden colour, corresponding to the four whorls of organs: sepals, petals, stamens, carpels (Fig. 4.6). These colours are themselves built up from a combination of three basic colours, called *a, b* and *c*. The outermost ring is

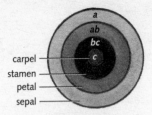

Fig. 4.6 Concentric rings of hidden colour, corresponding to four organ identities in a normal flower.

coloured a, the next ring in is coloured with the combination $a + b$, third in is $b + c$, and finally c is in the centre. These basic colours and their combinations therefore give a different colour signature to each whorl. Starting from the outer whorl and moving towards the centre, the combinations are: a, ab, bc, c, representing the identities sepal, petal, stamen, carpel respectively.

Now the key feature of the model is that if you remove one or more colours, the identity of the organs will change to a default determined by the remaining colours. Suppose, for example, that colour b is missing (Fig. 4.7). Instead of the colours being a, ab, bc, c, the flower will now have colours a, a, c, c. Since colour a alone corresponds to sepal identity, and c alone signifies carpel identity, a flower with rings a, a, c, c will have sepals in the outer two whorls and carpels in the inner two, giving the formula *sepal, sepal, carpel, carpel*. This is essentially what the mutant flowers belonging to the b class look like. The model has been expressly designed to account for the b class of mutants in terms of the loss of a particular hidden colour: b.

The a and c classes of mutants can be explained in a similar manner, through loss of their respective colours. In this case, though, there is an additional complication. To predict the correct pattern of organ identities, we must assume that the a and c colours are not completely independent but oppose each other

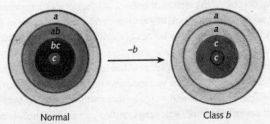

Fig. 4.7 Effect of losing the b hidden colour on organ identity.

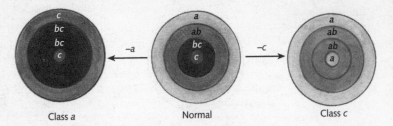

Fig. 4.8 Effect of losing the *a* hidden colour (*left*) or the *c* colour (*right*) on organ identity.

in some way. If for some reason colour *a* is missing, then the *c* colour appears in its place. Similarly, if *c* is missing, the *a* colour will substitute. Thus, in a mutant that lacks *a*, the *c* colour appears in all rings but the *b* colour is not affected, giving the colours *c, bc, bc, c* (Fig 4.8, left). This would signify a flower with the formula *carpel, stamen, stamen, carpel*, agreeing with the appearance of class *a* mutants. On the other hand, if we take *c* away, the *a* colour appears everywhere and we get *a, ab, ab, a*, signifying a flower that is *sepal, petal, petal, sepal*, as observed with class *c* mutants (Fig. 4.8, right). These rules may seem rather arbitrary, but remember that at this stage they have simply been devised to account for the appearance of the mutants. We shall return to how hidden colours can actually oppose each other in a later chapter.

The model therefore gives us a set of rules for predicting what type of organs will be made when a distinctive regional quality, symbolised by a colour, is lost. We can even predict what would happen if two hidden colours were missing. Suppose both colours *b* and *c* are absent: the flower would only be left with *a*, and because there is no *c* to oppose it, *a* will appear in all rings, predicting a flower that only consists of sepals. This is precisely what is seen when class *b* and *c* mutations are combined in the same plant.

Identity genes

So far I have described the effects of hidden colours in a rather negative sense, by showing what happens when they are removed. This is because of the reverse way in which we learn the DNA language through mutations, looking at what happens when a particular gene is defective. From a positive viewpoint, we could say that there are a specific set of genes in the plant, what I will call *organ identity genes*, that are dedicated to producing the set of *a, b* and *c* colours. The positive significance of these genes is to ensure that particular colours are made. Mutations that render one of these genes ineffective result in the loss of a colour, and so change the identity of the whorls of organs that develop.

It is important to emphasise that neither these genes nor the colours they produce represent instructions for how to construct a particular type of organ. They simply provide distinctions between regions. It might be thought, for example, that because $a + b$ results in an organ developing with the identity of a petal, then this colour combination specifies how a petal should be made. To see why this is not the case, look at Fig. 4.9, which compares flowers from *Antirrhinum* with *Arabidopsis*. The basic organisation of the two types of flower is the same: they both consist of concentric whorls of sepals, petals, stamens and carpels. This reflects a similar distribution of a, b and c hidden colours in concentric rings. Nevertheless, the structure of the various organs is quite different, allowing us to distinguish the two species quite easily. For one thing, the *Antirrhinum* organs are much larger, being about ten times the size of *Arabidopsis* in the linear dimension (for size comparison, see the tiny *Arabidopsis* flower within the circle in Fig. 4.9). But even adjusting for size, the organs obviously have a different structure. The five petals of an *Antirrhinum* flower are united together for part of their length to form a tube. At the end of

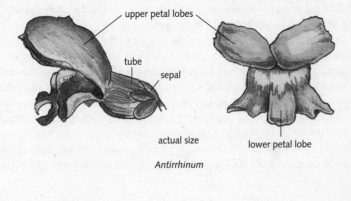

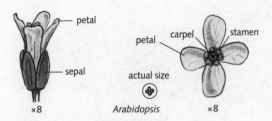

Fig. 4.9 Comparison of *Antirrhinum* and *Arabidopsis* flowers, each shown in side and face view. For size comparison, look at the smaller *Arabidopsis* flower inset within the circle, which is drawn to the same scale as the *Antirrhinum* flower (natural size).

the tube, the petals are more separate, forming five lobes, the lower ones providing a platform for bees to land on and prise open the flower, as shown in the side view of Fig. 4.9. In contrast, the petals of *Arabidopsis* are more spoon-shaped and are entirely separate from each other. Together, they form a symmetrical cross (hence the name Cruciferae, for the family of plants this species belongs to). Similar comparisons could be made for the sepals, stamens and carpels: in each case there are numerous differences in anatomy and shape that distinguish corresponding organs of *Antirrhinum* from *Arabidopsis*. So even though the identity of the organs in both species depends on a similar set of hidden colours, the detailed structure of the organs is different.

The point is that if the *a*, *b* and *c* hidden colours were giving precise instructions on how to make each type of organ, the organs should be identical in both species. If the details of how to make a petal were specified by the *a* + *b* combination, a petal of *Antirrhinum* should look the same as one from *Arabidopsis*. Clearly the colours are not giving instructions of this sort. They merely provide a distinction between different regions, allowing organs with separate identities to develop. It is as if the colours provide a common underlying pattern, but how this becomes manifested in the final organs of a flower can vary greatly according to the species.

In some cases, this variety of forms may go so far as to contradict some familiar notions. We normally think of petals as being the largest and most attractive organs of the flower. Yet in some species, this is a feature of the outer whorl of organs, the sepals rather than the petals. In flowers of the genus *Hydrangea*, for example, the sepals are sometimes much more conspicuous than the petals, so the colourful display we enjoy in garden varieties is almost entirely due to the sepals (Fig. 4.10). Although the relevant genes from these species have yet to be studied, it is reasonable to suppose that they will have a comparable set of *a*, *b* and *c* hidden colours to those in *Antirrhinum* or *Arabidopsis*.

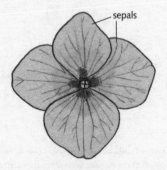

Fig. 4.10 Flower of *Hydrangea* with large showy sepals.

It is just that in the case of *Hydrangea*, this pattern of colours becomes manifested in a different way.

In the previous chapter, we also came across the notion of hidden colours that can be variously manifested. Recall that they were invoked as a way of accounting for the pattern of bristles in a fly. The fly was assumed to contain a patchwork of hidden colours, a set of regional differences that could not be seen. Although normally invisible, the patchwork could nevertheless be inferred from the way it was interpreted, resulting in one particular bristle pattern being displayed out of many other possibilities; just as someone flying over a football stadium and seeing various messages might infer that the people down below were responding to a grid of colours on the ground. The hidden colours that make up this patchwork are similar in kind to those I have been describing in this chapter as *a*, *b* and *c*. It is just that in the flower example, the patchwork has the configuration of four concentric rings.

I have been approaching hidden colours from two different viewpoints. In the previous chapter, a pattern of hidden colours was inferred from mutations that changed the way they were interpreted by a particular gene (*scute*). The hidden colours were left unchanged by the mutations; all that was affected was their interpretation. In this chapter, it is the hidden colours themselves that are proposed to change in order to account for the appearance of various mutants. In this case, the mutations are affecting a specific set of genes, the organ identity genes, needed for producing the *a*, *b* and *c* hidden colours. Each mutation results in the loss of a colour, changing the identity of the organs in some way. So in one case the mutations affect the *interpretation* of the hidden colours; whereas in the other case the mutations influence the *production* of the hidden colours. In both cases, the hidden colours provide a frame of reference that can be variously interpreted, eventually becoming manifested in the visible structure of the organism that develops; the colours are not providing a set of instructions for how to construct a particular organ or a bristle pattern. I realise that all this discussion of hidden colours and their interpretation may sound rather abstract at this stage, but I would ask the reader to be patient and eventually their meaning and utility for understanding development will become clear.

A change in outlook

Looking back on Goethe's views from our present perspective, we can see that many of his ideas turned out to be penetrating. The idea that the different organs of a plant might be variations on a theme has a modern resonance with the various hidden colours that confer distinct organ identities (I have been mainly concerned with the variations but will return to the nature of the theme in chapter 15). In my view, though, Goethe's greatest insight was his clear perception of how the study of abnormalities, what we now call mutants, could be used to understand the normal course of development. As he stated in his essay on plant metamorphosis:

> From our acquaintance with this *abnormal* metamorphosis, we are enabled to unveil the secrets that *normal* metamorphosis conceals from us, and to see distinctly what, from the regular course of development, we can only infer. And it is by this procedure that we hope to achieve most surely the end which we have in view. [my italics]

He clearly saw that this reverse form of logic, arguing from the abnormal to the normal, was a valid and important way to proceed in unravelling development. Perhaps it was Goethe's breadth of mind, his desire to understand the underlying unity of nature without too much concern for experimental details, that led him to this remarkable insight. I do not wish to imply that everything Goethe said about plants was gospel. Some of his ideas, like his notion that organs change in appearance due to a sap being gradually purified as plants develop, are of little modern significance. But his clear appreciation of the significance of abnormalities was certainly ahead of its time.

Goethe's perspective only came to experimental fruition in the twentieth century, as mutations affecting development started to be investigated in detail. The unravelling of the *abc* model is a good example of how the outlook underwent a change. Given its basic simplicity, it seems quite remarkable that the *abc* model for flower development was only proposed in the late 1980s, even though the experimental approach that lay behind it, the production and classification of mutants, had been well established for many decades before this. The advance had more to do with a change in the way that flowers were being looked at than in the development of new technology. I remember, when we had first obtained one of the class *a* mutants (*carpel, stamen, stamen, carpel*), going home in the evening after having spent some time looking at its flowers. It was clear that the outer whorl of sepals had been replaced by female organs, but it was less obvious what had happened to the next whorl, where petals normally form. It seemed that these organs were narrow and strap-like with abnormal structures at the ends. As I considered various models at home, it occurred to me that if the strange strap-like structures were due to a transformation of petals towards male organs, the stamens, a simple model could account for the various classes of mutant we knew about. The next morning, I rushed into the greenhouse to look at the mutant flowers again. To my delight the strap-like organs did indeed have some tell-tale features of stamens that I had overlooked the previous day. Later on we obtained some much clearer examples of this type of mutation where there could be little doubt that stamens had replaced petals, but the earlier anticipation of the result has remained with me as a striking example of how observations and descriptions are influenced by what you are looking for. In the 1980s we had started to look at flowers in a different way. At the back of our mind we had the notion that genes might act in combination to confer distinctions in identity. And one of the most important contributions to this new outlook on flowers came from studies on a quite different organism: the fruit fly, *Drosophila melanogaster*.

Segment identity in flies

About one hundred years after Goethe wrote his treatise on plants, the zoologist William Bateson described a comparable set of abnormalities in the animal world. Bateson was convinced that the only way to understand evolution was by studying how biological forms vary, and he therefore set about cataloguing the principle types of variation in his book *Materials for the Study of Variation* of 1894 (Bateson was later to become one of the founders of the science of *genetics*, a term he coined). Among the variations he described were some striking abnormalities in insects and crustaceans. Like plants, these animals are divided into parts that seem to be fundamentally equivalent. Look at a shrimp or a fly and you can easily see that they are made up of repeating units or segments. Almost the entire body of these animals seems to be based on segments that can be modified in various ways: some bear legs, some can have wings, while others have no appendages. Even the head, which bears antennae, eyes and various mouthparts, can be considered as being made up of segments.

Bateson came across abnormalities in which part of a segment seemed to be converted or transformed into something typical of a quite different segment: an insect with a leg at a position normally occupied by an antenna, or a crab with an antenna instead of an eye. It seemed that one sort of appendage was replacing another. What was surprising in all of these cases, just like in those of the flower, was that you didn't end up with a complete mess. The extra antennae or legs seemed to be remarkably normal even though they were growing in the wrong place. Bateson realised that these abnormalities were of great significance: 'Facts of this kind, so common in flowering plants, but in their higher manifestations so rare in animals, hold a place in the study of Variation comparable perhaps with that which the phenomena of the prism held in the study of the nature of Light.' He coined a special term, *homeosis*, for this particular type of variation, in which one member of a repeating series assumes features that are normally associated with a different member.

Yet, in spite of Bateson's prophetic words, the real significance of these *homeotic* mutants was overlooked for a very long time: they were treated as no more than curiosities. Strangely enough, it was their dramatic effects that discounted them in most people's eyes. Most of us have had the experience of desperately kicking or banging a machine after it has refused to perform properly. Occasionally, such acts of desperation can jolt the machine into working again but, although we may be pleased with the outcome, we do not imagine that we understand the machine any better after this. If anything, it seems even more mysterious to us when it responds to acts of frustration. This was the early view of homeotic mutations: that they were genetic jolts that caused a major change in development but were not themselves informative about the underlying mechanisms. Development was thought to be so complicated and subtle that major flips of this type were unlikely to be revealing. It was only during the 1960s and 70s, with the work of Ed Lewis at Caltech on segments in fruit flies, that the central importance of these homeotic changes began to emerge.

Before explaining the results Lewis obtained, I need briefly to describe a normal fruit fly. We can consider the main body of the fruit fly as being made of 14 segments, numbered starting from the head end (Fig. 4.11). The *head* contains three segments (0, 1 and 2). The next region of the body, called the *thorax*, comprises another three segments (3, 4 and 5), each of which carries a pair of legs. Segment 4 also bears a pair of wings whilst segment 5 has a pair of small appendages, called *halteres*, which are thought to help balance the fly during flight. The rest of the fly, the *abdomen*, is made of eight segments (6–13).

In the 1940s, Lewis had been searching for mutants that might be useful for studying the structure of genes, and he came across a mutant called *bithorax*. In the *bithorax* mutant, the halteres, the tiny balancing organs on segment 5, seemed to have been partly transformed into wings. It was as if segment 5 had become more like segment 4, where wings normally develop. As Lewis worked intensively on *bithorax* mutations, he started to realise that he was dealing with

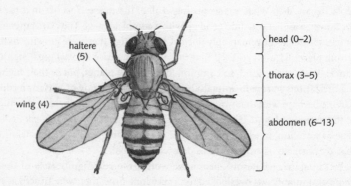

haltere (5)

wing (4)

head (0–2)

thorax (3–5)

abdomen (6–13)

Fig. 4.11 Adult fruit fly, with segment numbers indicated in brackets.

a complex of several genes, arranged next to each other along the DNA chromosome. Every gene corresponds to a small stretch of DNA, so in this case Lewis was dealing with several of these short stretches, lined up one after the other. He called this cluster of genes the *Bithorax Complex*.

Lewis discovered that many of the mutations in the Bithorax Complex affected only part of the haltere. The original *bithorax* mutations, for example, appeared to transform only the front half of the haltere into the corresponding half of the wing (Fig. 4.12, middle specimen). Another type of mutation gave the complementary result: only the rear half of the haltere was transformed into rear wing (Fig. 4.12, right specimen). As he recalled: 'You see, *bithorax* mutants were said to transform the haltere into wing. That was wrong. They only transformed the anterior part. So it was *very* exciting to find a mutant that did the complementary thing.' It was as if you could transform different parts of the haltere separately.

He then wanted to see whether the effects of the two sorts of mutation could be added together, to simultaneously transform both the front and rear halves of the haltere into wing. To test this, he crossed flies with the different types of mutation with each other, and eventually managed to combine both mutations in the same individual fly. The two types of transformations did indeed add up precisely, to give halteres entirely replaced by wings, resulting in a four-winged fly. This was a key advance because it told him that these mutations could not simply be ignorant kicks of the system. Why would one type of jolt consistently transform one half of the haltere whereas another type of jolt would always transform the other half? And why should they add up so nicely when they were combined to give a complete and perfect transformation? Lewis realised that these mutations were giving important clues as to how flies *normally* develop. His interests started to shift from studying the mutations in themselves to

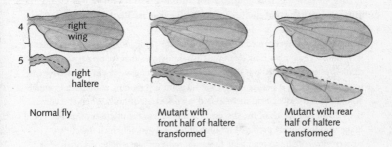

Fig. 4.12 Effects of two types of mutation in the Bithorax Complex, replacing the front half (*middle specimen*) or rear half (*right specimen*) of the haltere with the corresponding part of the wing. Only the right side of the fly is shown in each case.

thinking about what they might reveal about the normal pathway of development. Somehow the genes in the Bithorax Complex were critical for normal flies to establish separate identities for segments 4 and 5. This distinctive information was being lost in the mutants, so these segments became more similar to each other.

For a long time Lewis continued to collect and study many developmental mutations in different parts of the Bithorax Complex, each affecting the adult fly in different ways. Then, in the 1970s, he managed to get a mutation that removed a large chunk of DNA containing all of the genes in the Bithorax Complex in one go. How would a fly develop if it had *none* of the genes in the Bithorax Complex? The result was that the mutant died at a very early stage, just about the time that the larva was hatching from the egg case. To understand what was going on, Lewis therefore had to look at the early larvae rather than adult flies.

As with many insects, fruit flies spend their early feeding life as larvae, small grubs that eventually pupate and metamorphose into the adult flies. Larvae can also be divided into 14 segments, corresponding to the adult ones that will form later, but there are no appendages such as wings, legs or halteres to distinguish them. For that reason, larvae were considered to be rather uninformative, and nobody had taken much of an interest in looking at early larval stages of development. When Lewis saw that his deletion mutant died at such an early stage, he was forced to develop a method for cleaning and preparing the young larvae or embryos. A key step was treating these embryonic larvae with a chemical, lactic acid, that made it much easier to see their outer surface or skin, called the *cuticle*. Using this method, Lewis quickly saw how each segment on a larva could be distinguished from the others by its characteristic pattern of tiny thorn-like outgrowths, called *denticles*, on its outer surface. Look at the larva on the left in Fig. 4.13 which shows most segments of a normal individual identified in this way, according to their specific patterns of denticles.

Having developed this method for looking at larvae, Lewis could then examine the mutant that died early on. As he later recalled:

> Being lethal, we were forced to find out what the thing looked like. As a matter of fact, I don't remember how clean the first animal was that we prepared . . . I think it actually crawled out of the egg. Then I decided we've got to improve this method and make it permanent, it was no good just looking at one dead animal. So then I used a lactic acid method and it worked beautifully. When you come right down to it much of the success of bithorax was the discovery that you could make simple, quickly prepared mounts of the embryo, in which the cuticle pattern allowed you to read gene function, right off like a book. So often it's a little technique that allows you to make a big jump and the more I think about it, that was what really helped because we could study any mutation by putting a larva in a drop of lactic acid and alcohol.

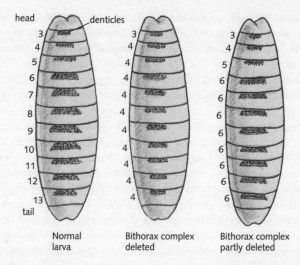

Fig. 4.13 Larvae from normal fly (*left*), mutant with all genes in the Bithorax Complex deleted (*middle*), mutant with part of the Bithorax Complex deleted (*right*).

By looking at the pattern of larval segments in his mutant with the entire Bithorax Complex deleted, Lewis made a remarkable discovery. The segments from 5–13 were no longer distinct from each other: they all resembled segment 4. In other words, the larva was normal at the head end (segments 0–4), but then it had nine rear segments that all looked like segment 4 (Fig. 4. 13, middle specimen). Remember that in the mutants I previously mentioned, it was only segment 5, the region that bears the halteres, that was transformed to resemble segment 4. But in the case of this new mutant, *all* of the segments from 5 onwards had assumed the same identity as segment 4. If such a larva could have survived to adulthood, it would have turned into a fly with 22 legs, the same number as an entire soccer team (it would also have had 20 wings).

Lewis went on to produce further deletions, removing only some of the genes in the Bithorax Complex. These gave similar types of result. For example, in one case the head and middle segments (0–6) were normal, but then the remaining segments behind this all looked like segment 6 (Fig. 4.13, right specimen). A general pattern was emerging from these mutants. The larvae might be normal from the head up to a certain point, segment 6 in this case, but then the segments behind this would simply reiterate the same identity.

To explain his results, Lewis proposed that the normal role of the Bithorax Complex was to produce a set of distinct 'substances' in each segment. The combination of substances could lead to each segment developing with a particular identity. The model can get rather complicated, so I want to present a simplified version in terms of hidden colours that nevertheless captures the essential features.

In my simplified model, the fly has only four segments, with distinct identities 1–4 (Fig. 4.14). The identities depend on a set of three overlapping basic hidden colours, denoted as *f, g* and *h*, distributed so that there are progressively more colours as you move towards the tail end of the normal fly: *f* is in three segments (2, 3, 4), *g* is in two segments (3, 4) and *h* is only in segment 4 (Fig. 4.14, left). The identity of a segment depends on the combination of hidden colours it contains. So if we remove the rearmost colour *h*, the identities of the first three segments would be unchanged but the fourth segment would now only have *f* and *g*, giving it identity 3 (mutant *h*, Fig. 4.14). Removing *g* as well as *h* would leave only *f* in three segments, giving them all identity 2 (mutant *hg*). Removing all three colours would give all segments identity 1 (mutant *hgf*). In other words, with this model we can account for the various mutants Lewis observed that were normal up to one point, but then reiterated the same type of segment, by proposing the loss of one or more hidden colours.

This is essentially how the genes in the Bithorax Complex confer distinctions between segments. We can say that the Bithorax Complex contains a set of *segment identity genes*. Each of these genes is needed for a particular hidden colour, and the combination of colours determines the particular identity of the segment. The normal role of these genes is therefore to generate hidden colours in their respective regions. Mutations that inactivate or remove one or more of the segment identity genes result in a loss of colour, and hence a change in some of the identities. (You will notice that in my simplified model I only mentioned three hidden colours, *f, g* and *h*, to distinguish four segments. In the real fly, the genes in the Bithorax Complex have to discriminate between the nine segments from 5–13. You might expect, as Lewis did, that this would involve eight separate substances or colours. However, it appears that only three

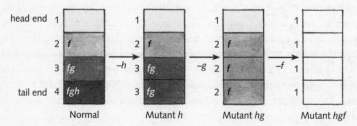

Fig. 4.14 Simplified model for conferring four segment identities (1–4) with a combination of three basic hidden colours, *f, g* and *h*.

major colours are involved as some of the differences between segments 5–13 are due to variation in the intensity of colour as well as the type of colour.)

Although the genes in the Bithorax Complex confer distinctions between segments 5–13, we still have to explain how the segments nearer to the head (0–4) get their distinctive identities. It turns out that there is another cluster of segment identity genes, called the *Antennapedia Complex* (Antennapedia is so named because a mutation in one of its genes results in legs growing in place of antennae). These genes act in a similar way to those of the Bithorax Complex, providing a further set of five hidden colours (which we can call *a, b, c, d* and *e*) that distinguish the segments at the head end (the role of the *a* and *b* colours is more restricted than the others). From head to tail, the distinctive segment identities in a larva, and the fly that develops from it, therefore depend on the combined action of eight hidden colours *a–h*. The first five, *a–e*, depend on identity genes in the Antennapedia Complex, whereas the last three, *f–h*, depend on those in the Bithorax Complex.

As with previous examples, these hidden colours are not instructions on how to make a particular type of structure, such as a segment bearing legs and wings; they simply provide distinctions between regions. For instance, a similar set of hidden colours to those in fruit flies has now been documented in many different types of insect, including beetle and butterfly species. Even though these species look very different outwardly, they have a common set of underlying hidden colours that provide distinctions between their segments from head to tail. Many of the obvious differences between these species most likely have to do with the variety of ways in which this common pattern can be interpreted and becomes visibly manifest. The hidden colours do not represent detailed instructions, they just provide a frame of reference.

Flowers and flies

Monstrosities in flowers and flies, in which parts of the organism appear to adopt mistaken identities, can be explained in very similar ways. In each case we can infer a pattern of hidden colours that provide distinctions between different regions of the organism, formalised as concentric rings in flowers or a linear array in flies. These regional differences in colour can be interpreted to give organs or segments with distinct features. The colours themselves depend on a particular set of genes, called organ or segment identity genes. Take away or mutate one of these genes and colours are lost from the pattern, changing the identities of some of the structures that develop.

I have chosen these examples from plants and animals to show how hidden colours depend on genes. In the previous chapter I showed how genes are also involved in interpreting these colours, responding to them in an informed way.

I now need to show how these two aspects can be brought together: how hidden colours are produced and how genes are able to interpret them. To answer this satisfactorily, we have to know more about the molecular properties of the genes and proteins involved. We have to turn from what these genes signify to how they are used.

CHAPTER 5

The internal world of colour

So far I have described how organisms can be thought of as containing a patchwork of hidden colours: a frame of reference that can be interpreted by genes. In this chapter I want to reveal these hidden colours more directly and show precisely how genes manage to interpret them. Because I am using colours in a more abstract sense than usual, I want to begin by looking more closely at our notion of colour and what we might mean by revealing an invisible colour. To help do this, we shall first look at colours from a different perspective, a bee's viewpoint. This will provide a useful stepping stone for approaching the more abstract hidden colours that underlie the development of organisms.

A bee's eye view

Bees do not see the world in the same way as we do. For one thing, they respond to a different range of colours. In experiments in which bees were trained to visit a glass of sugar water lit with a particular colour, they proved to be blind to red but could distinguish other colours such as blue, green or yellow. Bees make up for this deficit at the red end of the spectrum by being sensitive to a colour at the other end that is invisible to us: *ultraviolet*. As its name suggests, ultraviolet is a form of light that lies beyond violet in the spectrum. A rainbow viewed by a bee would have about the same overall width as one viewed by us, but would be missing red from one end and have ultraviolet tagged on at the other. This difference between bees and humans has to do with how cells in their eyes detect light. Humans have three types of colour-sensitive cells at the back of their eyes, each of which is most sensitive to one region of the spectrum: red, green or blue. The combination of these three primary colour sensors allows our brain to distinguish the full range of colours in the visible spectrum. Bees also have three types of colour-sensitive cells in their eyes. As in humans, two are most sensitive to blue and green, but the third is sensitive to ultraviolet rather than red.

This means that when a bee looks at a garden full of flowers it detects a different range of colours from us. To a bee, the red of a poppy is invisible. But poppies may nevertheless stand out because they reflect ultraviolet light, allowing the bee to detect a distinctive colour that we are oblivious to. One way of getting a better idea of what a bee can see is to take photographs using an ultraviolet camera (a camera fitted with a filter that only passes ultraviolet light, and loaded with a film that is sensitive to ultraviolet). Figure 5.1 shows several flowers photographed in this way. The pale areas are those that reflect ultraviolet light whereas the black areas do not. This is because the black areas contain specific molecules that absorb ultraviolet light. These striking patterns of contrasting ultraviolet colour are thought to provide visual guide marks to the bee, showing it the way to nectar and pollen within the flower. The ultraviolet camera reveals a world of colour patterns that is normally invisible to us but provides important cues to a bee.

Apart from the difference in colour sensitivity, there is a more profound sense in which bees do not see the world as we do. Our image of the world depends on the way our brains process visual information. The colours and patterns we experience when looking around depend on a very complex series of mental events. Defining a particular colour *experience* is therefore much more difficult than distinguishing a colour. Most people can distinguish the colour red, but it is much more difficult to say what the mental experience of redness is. We might agree that grass is green but I cannot tell for sure that your experience of green is the same as mine because I have no objective way of defining what a green experience is. Nevertheless, because our brains are structured in a similar way it is reasonable to assume that two people's experience of particular colours is comparable (assuming that neither is colour-blind). What seems reasonable when comparing experiences between two humans becomes unreasonable when comparing humans with bees, species with very different sizes and types of brain. Although both bees and humans can distinguish yellow from blue, the bee's experience of each colour is likely to be very different from ours. We do not know whether a colour sensation for a bee is anything like what we feel when experiencing a colour. We know that bees can see colours, because they

Fig. 5.1 Ultraviolet marks of flowers. (1) Golden cinquefoil (*Potentilla aurea*). (2) Marsh felwort (*Swertia perennis*). (3) White bryony (*Bryonia dioica*).

can be trained to visit particular colours and have the visual apparatus to detect and discriminate between them, but this does not mean that their experience of colour resembles ours.

The ultraviolet flower patterns a bee sees are twice removed from us. First, because they display a colour that lies outside our range of detection and is therefore unimaginable to us—we cannot imagine an extra colour on top of our normal visible range, because we can only picture the colours we are familiar with. Secondly, even if our eyes could detect ultraviolet light by having an extra type of visual cell that was sensitive to it, we would still not know how this would be related to a bee's experience of ultraviolet. When we look at pictures of the flower patterns in Fig. 5.1, we do not see what the bee sees. No illustration can show this to us. What the pictures reveal is more limited: a set of regions in the flower that absorb ultraviolet light to different degrees, which the bee can respond to because of its sensitivity to this part of the spectrum. Nevertheless, we can use the patterns as a way of understanding the underlying colours of flowers and how bees respond to them, without trying to enter into the mind of a bee. To account for the behaviour of bees it is still legitimate for us to talk about them seeing ultraviolet colours, as long as we are aware that we are using colour in a more abstract sense than usual.

The abstract nature of these colours can be clearly appreciated by imagining the following scenario. Suppose a naturalist spends many hours investigating bee behaviour but has no notion of ultraviolet light. From extensive studies on how bees behave on various flowers, the naturalist might come to the conclusion that there are some hidden marks on the flowers to which bees can respond. The naturalist might even propose a theory that many flowers exhibit 'invisible colours' that bees can see but that we are oblivious to. Now such a notion would sound very abstract, or even fanciful, to anyone without a knowledge of ultraviolet light and the mechanisms of bee vision. They would wonder what these mysterious invisible colours might be and how bees could possibly respond to them. The mystery would start to disappear, however, if someone came along with an ultraviolet camera and actually revealed the hidden patterns. And once the process of bee vision is explained in terms of ultraviolet-sensitive cells, the notion of an invisible colour that bees can respond to makes perfect sense. An abstract concept, such as an invisible colour, can be rendered quite comprehensible once we know something about the underlying mechanisms involved.

There is a parallel between this scenario and the study of development. In the previous chapters, we saw how hidden colours were proposed in order to account for the properties of particular mutants with altered patterns or identities. These hidden colours were thought to provide distinctions between regions which were responded to or interpreted by genes. The hidden colours had a comparable status to the invisible colours inferred by the naturalist studying bee behaviour. In the case of development, the distinctive territories of hidden colour were responded to by genes; for the naturalist, the regions of invisible colour were responded to by bees. To get a more concrete understanding of what is going on in development, I therefore need to describe the equivalent of an ultraviolet camera: a method that directly shows where the hidden colours are located. For each hidden colour, we would like to have an image analogous to the ultraviolet flower patterns, showing regions of the organism that have greater or lesser amounts of a particular hidden colour. Furthermore, we need to know what these hidden colours are and how genes can actually respond to them.

To address these issues in a satisfactory way, I shall need to go into some of the molecular properties of the genes and proteins involved. This is the aim of this chapter, and by the end of it we should have established a firm foundation for pursuing many of the fundamental issues concerning development that we shall encounter later on.

Probing the interior

I will begin by defining two states for a gene: *on* and *off*. Remember that a gene is a stretch of DNA that typically includes a region coding for a protein. Essentially every cell in an organism carries the same set of genes in its DNA. For a protein to be made, the coding region of a gene is first transcribed into RNA copies which are then translated to make a protein of a particular type: information flows from DNA to RNA to protein. So long as this process of making RNA and protein from a gene is happening in a cell, the gene can be said to be in the *on* or *expressed* state. When a gene is in the *on* state, the corresponding RNA and protein molecules will therefore accumulate in a cell. In some cells, however, the transcription of a gene's coding region into RNA may not occur, so that no RNA or protein molecules from that gene are made. The gene is still there in the nucleus of the cell, but in this case it is said to be in a *silent* or *off* state. Whether a gene is *on* or *off* in a cell is not a matter of chance but is a highly regulated affair. At any time during the development of an organism, a gene may be *on* in some cells and the same gene *off* in others. There may be cells in some parts of the organism, say those in the liver, in which a particular gene is being expressed, and cells in other parts, for example those in the brain, in which that same gene is silent. We can refer to this regional pattern of gene activity as the *expression pattern* of the gene. This is a central concept that I shall refer to many times, so it is important to remember that the expression pattern of a gene refers to where that gene is *on* or *off* in the organism at any given time.

The easiest way to determine whether a gene is *on* or *off* in a cell is to find out whether the particular RNA or protein molecules derived from the gene are present in the cell or not. If the gene is *on*, the corresponding RNA and protein should be present in the cell; whereas if it is *off*, these products should be absent. Identifying the products of a specific gene is rather a daunting task because there are tens of thousands of genes in the DNA of every cell, each coding for a different protein. How are we to distinguish the RNA or protein products of one specific gene from those derived from all the other genes that are in the cell?

The problem can be overcome by using a very specific molecular *probe* that can distinguish the products of one gene from all the others; it provides the equivalent of a specific light filter on a camera, separating one contribution from everything else. Two types of probes can be used: those that recognise particular RNA molecules and those that recognise particular proteins. They both work by matching the shape of RNA or protein molecules in the cell. The molecular shape of the probe is complementary to that of the RNA or protein it is designed to detect, and therefore only sticks to its matching counterpart, distinguishing it from all others—much as a shoe was used to identify Cinderella. By incubating a slice of tissue with a probe under the right conditions, the probe will only stick to those cells which contain the matching RNA or protein. The location of the probe can then be revealed because it is designed in such a way that it can be visually detected wherever it goes.

A probe effectively provides a very specific visual *stain*, revealing cells that contain a particular type of RNA or protein. We can compare this to revealing a message written on some paper with invisible ink. By treating the piece of paper with a special chemical (the probe) which sticks or reacts with the invisible ink, the message becomes visible. The chemical has stained the paper only where the invisible ink was located. You might have various types of invisible ink, each requiring a different chemical to reveal it. In the same way, each gene needs a different matching probe to detect its RNA or protein products. This means that for every gene, you need a different probe to stain for its expression pattern.

The trickiest bit in this whole procedure is getting the probe. Once you have a probe that can detect a particular type of RNA or protein molecule, the rest is cookery: mixing things together under the right conditions. A key advance in obtaining probes came with the advent of gene cloning in the 1970s and 80s. Once a particular gene has been *isolated* (the specific stretch of DNA that includes the gene has been multiplied in bacteria) it is not too difficult to make a probe that can detect the RNA or protein products from that gene. There is no need to get bogged down in the technicalities of how this is done, but to avoid these probes seeming too mysterious, I should mention that probes are molecules you are already familiar with: RNA or protein. Probes for detecting RNA are themselves RNA molecules (the sequence of bases in the probe is complementary to that in the RNA being detected). Similarly, probes for revealing proteins are themselves proteins (they are generated by injecting a rat or rabbit with a protein and waiting for the animal's immune system to produce an antibody—a protein that matches the one injected).

Expression of identity genes

We are now in a position to see how the previously inferred hidden colours can be revealed. In the previous chapter I described how hidden colours can depend on a set of *identity genes*. In the case of a flower, these genes were needed to produce concentric rings of hidden colour, resulting in four distinct identities of floral organs: sepals, petals, stamens or carpels. In the case of a fly, identity genes were needed for a set of colours that varied from head to tail, resulting in separate identities for the various segments. There were eight different colours in the fly, *a–h*, ordered alphabetically according to the segments they affect: *a* being nearest the head and *h* nearest the tail. The identity genes were recognised by virtue of mutations in them that gave organs or segments with mistaken or inappropriate identities, accounted for by a loss in one or more hidden colours. In all these cases, the hidden colours were inferred indirectly as a way of accounting for these various types of mutation, much as a naturalist might have inferred the existence of invisible colours by closely watching bee behaviour.

To reveal the hidden colours more directly, we will need to look at the *expression pattern of the identity genes* (remember that the expression pattern of a gene means where that gene is *on* or *off* in the organism at any given time). This means we need to have probes that can detect the expression of each identity gene. As I have mentioned, to make these probes you need to have isolated the relevant genes. This was first achieved for the identity genes affecting fruit fly segments in the early 1980s by three groups: Welcome Bender, Pierre Spierer and David Hogness and colleagues at Stanford, California; Matthew Scott, Thomas Kaufman and colleagues in Bloomington, Indiana; and Rick Garber, Atsushi Kuroiwa and Walter Gehring in Basel, Switzerland.

This was a major breakthrough because once these groups had isolated the identity genes, it became possible to make probes that detected where the genes were expressed during the development of a fly. The results showed that many of the identity genes were expressed at a very early stage in development: a few hours after egg laying, when the fly embryo is an elongated ball of cells about one-half of a millimetre long. Each probe stained a region of the embryo, showing where the identity gene was being expressed, where it was *on*.

Most importantly, the stained region corresponded to where the relevant hidden colour had previously been inferred to be. For example, based on the effect of mutations that remove colour *e*, this colour had been inferred to be in segments 4 and 5. When Michael Levine and Ernst Hafen in Walter Gehring's laboratory probed embryos to reveal the expression of the identity gene needed for colour *e*, staining was observed only in a stripe or belt around the embryo, where segments 4 and 5 would later develop (Fig. 5.2, left). In other words the pattern of gene expression mirrored that of the previously inferred hidden colour, as if the hidden colour was being directly revealed in front of their eyes! It would be like our naturalist, who had inferred the location of invisible colours purely based on bee behaviour, suddenly being shown a picture of a flower taken with an ultraviolet camera: directly revealing the distribution of the colours that had been the subject of his or her imagination for so long.

Similar results were obtained with identity genes for other hidden colours. Colour *f*, for example, had been inferred to start slightly further back than *e*, extending from segments 5–13; and sure enough, when Michael Akam and Alfonso Martinez-Arias in Cambridge stained for the gene needed for colour *f*, it was found to be expressed in these segments (Fig. 5.2, middle). The gene for the next colour in the alphabet, *g*, was found to be expressed further back still, in segments 7–13 (Fig. 5.2, right). As the genes needed for each of the eight hidden colours *a–h* were studied in this way, their staining pattern revealed a set of regions or zones in the fly that progressed from head to tail, with *a* nearest the head end and *h* towards the tail. It was as if a hidden map of colours was actually being seen in the flesh.

A similar result was obtained for the genes affecting flower organ identity. In 1990, some of these genes were isolated by Hans Sommer, Zsuzsanna Schwarz-Sommer and colleagues working on the *b* colour of *Antirrhinum* in Cologne; and by Martin Yanofsky, Elliot Meyerowitz and colleagues working on *c* from *Arabidopsis* at Caltech. Probes were then made to stain for the expression of these genes. Staining was observed at a very early stage of flower development: when the flower-bud was of the order of one-tenth of a millimetre wide. At this stage, the bud is a dome-shaped group of cells with a series of small bulges or outgrowths on its periphery (Fig. 5.3). These bulges will grow to form the outer

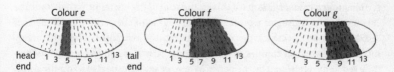

Fig. 5.2 Schematic representation of fruit fly embryos in side view showing regions expressing identity genes for hidden colours *e*, *f* and *g*. The dotted lines indicate the regions from which the various segments in the fly will develop.

Colour *c*   Colour *b*

Fig. 5.3 Schematic representation of very young flower-buds cut in half and stained to reveal the expression pattern of identity genes needed for hidden colours *c* and *b*. At this stage of development the flower-bud comprises a central dome with sepals (s) bulging out on the periphery. The regions of the dome that will form petals (p), stamens (st) and carpels (c) are indicated.

whorl of organs (sepals); the other three whorls (petals, stamens, carpels) will form later on from the dome.

As with the fly, the expression pattern of the flower identity genes was similar to the previously inferred distribution of hidden colours. Recall that three basic colours had been proposed (*a*, *b* and *c*) to give four concentric rings of colour, *a*, *ab*, *bc*, *c*, corresponding to *sepal, petal, stamen, carpel* identities. According to this scheme, colour *c* is located in the central two rings, corresponding to the stamen and carpel whorls. When early flower-buds were probed to reveal the expression of the identity gene needed for *c*, staining was observed only in the central region of the dome, precisely where the stamen and carpel organs would later emerge (Fig. 5.3, left). The staining pattern coincided with the region previously inferred to have colour *c*. Similarly, staining for the gene needed for *b* revealed expression in a region of the dome that would form petals and stamens, precisely where colour *b* was thought to play a role (Fig. 5.3, right). The abstract rings of hidden colour were being directly revealed as territories of gene expression in the developing flower-bud.

We may conclude that the hidden colours of flowers and flies correspond to regions in which identity genes are being expressed. When a given identity gene is *on* in a cell, producing its encoded protein, it gives that cell a particular hidden colour. In other words, *the protein produced by the identity gene is responsible for the hidden colour*. Compare this to the role of pigments in visual colours. If you colour a region on a piece of white paper with red paint, the colour you see depends on specific types of molecule in the red paint. These pigment molecules absorb all colours in the visible spectrum except red, ensuring that only red light is reflected by the paper. When we say that the region on the piece of paper is red, we are monitoring a particular type of molecule in that region, with a property that we happen to be sensitive to (the absorption of all colours except red). Similarly, when a bee sees ultraviolet markings on a flower, it is responding to a region containing molecules that absorb colours other than ultraviolet. Nevertheless, it is still convenient to name these molecules after the colour they confer: red or ultraviolet pigment molecules, as the case may be. In a similar way, each hidden colour depends on the presence of a particular type of molecule: a protein encoded by an identity gene. We can name the protein after the hidden colour it is responsible for: protein-*f* gives hidden colour *f*, protein-*g* gives hidden colour *g*. These proteins are not, however, responded to by the visual system of animals, but are responded to by genes.

Let me summarise the story so far. Based on the study of various mutants, flowers and flies can be inferred to have a series of hidden colours, conferring distinctions between their various organs or segments. These hidden colours are produced from a particular set of genes, identity genes. When the identity genes were isolated, it became possible to make probes that detected where each of them was expressed: the region of the organism which contained the RNA and protein derived from each gene. This allowed the location of each hidden colour to be directly revealed, just as a filter on a camera can be used to locate a previously invisible colour. Once this was done, it became apparent that hidden colours were equivalent to proteins being produced from the various identity genes. These proteins are responsible for hidden colours in the same way that pigment molecules are responsible for visual colours.

I now want to deal with the problem of how the hidden colours are interpreted. Whereas visual colours are responded to by animals detecting light, hidden colours are responded to by genes. The mechanism by which genes interpret hidden colours is rather a complicated story, so I will need to tell it in several stages. The first will be to describe the major properties of the proteins responsible for hidden colours—what the shape of these proteins allows them to do in a cell. I will then go on to show how genes can respond to these proteins according to some simple rules. The final step will be to elaborate the rules further, to give a more realistic feel for what is going on.

Master proteins

There is an important feature that is common to all the proteins responsible for hidden colours: they are able to recognise and bind to small stretches of DNA sequence. Recall that what a protein does depends on its shape. In this case, all of these proteins have shapes that match small sequences of DNA. For example, one protein giving a hidden colour has a shape that fits the sequence of bases TTATTG very nicely. This protein will therefore stick to a stretch of DNA that contains this sequence: it will bind to the particular *site* in the DNA that has the sequence TTATTG. We can refer to this six-base sequence as the *binding site* that is recognised by this protein. You can think of the binding site as a sort of lock, and the matching protein as the key that fits it. Just as differently shaped keys will fit distinct locks, proteins giving different hidden colours can have slightly different shapes, leading them to recognise different DNA sequences or binding sites.

The reason that binding to DNA is so important is that it allows these proteins to influence the expression of genes. They can affect whether a gene is in the *on* or *off* state by directly binding to sites within it. Because of their ability to influence the activity of genes in this way, these DNA-binding proteins will be called *master proteins* (the sense in which these are and are not true masters will eventually become apparent). Thus, *hidden colours correspond to master proteins that can bind to particular sites in DNA*.

Responding to master proteins

To see how these master proteins work, I will need to distinguish between two sorts of genes: those that are needed to *produce* hidden colours, and those that can *respond* to them. So far in this chapter I have been describing the expression pattern of one set of genes: the *identity genes* needed to produce hidden colours. I shall now need to refer to the second class of genes, those that are responding to these hidden colours. I will call these *interpreting genes*. An interpreting gene can respond to or interpret the hidden colours produced by identity genes. The identity genes produce a frame of reference, whereas the interpreting genes respond to it in particular ways. It is important to bear in mind that these two types of gene also produce two different sorts of protein. On the one hand, there are the master proteins produced by the identity genes. On the other, there are the proteins produced by interpreting genes, which may have various other roles in the cell. (The distinction I have made between identity and interpreting genes, although convenient for our present purposes, is somewhat over-simplified. We shall return to this issue in Chapter 9.)

Like the identity genes, each interpreting gene can be in an *on* or *off* state. When the interpreting gene is *on* in a cell, its corresponding protein is made; whereas when the gene is *off*, no product accumulates. What determines whether an interpreting gene is *on* or *off*?

Look at the left panel of Fig. 5.4. It shows that an interpreting gene is divided into two regions: a *coding region* which carries the information to make a protein, and a new region that I have not mentioned before, called the *regulatory region*. (For convenience, the regulatory region is shown only to one side, but regulatory regions may be in other positions as well.) The regulatory region is typically several thousand bases long and contains a series of binding sites, short stretches of DNA to which master proteins can bind. We can think of the regulatory region as a string of different locks (binding sites) which are

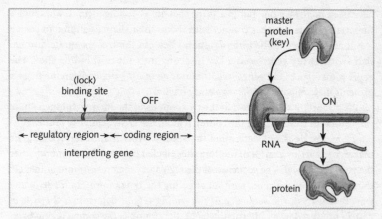

Fig. 5.4 Structure of an interpreting gene showing the regulatory region, which contains a binding site, and the coding region, which is used to make RNA (which in turn leads to protein being made). In the left panel, no master protein is present and the interpreting gene is *off*. In the right panel, a master protein is present and it binds to the regulatory region: this encourages the coding region to be transcribed into RNA so the gene is *on*.

recognised by various types of key (master proteins). For simplicity, I have only shown one of these binding sites (one lock) in the figure.

To summarise the story so far: identity genes produce particular types of protein that can bind to DNA, called master proteins (hidden colours). Each identity gene can produce a different master protein (a different hidden colour), which may recognise and bind to a distinctive sequence of DNA, called its binding site; much as each key fits its own type of lock. The binding sites are located in the regulatory region of interpreting genes that respond to the master proteins.

I now want to try and bring some of these ideas together to show how an interpreting gene can be influenced by master proteins. Remember that an interpreting gene may be in one of two states, *on* or *off*. When it is *on*, its coding region is being used to make protein, whereas when it is *off*, no protein is being made. Now which of these states the gene is in depends on which master proteins are bound to its regulatory region. To give a simple illustration, look again at Fig. 5.4. In the left part, no master protein is bound to the regulatory region, and the interpreting gene is in the *off* state. In the right part, a master protein is bound to the gene, leading to the *on* state. By being bound to the interpreting gene, the master protein encourages or facilitates transcription of the coding region, resulting in the gene being *on*. (One way that master proteins may do this is by making physical contact with the proteins that transcribe DNA into RNA, influencing their shape. These transcribing proteins may originally have been there all along but in a shape that did not allow them to work. Once the master protein makes contact with them, their shape is altered in such a way that they start guiding transcription.) You can think of the master protein that switches the gene *on* as a key inserting into a lock (binding site). The expression state of the gene—whether it is *on* or *off*—depends on the master proteins that are bound to its regulatory region.

It is important to be clear that the two regions of the interpreting gene have very different roles. Unlike the coding region, the DNA sequence of the regulatory region does not determine the type of protein an interpreting gene makes, but carries a series of binding sites (locks) that can determine whether the gene is *on* or *off*. The sequence of the regulatory region therefore influences the *expression* of the gene, without affecting the type of protein made in any way. By contrast, the sequence of the coding region determines the *type of protein* produced by an interpreting gene. For example, the coding region of an identity gene will give rise to a particular master protein; whereas the coding region of an interpreting gene will produce a protein with a different function in the cell.

In my simplified example, whether the interpreting gene is *on* or *off* depends on whether or not the relevant master protein is around in the cell. What determines if the master protein is present? Recall that the master protein responsible for a hidden colour is encoded by an identity gene that itself can be *on* or *off* (Fig. 5.5). Whether or not the master protein is present therefore depends on the expression state of the identity gene. There is a linear chain of events: if the identity gene is *on*, the master protein will be made and hence the interpreting gene will also be *on* (Fig. 5.5, left); if the identity gene is *off*, no master protein will be made and so the interpreting gene will be *off* (Fig. 5.5,

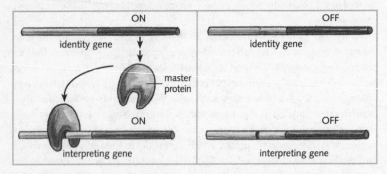

Fig. 5.5 Example of a relationship between the expression of identity and interpreting genes. When the identity gene is *off* (*right*), no master protein is made and so the interpreting gene is also *off*. When the identity gene is *on* (*left*), master protein is produced, which binds to the interpreting gene, leading to its being *on* as well.

right). In other words, in this example, the state of the interpreting gene simply follows whatever state the identity gene happens to be in. The expression patterns of the two genes are coupled together via the master protein. For the present purposes I am ignoring the important problem of what determines whether the identity gene is *on* or *off*; this is a more complex question that will be dealt with in later chapters.

A simple game

So far I have given the basic principles of how the expression pattern of two types of gene (identity and interpreting gene) can be coupled together, but to see how this is related to the interpretation of hidden colours, I need to place this process in the context of a developing organism. To do this, I will use the eight hidden colours of flies, *a–h*, as an example. Each of these hidden colours corresponds to a particular type of master protein. We can now play a game of predicting how interpreting genes will respond to this pattern. To begin with, I will make three simple assumptions or rules:

Rule 1: Each type of master protein binds to a distinctive DNA sequence (i.e. each key fits its own unique type of lock). The binding site can be symbolised by a corresponding capital letter. Thus master protein-*a* binds to a DNA sequence called an A-site. Similarly, master protein-*b* binds to a B-site, protein-*c* to a C-site and so on, up to protein-*h* binding to an H-site.

Rule 2: The regulatory region of an interpreting gene may contain only one of these sites, A or B or C or . . . H (i.e. each interpreting gene carries only one lock).

Rule 3: An interpreting gene will only be *on* if a master protein is bound to its regulatory region; if no master protein is bound, the gene will be *off* (i.e. the gene will be active only if there is a key in the lock).

With these rules in mind, we are now in a position to make some predictions. Suppose that the regulatory region of an interpreting gene contains an E-site. Because the gene will only be *on* if a master protein is bound, it will be expressed solely in those cells having master protein-*e*. We know that the identity gene for protein-*e* is not expressed everywhere in the organism, but only in a belt of cells that form segments 4 and 5. So this is precisely the region of the organism where our interpreting gene will be expressed as well (Fig. 5.6). The expression pattern of the interpreting gene will simply follow that of the identity gene: it will be in the same belt of cells. Note that this system determines *when* as well as where the interpreting gene will be expressed. Clearly the interpreting gene will not be *on* in the embryo before protein-*e* appears. Similarly, if protein-*e* were to disappear later on in development, the interpreting gene would also get switched *off*.

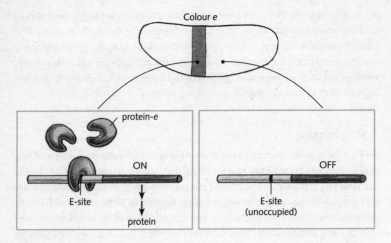

Fig. 5.6 Expression state of an interpreting gene with an E-site in its regulatory region, in various parts of the embryo. In the band of cells that contain protein-*e*, the interpreting gene is *on*, whereas in other regions of the embryo the gene is *off*.

An interpreting gene with a different site in its regulatory region, say an F-site, would be expressed in a different territory of the animal, in this case towards the rear where protein-*f* is located. Again it would only be expressed at the time that protein-*f* was around.

Altogether, there are eight different ways that interpreting genes can be expressed in this system, corresponding to the eight patterns of hidden colour, *a–h*. In other words, eight different interpretations or responses can be made to this framework of hidden colours, according to the rules of this game. Any particular interpreting gene will only be expressed in one way, make one interpretation, according to the binding site in its regulatory region. Suppose, for the sake of argument, that there were 80 different interpreting genes, each with its own regulatory region and coding region. Ten of these might have an E-site in their regulatory region and would therefore be expressed in the belt of cells where master protein-*e* is located. Another ten might have an F-site, resulting in them being switched on wherever master protein-*f* is. Continuing in this way, each territory of hidden colour would have its own set of ten interpreting genes that were being expressed there.

What has all this achieved? Remember that the characteristics of every cell depend on the types of protein it contains (Chapter 2). The types of protein in a cell will affect its shape, size and all the chemical reactions that occur within it. This means that the proteins produced from interpreting genes can influence various cell characteristics. Suppose, for example, that the set of interpreting genes with an E-site in their regulatory region produce proteins that confer a certain feature to a cell, say making a hair grow out from its surface. Any cell that expresses this set of genes will therefore produce a hair. But these genes are only expressed in the territory of cells where master protein-*e* is located, so hairs would specifically appear in this part of the animal: the fly would develop with a hairy belt of cells in the region containing hidden colour *e*. By responding to the hidden colour, the interpreting genes have conferred a distinctive feature to one region of the animal, the production of hairs. If the same set of interpreting genes had a different site in their regulatory region, say an F-site, the hairs would appear in a different part of the animal, where master protein-*f* is located. The hidden colours or master proteins are providing a frame of reference that can be interpreted by genes, leading to various parts of the animal having distinctive features. I have given the production of hairs as one example, but the same would apply to the many other features of a cell. The system allows cells in each of the eight territories of hidden colour to have distinctive properties.

Playing with the rules

According to the rules I have given so far, there are only eight possible ways that interpreting genes can respond to the eight hidden colours *a–h*. However, more possibilities can be opened up by modifying some of the rules. The rules themselves depend on the DNA sequence of the regulatory region and on the structure of the master proteins. So by playing with the rules, all I am doing is extending or modifying these properties in various ways.

Say, for instance, that instead of only allowing one binding site in each regulatory region (rule 2), two sites are now allowed (two locks per gene). There is plenty of room in each regulatory region to accommodate this because each site is only a very short stretch of DNA. It now becomes possible for an interpreting gene to be expressed in two hidden colour territories, giving a *combined* response. An interpreting gene with an E-site and an F-site, for example, would be expressed in two zones of the animal, corresponding to where the protein-*e* or protein-*f* are located.

A new complication has, however, been introduced if the zones of master protein overlap. In such overlapping regions more than one master protein (e.g. both protein-*e* and protein-*f*) will be bound at the same time to a regulatory region. To deal with this we can modify one of the rules. One possibility is that instead of saying that an interpreting gene will only be *on* when one master protein is bound (rule 3), we could change this to the gene being *on* when *one or more* master proteins are bound (i.e. when one or more locks are occupied). In other words, the master proteins do not interfere with each other when they are bound at the same time. This means that the interpreting gene can be expressed in both hidden colour territories, including where they overlap.

According to these new rules, there could be 28 extra ways of responding to the eight hidden colours. This is because 28 different pair-wise combinations can be made from the eight sites A–H. Taken together with the eight possible cases with a single binding site, we end up with 36 different ways of responding to the eight original expression patterns of the identity genes. Each interpreting gene will only respond in one way out of a possible 36, according to the particular combination of binding sites in its regulatory region.

There are other rules we could propose that would lead to yet other possible interpretations. For example, suppose we change rule 3 so that *two* master proteins (rather than one or more) have to be bound at the same time for an interpreting gene to be *on*. In other words, one master protein is no longer enough to switch the gene *on*, it now needs two master proteins to be there. It would be like having to turn two locks to open a door. With this new rule, an interpreting gene will *only* be expressed in regions where the territories of master protein overlap. A gene with an E-site and an F-site will only be *on* where the zones of protein-*e* and protein-*f* overlap. This represents another set of possible patterns.

The important point is that even with just eight hidden colours and two binding sites per interpreting gene, it is possible to produce quite a range of interpretations with just a few basic rules.

The molecular antenna

This range of possible responses is only scratching the surface. So far I have only mentioned eight master proteins in the fly, the eight colours *a–h* that are involved in distinguishing between segments. It is not yet known how many types of master protein there are in the fly altogether, but many more than eight have now been discovered—a reasonable estimate is that there are of the order of one thousand. Each of these proteins is responsible for a particular hidden colour distributed at certain times and places within the organism. The overall result is a complex internal patchwork of colours. Furthermore, there are usually more than two binding sites in each regulatory region of an interpreting gene: ten binding sites is more realistic. This means that the number of possible interpretations is unimaginably large (of the order of many millions of billions of billions). It follows that the particular combination of binding sites in the regulatory region of a gene will result in one response out of an enormous repertoire of possibilities.

We might summarise by saying that the combination of binding sites in the regulatory region of a gene determines how that gene responds to or interprets the pattern of hidden colours. The regulatory region provides a very specific chemical antenna that will only be activated at certain times and places, depending on the set of binding sites it contains. Each interpreting gene carries its own antenna (regulatory region) precisely tuned to respond to only a specific subset of master proteins. The beauty of this system is that once you have a patchwork of hidden colours to refer to, any gene with a regulatory region can make its own interpretation. A gene with one set of binding sites in its regulatory region will make one interpretation, a gene with a different set will make another interpretation.

To get an overall feel for the number of genes involved, we will suppose, for the sake of argument, that one thousand genes in fruit flies code for different types of master proteins. Now the interpreting genes essentially comprise *all* the remaining genes in the DNA of the fly, say ten thousand genes. These interpreting genes can respond to the hidden patchwork through their regulatory regions. The hidden colours therefore provide a universal frame of reference that is open to all of these genes to interpret, allowing each gene to be expressed in particular regions of the organism.

I have glossed over some additional complications that I should briefly mention. I have assumed that the binding of a master protein to an interpreting gene always leads to the gene being switched *on*. However, it is also possible that by binding to the regulatory region, the master protein could switch the gene *off*, by interfering with, or blocking, the transcription of its coding region into RNA. In practice, the way that a gene responds to a particular master protein (i.e. whether it is switched *on* or *off*) is a rather complex affair that depends on how that protein interacts with other types of master protein that are also bound to the regulatory region. This means that each master protein effectively monitors more than the DNA sequence it directly binds to because of its interactions with other master proteins. Thus the specific features of an E-site may include more than the region of DNA directly contacted by protein-*e*. Furthermore, the way an interpreting gene responds to hidden colours can be more subtle than I have indicated. Genes can have more states than simply *on* or *off*, like variable light switches that allow a continuous range of intensities. This is because the rate at which a gene is transcribed into RNA can vary according to the master proteins bound to its regulatory region, so a graded or quantitative response to the hidden colours is possible.

To summarise: studies on mutants with altered patterns or identities led to the idea that multicellular organisms contain a series of abstract territories, symbolised by hidden colours, that can be interpreted by genes. By isolating and studying many of the genes involved, it became possible to get a more concrete picture. The hidden colours correspond to master proteins, produced in particular regions of the organism by identity genes. The pattern of master proteins is then responded to by interpreting genes. Each interpreting gene carries a molecular antenna, a combination of binding sites in its regulatory region. The expression pattern of the interpreting gene depends on how its molecular antenna interacts with the pattern of master proteins in the organism. This allows many thousands of interpreting genes to respond in various ways to a common frame of reference, the patchwork of hidden colours.

I have spoken throughout of genes *interpreting* hidden colours. But it may seem that I could equally well have said that the hidden colours are *dictating* to the genes, telling them when to be *on* or *off*. From a logical point of view, the hidden colours, or master proteins, would appear to come first and then act to influence genes. Why not say that the colours are driving or dictating to the genes, rather than the genes interpreting the colours? I want to deal with this issue in the next chapter.

CHAPTER 6

Evolution of locks and keys

You are working in a building with lots of rooms. There is something slightly unusual about the rooms because each has a door with several locks on it, designed in such a way that to open a door you need a key that fits at least one of its locks. It so happens that you mainly use three of the rooms and have a key that matches a lock on each of their doors. But you now need to enter a new room, room 23, which your key doesn't fit. Frustrated, you try and change your key, filing off some bits here and there. Eventually you manage to modify your key so that it now fits one of the locks on room 23, allowing you to enter. The only trouble is that you are now locked out of the other three rooms because your key no longer fits them. In adapting the key to fit a different type of lock, you have sacrificed its ability to work on the others.

There is a much simpler way of solving the problem which avoids all this: don't mess around with the key but get the locks on room 23 changed. You ask the locksmith to install an additional lock on room 23 that matches your key. By installing the new lock, the ability of your key to work on the other doors is not affected, so you are now able to open the new room as well as the original three. Similarly, if it happens that a door currently opened by your key, say on room 12, needs to be closed to access by you (perhaps for security reasons), it would be a mistake to change your key because then it would no longer work on any of the doors you still need to enter. It would be better to simply remove the lock on room 12 that your key fits. In other words, if you want to modify a system involving components that recognise or match each other, like locks and keys, it can make a big difference which way round you change the components. In this case, it is much easier to change the locks than the keys.

If we imagine the locks being continually modified in this way over a period of time, with some locks being added or taken away from certain doors, we will end up with a building that has a particular pattern of locks on its rooms which can be opened with a set of matching keys. A particular key may open a set of doors, say on rooms 5, 10 and 23. Now without knowing the history of the building, it may seem that the key has been designed to open just these doors.

After all, you need the key to open or close a door, so it seems to have more control than the lock. But when we know the building's past, it is clear that the keys have only played a rather passive role in the design. It is the locks that have been changed and modified, not the keys. The keys have in a sense been at the mercy of where the locks they fit have been placed, rather than the other way round. We might almost say that by acquiring a matching lock, certain doors have allowed themselves to be opened by a key, rather than the key being able to dictate which doors it can open.

Evolution of interpretations

In the previous chapter, we saw how a master protein (hidden colour) can bind to a particular site in the regulatory region of an interpreting gene, switching the gene on or off, much as a key fits into a lock. Now, just as it was easier to change the locks than the keys in our building full of rooms, much of biological evolution has involved changes in the binding sites within regulatory regions (locks) rather than in the master proteins themselves (keys). Because a typical master protein might bind to as many as one hundred different interpreting genes, an enormous constraint is imposed on the extent to which the shape of this master protein can be modified during evolution: any significant change may jeopardise the expression of all one hundred genes it normally binds to, most likely with disastrous consequences for the development and survival of the organism. In contrast, by changing a binding site in the regulatory region of a gene, only the expression of that gene will be directly affected. For this reason, evolutionary changes are often likely to involve mutations in the sites within regulatory regions, rather than alterations in the regions coding for the master proteins themselves.

An example of a change in a regulatory region might be the creation of a new binding site. Binding sites are quite short stretches of DNA, typically six to ten bases long. In a regulatory region a few thousand bases long, it is not too improbable that a chance mutation altering one or two bases in the DNA could create a new binding site for a master protein, or at least something that came reasonably close to a new binding site. This sort of mutation might start to couple the interpreting gene to a different master protein, modifying the gene's pattern of expression. If this new pattern proved advantageous for the organism, further mutations in the regulatory region might then be selected for to improve the match, creating an even better binding site. A new binding site (lock) has evolved to match a master protein (key) that was already around. It is also easy to see how a mutation could lead to the loss of a binding site in a regulatory region. A change in just one base in the DNA sequence of a binding site could mean that the master protein that normally recognises it can no longer bind. The interpreting gene would then no longer respond to this particular master protein.

Thus, the combination of binding sites in the regulatory region of a gene is something that has gradually evolved. During the course of evolution, particular binding sites have arisen or been lost, changing the way interpreting genes respond to the patchwork of master proteins.

Master proteins are only masters in the sense that they can influence the activity of many genes, just as a key might open many doors. They are not masters in the sense of dictators, having evolved all the information that decides which interpreting gene should be on or off. This is because of the way the system has evolved, through genes modifying their response to the master proteins, rather than the master proteins evolving more and more complex shapes that allow them to dictate to more and more genes.

I do not want to give you the impression that master proteins never change. At the early stages in the evolution of a master protein, when it may bind to just a few genes, there may be quite a bit of room for change. But as more interpreting genes evolve suitable binding sites and come under the influence of the master protein, the possibilities for change become more limited. It then becomes increasingly more likely that altered patterns of gene expression involve changes in interpretation, rather than changes in the master proteins themselves.

We can now see that genes *interpret* hidden colours, in the sense that interpretation was defined in Chapter 3: (1) The hidden colours provide a *frame of reference*, a distribution of master proteins of various types. (2) Each gene responds to this pattern *selectively*, being expressed at various times and places in the organism according to the set of binding sites in its regulatory region. A different combination of binding sites leads to a different pattern of expression. (3) The particular selection made in each case is *historically informed*, depending on a series of historical events that have led to one set of binding sites in the regulatory region rather than another.

Families of colour

Although the evolution of master proteins is constrained, there is one very important way in which these restrictions can be partially overcome: through a process called *gene duplication*. This occurs when a mistake is made during the copying of DNA. Remember that DNA is normally copied once every time a cell divides. Occasionally an error is made in the copying process such that one stretch of DNA ends up being copied twice instead of once. The details of how this occurs need not concern us here: what matters is that it sometimes results in an extra copy of a gene being incorporated in the DNA. This means that if

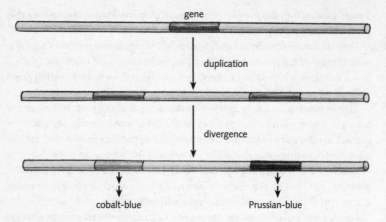

Fig. 6.1 Duplication and divergence of a gene coding for a master protein, resulting in genes for two different types of blue hidden colour.

we have an organism with one gene for a master protein, very occasionally a descendant will be produced with an extra copy of the gene. The descendant now has two copies of the gene for a hidden colour (Fig. 6.1, upper part). These two copies will have exactly the same sequence of bases in their DNA: they will be 100% identical.

Duplication seems to be of little consequence at first, but in the longer term it can provide greater evolutionary flexibility as the duplicate copies *diverge*. This is for the same reason that if you have two copies of a key, you can tinker around with one of them without jeopardising your ability to open doors, because the other copy acts as a backup. If you have two copies of a gene, some mutations that might normally be detrimental to the organism could be allowed because the genes act as backups for each other. Mutations would be expected to accumulate over a period of evolutionary time, so that the duplicate genes eventually start to diverge in sequence, indicated by the different shading of the genes in Fig. 6.1. The duplication has been followed by divergence between the copies, so the DNA sequences of the duplicate genes are no longer 100% but, say, only 90% identical (that is, out of every ten bases of DNA sequence, there is now on average one difference, just as the word *convection* only differs in one out of ten letters from *conviction*). As the DNA sequences diverge, so the proteins encoded by each of the duplicates may also start to diverge. Perhaps this would lead to their shapes becoming slightly different in some way, so that they now bind to regulatory regions with a slightly different specificity. Having started off with one hidden colour, two different versions have evolved. The two hidden colours will be closely related to each other because they are both derived from the same original gene. A convenient way to indicate this relatedness is to symbolise them by similar colours, say *cobalt-blue* and *Prussian-blue*. Both are in the *blue* family of colours. We could imagine the process of duplication and divergence repeating itself so that eventually you end up with a large family of master proteins, each with a slightly different shape, all being symbolised by distinctive types of blue.

This process of divergence between gene copies does, however, eventually reintroduce the same old constraints. As the duplicate master proteins start to recognise different sites, they no longer act as backups for each other: they are no longer pure duplicates. Thus even though duplications can allow some extra flexibility initially, constraints eventually build up again as the duplicates diverge.

Many of the hidden colours affecting identity in flowers and flies are thought to have arisen by duplication and divergence. This became very clear when the identity genes needed for the hidden colours were isolated in the early 1980s. Once a gene has been isolated, the sequence of its DNA and encoded protein can be determined. By comparing the sequences of different identity genes, it soon became apparent that they had arisen by gene duplications. For example, the eight fly genes needed for colours *a–h* (page 77) all have similar DNA sequences. The percentage similarity between the eight genes is particularly high in one stretch of their DNA, about 180 bases long, named the *homeobox*.* The homeobox provides a sort of common signature, showing that the eight genes all started as duplicates of each other. Because of this, the identity genes of the fly are also sometimes referred to as *homeobox genes*, as they all share this region of similarity in their DNA. This does not mean that the homeobox region is identical in the different genes; they each have slight differences in this region due to divergence. It is just that the homeobox is the region of greatest similarity between the genes.

The reason that the homeobox is thought to be so well conserved between duplicates is that it codes for the part of the master protein (called the homeo-domain) that makes direct contact with the DNA: the region of the master protein that fits into the binding site (equivalent to the part of a key you insert into a lock). Any major alterations in the sequence of this region are likely to disrupt the ability of the master protein to work at all, and alterations have therefore been selected against during evolution.

Because the master proteins encoded by these genes are similar or related to each other, we can symbolise them as a family of related hidden colours, say various types of *green* (Fig. 6.2). We can replace the colours *a–h* with types of

---

*Homeo derives from homeosis, the term originally used to describe mutants with mistaken identities, and box is appended because the DNA sequence could be highlighted by drawing a box around it.

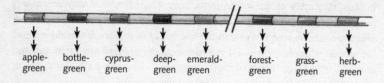

Fig. 6.2 Cluster of identity (homeobox) genes in fruit fly DNA with the corresponding hidden colours they code for.

green that begin with the same letter: colour *a* becomes *apple-green*, colour *b* is now *bottle-green*, *c* is *cyprus-green*, *d* is *deep-green*, *e* is *emerald-green*, *f* is *forest-green*, *g* is *grass-green* and *h* is *herb-green*. The fruit fly is divided up into territories coloured with various types of green, starting with apple-green at the head end and finishing with herb-green at the tail end. Each type of green territory corresponds to a region of cells where a particular type of master protein is made.

As shown in Fig. 6.2, the identity genes needed for the various green colours are arranged in two clusters in the DNA: five genes are in one cluster and three in the other. You may notice that the order of the genes in the clusters is the same as the order of hidden colours from head to tail in the organism. For example, looking at the cluster on the left, the gene for apple-green is followed by the gene for bottle-green, followed by the gene for cyprus-green and so on, paralleling the order of the corresponding territories of hidden colour from head to tail in the animal. It is still not clear why the order of these genes in the DNA should correspond so nicely with their order of expression in the organism, but it most likely has to do with the way the duplications have evolved.

Many of the identity genes affecting whorls of flower organs have also arisen by duplication and divergence. These genes each contain a similar stretch or signature in their DNA sequence. This signature is not the same as the one in the fly identity genes, and is therefore given a different name: the *MADS-box.**

Instead of greens, we could represent the set of master proteins encoded by these genes as *reds*. The hidden colours of the flower previously referred to as *a*, *b* and *c* can now be replaced by *amarone-red*, *burgundy-red* and *claret-red* respectively (fortunately the wine trade has provided us with many names for reds). The flower-bud can be thought of as containing concentric territories of red colour: starting with amarone-red in the outermost whorl (sepals), then amarone-red + burgundy-red (petals), then burgundy-red + claret-red (stamens), and ending with claret-red in the centre (carpels). This pattern of

*MADS is an acronym based on the names of some of the earliest described members of this family.

hidden colours is what gives a separate identity to the various whorls of flower organs.

Let me summarise the main points so far. Flies and flowers contain a set of identity genes that are expressed in various regions of the organism to produce master proteins. This distribution of master proteins is equivalent to a map or patchwork of hidden colours. Many of the master proteins are related to each other because the various identity genes arose by duplication and divergence, and this can be symbolised by the red (flowers) or green (flies) families of hidden colour. The map of hidden colours provides a frame of reference that can be interpreted by many genes through their regulatory regions. The combination of binding sites in a regulatory region acts like a specific molecular antenna, responding to the pattern of hidden colours in such a way that each of these genes comes to be expressed at certain times and places in the organism.

Genes and language

It is useful to compare this view of genes with the way our own language works. Each *gene*, made up of a sequence of DNA bases, is often compared to a *word* comprising a series of letters. The equivalent of all the thousands of genes in the *total DNA* of an organism might then be a large *dictionary* with a vocabulary of thousands of words. There is, however, a fundamental difference between dictionaries and DNA when it comes to expressing their contents. To express a word from a dictionary, someone needs to look up the word and pronounce it. The word itself does not carry information that tells you whether to say it or not—this comes from the reader who is using the dictionary. A gene, however, does carry information in its regulatory region that determines when and where it is expressed. The gene contains a molecular antenna, a series of binding sites, ensuring that it is expressed in some cells and not others. It would be as if each word had a large prefix that ensured it was pronounced at certain times and places.

Although analogies with the written word break down here, there is in my view a better type of linguistic comparison: with the way we use words in our head. When you talk, or experience a train of thought, the words seem to come automatically. Suppose you have a thought like 'I wonder how bees look at flowers.' You do not look up each word, like 'I', then 'wonder', then 'how', in a mental dictionary, because to do so you would first have to know what words you wanted to look up: to look up 'wonder' in your head you would already have to know that 'wonder' is the word you wanted, defeating the whole point of the exercise. The words we use are in a sense stored in our brain, but they *occur* to us under particular conditions rather than being something we look up in a mental reference library. We are most conscious of this when a word is on the tip of our tongue and we have difficulty in recalling it. We have to wait until the word comes to us almost of its own volition. Thoughts are not something we plan and then execute by looking up the appropriate words, we just have them. We can of course plan to think about something, like 'I am going to spend the next hour thinking about bees'; but we do not plan the thoughts we will then have about bees and retrieve the words accordingly, because to do so would mean that we had already had the thoughts. Rather, we might start by contemplating some aspect of bees, and this would lead to other thoughts and words coming to mind. We experience a wandering train of thought rather than a planned series of events.

By analogy with genes, we might notionally divide each word in our brain into two parts. One part has to do with what gets expressed as the word occurs to us, and is responsible for how the word 'sounds' in our head. By 'sound' here, I mean the experience of having the word in our conscious mind, irrespective of whether we say it aloud or not. This would be equivalent to the coding region of a gene producing a particular protein. The second part of a word would determine when the word occurs to us, ensuring that each word comes to our mind under certain conditions. This would correspond to the regulatory region of a gene. It is as if each mental word carries information that leads to its being expressed or manifesting itself in our consciousness according to the conditions in our brain, rather than just being an entity that we retrieve. Words in our mind are not the same as those written down on a page; they are networked or locked into the thinking process. Of course the way they are locked in is not immutable: it can change as our experiences and mental processes develop. At any one time, the 'regulatory part' of each word is historically informed, depending on our previous learning experiences.

I am not saying that mental words are as simple as a linear sequence of subunits in a gene. We do not yet know how words work in our mind, but they most likely reflect a complex set of interactions between cells in our brain. It may be that these interactions would defy being simply broken down into the equivalent of regulatory and coding parts of a gene. My reason for drawing this comparison between mental words and genes is not to give an oversimplistic view of the mind, but to give us a better sense of how genes work than is implied by the notion of a dictionary. There is no independent reader dipping into the gene volumes held within each cell. Genes carry information that leads to their being expressed at certain times and places.

Now of course the whole process of thinking is remarkably interactive. Every word or thought that occurs to you leads to new words coming to mind. There is an ever changing state of mind in which each word or thought feeds off the previous ones. I have begun to show that this is also true for genes. Genes come to be expressed by interpreting hidden colours; and these hidden colours or master proteins themselves depend on a set of genes (identity genes). Genes feed off each other much as words do. But I have yet to explain how the genes coding for the hidden colours themselves get to be expressed in a pattern. It is all very well saying that genes interpret a complex patchwork, but what sets up the patchwork to begin with? I have started my explanations in mid-stream, as it were, assuming that the pattern of hidden colours is already given. As we shall see, the production of this pattern depends on further interactions between genes and proteins. Before dealing with this, however, I want to address another issue in the next chapter: the hidden colours of humans.

CHAPTER 7

The hidden skeleton

> As Gregor Samsa awoke one morning from uneasy dreams he found himself transformed in his bed into a gigantic insect. He was lying on his hard, as it were armour-plated, back and when he lifted his head a little he could see his dome-like brown belly divided into stiff arched segments on top of which the bed-quilt could hardly keep in position and was about to slide off completely. His numerous legs, which were pitifully thin compared to the rest of his bulk, waved helplessly before his eyes.

This is how Kafka starts his short story, *Metamorphosis*, describing the life of Gregor after he was transformed into an insect. Kafka's choice of an insect as the vehicle of his nightmare was not accidental. Insects live within their skeletons: they carry their supporting framework on the outside of their body, conjuring up a claustrophobic image of being helplessly trapped and entombed within a hard casing. They seem to have an imprisoned and alien existence, providing useful fodder for science fiction and horror stories. Nevertheless, there is a basic similarity between insects and ourselves. Like us, they have a head with mouth and eyes at one end, a body bearing limbs, and an anus at the tail end. This is what allows Kafka's transformation of a human into an insect to work so well: we can readily substitute human parts for corresponding parts of a beetle, head for head, main body for main body. We might take this as indicating that *vertebrates*, animals with internal bony skeletons such as ourselves, and *arthropods*, jointed animals with an outer casing such as insects, are formed on similar principles. Perhaps there is a common system that underlies the formation of these different types of animal. Another view would be that this similarity between vertebrates and arthropods is simply a trivial consequence of the limited number of ways that an animal can function. There are only so many ways that animals can operate. Having a head at one end, a main body with limbs, and a tail end, is a particularly convenient arrangement and it is not surprising to find it in different types of animal. In this view, a beetle and human are entirely different types of creature, any resemblance between them being a superficial consequence of limitations on the ways animals can function.

The question of whether vertebrates and arthropods are formed in a similar way or are built entirely differently was a burning issue in the early nineteenth century. It culminated in a famous confrontation in 1830 between two highly respected professors at the Muséum d'Histoire Naturelle in Paris: Georges Cuvier and Etienne Geoffroy Saint-Hilaire (for an excellent description of this debate, see *The Cuvier–Geoffroy Debate* by Toby Appel).

Form and function

The early part of the nineteenth century was the golden era of comparative anatomy, with the Muséum d'Histoire Naturelle in Paris being at the heart of many of the most exciting new discoveries. Scientists were busily dissecting, describing and comparing many different types of animal for the first time, trying to discover the secrets of their internal anatomy. The legacy of this era can still be observed today in the Paris Museum's Gallery of Comparative Anatomy. There you can see what looks like a skeleton march: a stunning display of skeletons from all sorts of animals, arranged so that they all appear to be walking in the same direction (Fig. 7.1). The skeletons bear witness to a period in which the study of anatomy was at the cutting edge of biological research. Based on these sorts of investigations, Cuvier and Geoffroy each thought they had uncovered the unifying principles that governed anatomy. Their approach was, however, very different.

You might classify musical instruments in two ways. One would be to classify them according to how they look: violins and cellos have a similar shape and are made of wood and string; trumpets and horns look similar and are made of metal. This is a classification based on the *form* or structure of the instrument. Alternatively, you could classify them according to how they make sounds: you have to pluck or bow a stringed instrument; a woodwind instrument is sounded by blowing, either directly or through a reed; to play a brass instrument you have to press your lips against a mouthpiece so that they vibrate when you blow; finally, you bang or hit a percussion instrument. In this case, the classification is based on the way the instrument *functions*. These two approaches to classification—the form of the instrument or the way it functions—emphasise different aspects of the instruments. If you were to classify a saxophone based on *form*, you might place it with the brass instruments, whereas if *function* was your main criterion you would place it in the woodwind section because you play it by blowing through a reed (saxophones are normally classified as woodwinds for this reason). In spite of these differences, the two types of classification will often give similar answers because the structure of the instrument is obviously closely connected to the way you make it sound.

For Cuvier, the key to understanding the structure of an animal lay in the way it functioned. Each animal was beautifully designed to function in a

Fig. 7.1 Gallery of Comparative Anatomy, Muséum National d'Histoire Naturelle, Paris.

particular way and this dictated its form and structure. To Geoffroy, things were the other way round: the fundamental feature of animals was their unity of form. The specific way they functioned was a secondary matter. Following these two approaches, Cuvier and Geoffroy each arrived at a different formulation of the key rules underlying anatomy.

Cuvier and functional integration

Based on extensive animal dissections and studies, Cuvier decided that it was no good trying to understand organs or tissues in isolation: they only made sense when they were seen as parts of an integrated active individual. Why, for example, should birds have feathers? Feathers only make sense if they can be attached to a specialised type of forelimb, making a wing. A wing only makes sense if there is a certain type of collar- and breastbone for the wing muscles to attach to. The muscles in turn can only work if they can be provided with high levels of oxygen, requiring a particular type of chest and breathing system. Carrying on in this vein, the whole body plan of a bird could be deduced, starting from just a feather. Once told that an animal has a feather, we could work out that it must also have a certain type of collar-bone, and given a particular collar-bone we could infer that the owner had feathers. A feather without a bird, or a bird without feathers, just would not make any sense: the resulting animal would be illogical, unable to function properly and so could not exist. Cuvier applied and extended this principle to an enormous range of animals. Given a single fossilised bone he became expert at predicting what the rest of the skeleton looked like, what the animal ate and how it moved.

Cuvier was aware, however, that in practice many of the relations between parts had to be worked out retrospectively. If someone had never seen a bird and was given a feather, it is very unlikely that he or she would be able to deduce the structure of a bird from scratch. In practice, we cannot work everything out from first principles, but need to glean knowledge from the animals around us. Cuvier's deductions using fossilised bones were based on comparisons with bones of other known skeletons rather than purely logical deductions. He thought that this had more to do with our inadequacies in logical thinking than with any weakness in his theory. If we were intelligent enough, perhaps we could work out the structure of a bird from a feather without first needing to look at any examples.

It is a pity that Cuvier never met Sherlock Holmes, another great exponent of the power of deductive logic. This is how Holmes describes the art of deduction in *A Study in Scarlet*:

> From a drop of water, a logician could infer the possibility of an Atlantic or a Niagara without having seen or heard of one or the other. So life is a great chain, the nature of which is known wherever we are shown a single link of it. Like all other arts, the Science of Deduction and Analysis is one which can only be acquired by long and patient study, nor is life long enough to allow any mortal to attain the highest possible perfection in it.

To Cuvier, the logical principle that connected the different parts of an organism together was their functional interdependence: parts were inextricably linked because they relied on each other to work properly. To some degree this may seem rather obvious and it had indeed been stated by others before. Cuvier's achievement was to pursue this idea in a systematic way and to elevate it to a fundamental law which he called the *Conditions of Existence*:

> it is this mutual dependence of the functions and the aid which they reciprocally lend one another that are founded the laws which determine the relations of their organs and which possess a necessity equal to that of metaphysical or mathematical laws, since it is evident that the seemly harmony between organs which interact is a necessary condition of existence of the creature to which they belong and that if one of these functions were modified in a manner incompatible with the modifications of the others the creature could no longer continue to exist.

For a creature to exist at all, its various parts must function together properly. This means that if one part was altered so that it no longer worked well with the others, the organism would not be able to survive and exist. Functional integration was therefore a necessary condition for existence and was the guiding principle behind all animal designs. Cuvier took the harmonious arrangements observed in animal anatomy to be a reflection of God's wisdom: species had been created by God along logical principles with parts that worked well together.

Perhaps Cuvier's crowning achievement was the way he applied his principle to the overall classification of animals. After surveying the animal kingdom, he proposed that there were basically four types of animal, each organised along different lines: (1) *vertebrates*, animals with a backbone, (2) *molluscs*, soft bodied animals such as slugs and snails, (3) *articulates*, jointed animals such as insects and shrimps (i.e. arthropods), (4) *radiates*, radially symmetrical animals such as starfish and jellyfish. Any animal could be classified as belonging to one of these four categories, or *embranchements*, just as an orchestral instrument can be classified as belonging to strings, woodwinds, brass or percussion.

Coming up with this four-fold classification may not seem particularly astounding, but imagine you had the task of dividing up the entire animal kingdom in a sensible way. Where would you start and what would your criteria be? You might begin with divisions like birds, fishes, mammals, etc. The trouble with this sort of classification is that it is somewhat superficial and anthropocentric: it is biased towards animals we are more familiar with. When set in the context of the whole animal kingdom, birds, fishes and mammals are seen to be all organised along very similar lines. Cuvier came up with a less parochial classification that recognised fundamental features of animal function and structure. To Cuvier, the four embranchements represented qualitatively different types of functional organisation. In his system, mammals, birds and fishes all belonged to just one of the embranchements, the vertebrates. The remaining three embranchements were reserved for fundamentally different animals such as insects, slugs and starfishes. (This minority status of vertebrates is even more apparent in modern classifications which recognise about thirty-five major groups or phyla, of which the vertebrates are only one.)

Now according to Cuvier, although you could fruitfully compare animals belonging to the same embranchement with each other, comparisons between embranchements would be meaningless. It would be like trying to compare a violin with a trumpet. Humans, for example, belonging to the vertebrates, could not be compared with insects, belonging to the articulates, because they are organised in a totally different way. For Cuvier, the key to the animal kingdom was functional integrity, which he thought led inexorably to the four-fold classification of distinct types. It was this division of animals into qualitatively different categories that eventually led to Cuvier's conflict with Geoffroy.

Geoffroy and the principle of connections

Geoffroy approached the problem of animal structure from a quite different angle. Compare the way horses and humans walk. When a horse walks, the back legs appear to bend at the knee in such a way that the lower part of the leg swings forwards and up (Fig. 7.2). Try to do a similar thing with your legs and you'll end up in hospital because when you walk, the lower part of your leg swings

Fig. 7.2 Horse walking (from Leonardo da Vinci, Windsor Royal Collection, no. 12341).

backwards below the knee. The front legs of a horse seem to be a bit more comfortable to think about: the lower part of the leg swings back below the knee (Fig. 7.2). However, we should really be comparing the front legs of a horse to their real human equivalent, arms. If you let your arms fall to your side and bend them at the elbows, the lower arm swings forward, the opposite to the way a horse's front legs seem to bend.

To see why our joints appear to work in the completely opposite way to a horse's, we need to look at the internal anatomy of limbs. If you follow the arrangement of bones in your leg starting from the pelvis, you have a single thigh bone (femur), which is followed at the knee joint by two shin bones (tibia and fibula), which are in turn jointed at the ankle with bones in the foot. The back leg of a horse has a very similar arrangement but if you try to match its bones with ours, one for one, it turns out that what I have been calling its knee corresponds to our ankle, and its lower leg matches the middle bones (metatarsals) of our foot, ending in the middle toe (Fig. 7.3). To turn yourself into a horse, you have to imagine the middle bones of your foot growing to the length of your shin, and the nail on your middle toe enlarging to form a hoof. The same is true for the front limbs: you have to imagine the bones of your middle hand and finger growing to be about as long as your forearm. It is now easy to move like a horse. You swing the back legs forwards by bending your ankles and the front legs backwards by flexing your wrists. The sequence of

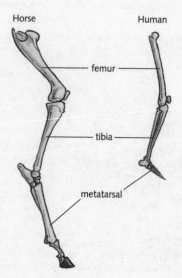

Fig. 7.3 Bones in the rear limbs of horse and human.

bones in a horse and the way they bend at the joints is similar to ours, it is just that the relative lengths of the various bones are quite different.

Now surely if you were designing horses and humans purely on functional grounds, it is unlikely that you would choose exactly the same arrangement of bones and be satisfied with just varying their relative lengths. Why not have different arrangements in each case that best suit the purpose they are put to? The problem gets worse as you look at other animals. To turn your hand into a bat's wing, you elongate your fingers, splay them out and cover them with a web of skin. A whale's flipper represents yet another distortion of the same arrangement of bones. It seems that Nature is remarkably unimaginative when she tries to make bodies and limbs, always seeming to use the same arrangement of bones. To Geoffroy, this indicated that function was not the primary consideration in animal anatomy. The real key had to be something else: something universal that transcended the particulars of each function, be it flying, running, swimming or holding.

Geoffroy pointed out that what was the same in all these cases was the way the bones were *connected* to each other. The order of connections was relatively invariant even though the length of each bone could change. It would be as if you had drawn a map of the various bones on a piece of rubber and stretched it in various ways. You might end up with skeletons with different overall shapes, but no matter how the piece of rubber was deformed by stretching or pulling, the order in which the bones were connected with each other would always remain the same. Geoffroy called this the *Principle of Connections*: bones maintain the same connections in different animals, irrespective of which particular functional use they are put to. In the 'Who's Who' of bones it is not what you do but who you are connected to that matters.

It followed from this that all mammals had a common map of bone connections: they all conformed to a *unity of plan*. This unified plan did not correspond to any particular animal but to a basic layout that was common to all. In a similar way, I could try to illustrate all stringed instruments by a generalised diagram showing their basic layout. The generalised diagram is not itself an instrument, it is a map that depicts the common arrangement of parts in a violin, cello or double-bass. The unified plan is an abstraction, a general property which is manifested in individual cases.

Conflict and controversy

Within certain limits, Geoffroy's principle of connections and unity of plan were acceptable and even welcomed by Cuvier. It clarified the well-known fact that different mammals display similar bone arrangements. It was only when Geoffroy started to claim true universality and primacy for his principles that trouble started to brew. First of all, he claimed that mammalian skeletons were not only similar to each other but also displayed the same basic connections as skeletons of fishes. On the face of it, this seemed absurd because there were many fish bones, particularly those in the head, that had no obvious counterparts in mammals. Undeterred by this, Geoffroy applied his principle of connections, ignoring superficial differences in the relative sizes and functions of bones, to establish what he thought were the more fundamental similarities.

Cuvier found some of Geoffroy's comparisons convincing, others less so, and if matters had stopped there, their disagreements might merely have been a question of emphasis. Geoffroy's next step was to change all that. In 1820 he proposed that not only did different vertebrates, such as mammals and fishes, have the same type of bone arrangements, but that they also shared the same basic layout with insects—animals that don't even have any bones! This flatly contradicted Cuvier's view that vertebrates and insects belonged to distinct embranchements. To Cuvier, comparing a human to a fly was simply not allowed: it was like comparing chalk with cheese.

Geoffroy supported his radical claim by pointing out that insects and vertebrates shared many features in their basic organisation. They both had a head with mouth and eyes, and the rest of the body was divided into two segmented regions: the thorax (the chest in humans) and abdomen. The main difference between them was that whereas vertebrates were supported by a bony skeleton from within, insects were supported from the outside by a hardened casing. Geoffroy proposed that these apparently different types of organisation were really two sides of the same coin; it was just that vertebrates were organised around a skeleton whereas insects lived within theirs. To turn yourself into an insect you have to imagine your skeleton expanding outwards, whilst your body remains the same size. Eventually the bones reach and enclose the skin, and all the soft tissues and vital organs end up interior to them, like a very elaborate bone marrow. He believed that once this was appreciated, detailed correspondences between insect and vertebrate structures fell into place according to his principle of connections. For example, the main trunk of the insect would correspond to the vertebral column (backbone), and its legs would be equivalent to our ribs. Not surprisingly, this view of insects as perambulating skeletons walking around on their ribs was greeted with disbelief by many: the detailed correspondences that Geoffroy drew between insects and vertebrates just seemed too fanciful. Nevertheless, Geoffroy stuck to his guns, convinced that the principle of connections revealed the deep unity shared by organisms. Sceptics, he thought, were simply being distracted by the more obvious superficial features.

For Cuvier, these comparisons between vertebrates and insects were both ridiculous and offensive, and for the next ten years he tried to undermine Geoffroy and his followers. As the historian Toby Appel summarises:

> the tension between the two naturalists was apparent to all, and weighed heavily not just upon the principals but upon the entire scientific community. In all their lectures, articles, books and reports at the Académie, Cuvier and Geoffroy continued to snipe at each other, but neither responded to the other head-on. Geoffroy accused Cuvier of denouncing him behind his back, rather than presenting a reasoned critique of his work. For ten years, beginning in 1820, he tried to lure his scientific opponent into a public confrontation, but Cuvier preferred the strategy of indirect attack.

Then, on February 15th, 1830, the bubble burst. The occasion was one of the weekly meetings of the Académie des Sciences in Paris where Academicians presented reports or papers. Two relatively inexperienced young naturalists, Meyranx and Laurencet, had submitted a memoir to the Académie pointing out that if a vertebrate was bent back on itself, its organs would be in similar positions to those of a cuttlefish, a mollusc. This was music to Geoffroy's ears because it implied that his universal type could now be extended to yet another of Cuvier's embranchements: the molluscs. When Geoffroy was asked to give a verbal report to the Académie on this memoir, he seized the opportunity not only to commend it but also to promote his own ideas. Surely the memoir illustrated how, by applying the principle of connections, real progress could be made to reveal the fundamental unity of animals. This was so much more exciting than the old-fashioned view, which he exemplified by a quote from one of Cuvier's classic works.

This was too much for Cuvier. He stood up at the end of Geoffroy's report and protested against both the scientific content and the personal slur in the report, promising to deal with the issues in detail at a later date. At the following week's meeting of the Académie, Cuvier arrived armed with diagrams of a cuttlefish and a vertebrate bent back on itself, prepared to do battle. He proceeded to show that many of the proposed similarities between molluscs and vertebrates simply did not hold up to close scrutiny. The debate, which was now starting to attract considerable public attention, continued vigorously for four more meetings, with the adversaries reduced to squabbling at one point over who should be allowed to speak first. The last of these meetings signalled the end of the verbal contest before the Académie, but by then it had caught the imagination of the scientific community and the public at large. Here was a fundamental issue about our place in Nature being disputed between two giants of the French intellectual establishment. Newspapers, journals and books publicised the debate, some siding with Cuvier, others supporting Geoffroy.

The Darwinian solution

About thirty years after the debate in the Académie, Charles Darwin thought he had provided a solution to the conflict in his book *On the Origin of Species* (1859). To Darwin, anatomical arrangements were simply a consequence of the way evolution worked. The *functional* aspects of anatomy that Cuvier emphasised could be accounted for as adaptations resulting from natural selection: organisms were made of parts that worked well together because such an arrangement was more likely to lead to survival and reproduction. In Darwin's words:

> The expression of conditions of existence, so often insisted on by the illustrious Cuvier, is fully embraced by the principle of natural selection. For natural selection acts by either now adapting the varying parts of each being to its organic and inorganic conditions of life; or by having adapted them during long-past periods of time.

To Darwin, functional integration was an adaptation that had gradually evolved over millions of years by selection. He acknowledged Cuvier's principle but instead of seeing it as a reflection of God's design, he explained it as an outcome of natural selection. Unity of *form*, as highlighted by Geoffroy, could also be accounted for by Darwin through common descent. The anatomy of horses, humans and all other mammals was similar as a consequence of their shared ancestry. Because evolution proceeded by gradual alterations, each step modifying what went before, ancestors and their descendants would be expected to retain certain features in common. In the case of vertebrates, Darwin pointed out that the sizes and shapes of bones were more likely to change than their arrangement, accounting for Geoffroy's principle of connections:

> The bones of a limb might be shortened and widened to any extent, and become gradually enveloped in thick membrane, so as to serve as a fin; or a webbed foot might have all its bones, or certain bones, lengthened to any extent, and the membrane connecting them increased to any extent, so as to serve as a wing: yet in all this great amount of modification there will be no tendency to alter the framework of bones or the relative connexion of the several parts.

Fins, wings and other types of mammalian limb had a similar set of bone connections because they evolved by gradual modification from a common ancestral appendage. Geoffroy's principle of connections and unity of plan were no longer fundamental laws but simply a historical consequence of common ancestry.

There was a further sting in the tail for Geoffroy's principle. Because the basic layout of an organism was featured in a common ancestor, this feature itself would be something that had evolved. For example, the common ancestor of mammals did not appear from nowhere: it also had an ancestry going even further back in time. Thus, according to Darwin, the layout that linked a class of animals, such as all mammals, was itself an adaptation that had arisen earlier on in their ancestry by natural selection. This meant that the unified plan (or unity of type, as Darwin referred to it) had its origins in the functional adaptations of the past. For Darwin, Geoffroy's law of form became subsumed in Cuvier's law of function:

> It is generally acknowledged that all organic beings have been formed on two great laws—Unity of Type, and the Conditions of Existence . . . in fact, the law of the Conditions of Existence is the higher law; as it includes, through the inheritance of former adaptations, that of Unity of Type.

Now if a basic layout is something that has arisen during evolution, you could imagine its having arisen more than once. For instance, the common ancestor of vertebrates could have evolved its basic layout quite independently of the common ancestor of arthropods (insects, crustaceans, etc.). There was no fundamental reason for supposing that layouts of insects and vertebrates had anything in common. Of course, if you go back far enough in evolutionary time, you might come across the ancestral stock from which both vertebrates and insects were descended. But this ancient organism might not have had much of a layout to speak of: it may have been a simple aquatic animal that had neither an internal nor an external skeleton, and therefore would lack a framework of parts that could be related to those of modern insects or vertebrates. And even if there was some sort of simple layout in this ancient animal, any vestiges of similarity between this and modern forms would most likely have been wiped out during the long periods of evolution. Given the many notable differences between the anatomy of insects and vertebrates, it seemed most likely that much of their basic layout had effectively arisen independently and had little in common. Modern insects and vertebrates were constructed along fundamentally different lines and could not be meaningfully compared. As far as this issue was concerned, Cuvier had won the day and Geoffroy's position became no more than a historical curiosity.

Vertebrate homeobox genes

I would not be going to the trouble of telling this story if it ended here. The first stirrings that were to rekindle the issue of how vertebrates were related to arthropods came in 1984, about one hundred and fifty years after the original dispute between Cuvier and Geoffroy. It was the time when the identity genes affecting segments in the fruit fly had just been isolated. Recall that these genes are needed for a set of hidden colours that define distinct regions in the fly. The genes code for a set of related proteins, symbolised by eight green colours, distributed from head to tail in alphabetical order: from apple-green at the head end to herb-green at the rear.

We might say that the arrangement of these hidden colours sets a basic layout for the fly from head to tail. Remember that the main body of a fly consists of a series of 14 segments with different identities: three head segments bearing appendages such as antennae, followed by three segments of the thorax bearing legs and sometimes bearing wings as well, and then eight distinct segments in the abdomen. These distinctions between the segments, the head to tail layout, depend on a family of green hidden colours. Mutations that remove one or more of the hidden colours result in a failure to distinguish between some regions, giving several segments with the same identity. If all the hidden green colours were missing, you would end up with a monotonous arrangement of 14 similar segments, losing the distinctive layout from head to tail. In other words, the anatomical layout itself depends on a deeper layout: the map of hidden colours.

Given the many differences in the anatomical layout of insects and vertebrates, we might expect that their map of hidden colours would also be quite distinct. This issue could be examined directly as soon as the genes for hidden colour, the identity genes, had been isolated from flies. Once you have isolated a gene from one species, it is possible to look for related genes in other species. So, shortly after the identity genes needed for hidden colours had been isolated from flies, Bill McGinnis and Michael Levine, in Walter Gehring's lab in Switzerland, started to investigate whether similar genes were present in other animals. Because of their distinctive anatomical layout, the expectation was that vertebrates would either not contain similar genes at all, or that they would have a different significance for vertebrates than for arthropods.

The method they used to look for related genes in other species is not too different, in principle, from the method I described in Chapter 5 for locating where genes are expressed. An isolated gene is used to make a molecular *probe*, but in this case the probe does not detect RNA or protein molecules, but a specific stretch of DNA sequence (the probe is itself a DNA molecule that has been chemically modified). Each organism contains many thousands of genes, each gene being a stretch of DNA. The probe can recognise a particular DNA stretch in this vast array: it sticks or *hybridises* only to this stretch, distinguishing it from the others. The probe that McGinnis and Levine made was derived from an identity gene of fruit flies and was used to recognise any stretch of DNA that was similar to it in sequence. If a similar or related gene happened to be present in the DNA of a different species, such as that of a human, the probe would locate or cross-hybridise with that gene. They tested their probe on DNA from a range of other organisms, including worms, as Bill McGinnis recalls:

Mike suggested we might as well test some worms, so my wife Nadine walked to the local fisherman's bait shop and got a variety of insects and segmented worms. I had the DNA isolated even before we identified what species they were. I put in some vertebrate DNAs hoping for some cross-hybridisation, but really thinking of them as negative controls. When I pulled the first blot out, there were obviously some strongly hybridising fragments in the human, frog and calf lanes. I was so excited my hands were shaking, but most were sceptical about the result and I was a bit disappointed.

The probe was recognising or cross-hybridising with stretches of DNA from vertebrates, indicating that these animals had genes that were similar or related to the identity gene from flies. McGinnis repeated the experiment and consistently got the same answer, so that eventually people became more convinced. Now as soon as genes from other species have been identified by a probe in this way, they can be isolated or fished out and studied in detail. Several laboratories therefore embarked on a series of molecular fishing expeditions, using probes based on the fly identity genes to isolate similar genes from vertebrates including frogs, mice and humans.

The vertebrate genes isolated in this way were remarkably similar in many details to the genes from flies.

The first point of similarity came from comparisons of DNA sequence. Recall that the eight genes affecting segment identity in flies arose by gene duplication and divergence (Chapter 6). Consequently, they all share a very similar stretch of sequence in their DNA called the homeobox, and because of this, these identity genes are sometimes referred to as *homeobox genes*. The homeobox is in the coding region, so these genes produce related proteins, symbolised by a family of green colours. Now the various genes isolated from vertebrates also contained a homeobox: *they were also homeobox genes*. This meant that the vertebrate genes coded for proteins that belonged to the same family as in flies: the family of green colours. This is perhaps not so surprising, as the vertebrate genes were after all isolated because of their similarity to the fly genes. The probes were designed to find genes in vertebrates that were related to the identity genes of flies, so it is not too unexpected that genes isolated from vertebrates in this way would contain similar stretches of DNA, such as the homeobox. What was more intriguing was the extent of correspondence between the different *types* of fly and vertebrate homeobox genes. Not only did the vertebrates contain genes for a family of green colours, but each type of green was also represented. For example, one of the vertebrate homeobox genes was most similar to the fly homeobox gene needed for the apple-green colour: its DNA sequence was clearly nearer to the gene for apple-green than to any of the other seven homeobox genes in the fly. Another vertebrate homeobox gene was most similar to the fly gene producing bottle-green. Yet another was

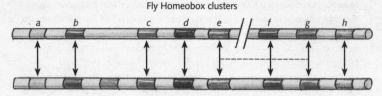

Fig. 7.4 Similar arrangement of homeobox genes in fruit flies and humans or mice. In flies the genes *a–h* are in two clusters (the Antennapedia and Bithorax Complexes), whereas in humans and mice the equivalent set of genes is arranged in one cluster (dotted lines indicate related genes that cannot be unambiguously matched up).

obviously closest to the gene for cyprus-green. Each vertebrate gene could be matched up as a counterpart to one of the eight fly homeobox genes. Thus, vertebrates had a series of homeobox genes that produced the equivalent of each type of green, from apple-green to herb-green.

The next point of similarity between fly and vertebrate genes came from looking at how the genes were arranged in the DNA. In flies, the eight homeobox genes—coding for the green hidden colours—are clustered together along the DNA. Five of the genes are in one cluster and three are in a separate cluster. I have illustrated this in the top part of Fig. 7.4, each gene being labelled with the initial letter of its corresponding hidden colour. Thus, genes labelled *a–e* are in one cluster and *f–h* are in the other. As the homeobox genes from vertebrates started to be analysed, it became clear that they were also clustered in a very similar way. Robb Krumlauf, who was working in London on the homeobox genes from mice, recalls:

> Walter Gehring in late '86 had the first homeobox workshop. There were about eight of us guys working on vertebrates sitting there talking to each other. At that time we all had small groups of individual homeobox genes. Where the difference came I think, is that Denis Duboule, who was in Pierre Chambon's lab in Strasbourg at that time, had a cluster of four or five genes like my cluster and we started talking to each other and trading sequence.

It was becoming apparent that the various homeobox genes in vertebrates were clustered together in a particular way. The arrangement of the homeobox genes in mice was then pieced together in detail by Robb Krumlauf and colleagues in London, and Denis Duboule and Pascal Dollé in Strasbourg. At about the same time, Eduardo Boncinelli and colleagues in Naples were working out the arrangement of human homeobox genes. The results from mouse and humans were essentially the same. There was a striking similarity in the way vertebrate and fly homeobox genes were organised: the genes were clustered together in a similar *order* in the DNA, from *a–h* (Fig. 7.4, bottom). In one way, the vertebrate arrangement was even simpler than that in flies because the complete set from *a–h* was clustered in a single region of DNA rather than in two separate ones.

The strong implication of this detailed correspondence was that insects and vertebrates had inherited their clusters of homeobox genes from a common ancestor. The ancestral aquatic animal that eventually gave rise to both the vertebrates and insects must have had a set of homeobox genes in its DNA arranged in the order *a–h*. These genes had then been preserved in pretty much the same arrangement throughout the evolution of vertebrates and insects. At some time in the evolutionary lineage that gave rise to flies, the cluster became split into two separate parts to give the two clusters *a–e* and *f–h* (this split has not happened in all insect lineages because some insects, such as beetles, still have a single cluster).

There are some additional complications I have glossed over. First of all, mammals have four copies of each gene cluster in their DNA. It seems as though a long stretch of DNA, including the entire cluster of homeobox genes from *a–h*, has been duplicated several times during the evolution of vertebrates, resulting in four copies of each cluster (giving a total of about 32 genes). Following these large-scale duplications, there has been some divergence between the four clusters, giving rise to slightly different versions of each type of gene: a human or mouse contains several slightly different versions of the gene for apple-green, several versions for bottle-green, etc. Secondly, there have been some extra duplications (and some deletions) of individual genes within clusters. This means that some clusters can have more genes than others (for example, the vertebrate cluster in Fig. 7.4 has nine genes rather than eight) and that some of the vertebrate genes cannot be matched up unambiguously with their counterparts in flies (this is indicated by the dotted line in Fig. 7.4). Even with these qualifications in mind, the degree of conservation between homeobox genes in insects and vertebrates is striking.

The hidden map of vertebrates

What significance might these homeobox genes have for vertebrates, given that the anatomical layout of vertebrate animals seems so different from that of insects? In flies, the homeobox genes are required to establish distinct territories in the animal, underlying the basic layout from head to tail. It could be that these genes play a comparable role in vertebrates, even though their anatomy looks so different from that of insects. Alternatively, the vertebrate homeobox genes might have nothing to do with the head-to-tail layout in vertebrates: they could have an entirely different significance for humans than for flies.

A good way to test these possibilities would be to look at when and where the homeobox genes are expressed in developing animals (recall that by expression of a gene I am referring to where it is switched on, producing its protein product, at any given time). In flies, each homeobox gene is expressed in a particular region along the head–tail axis of the early embryo. This gives a set of green hidden colours distributed in a particular order from head to tail, with apple-green at the head end through to herb-green at the tail end. If the vertebrate homeobox genes were also involved in the head-to-tail layout, they might also be expressed in a similar order of territories in the organism. To examine this, vertebrate embryos were stained to reveal where the various homeobox genes were being expressed: where they were on or off.

Before describing the results, I need to give you an idea of what vertebrate embryos look like. Figure 7.5 illustrates human embryos at various stages of development, from the second to the fifteenth week after fertilisation. They are shown at about their natural size. You can see that there is a distinct head and tail end to embryos even at the earliest of these stages. As development proceeds, various structures, such as the limbs, gradually appear. All mammalian species follow a very similar course of early development, although some may go through the stages more rapidly than others.

Now look at Fig. 7.6, which shows two mouse embryos stained to reveal the

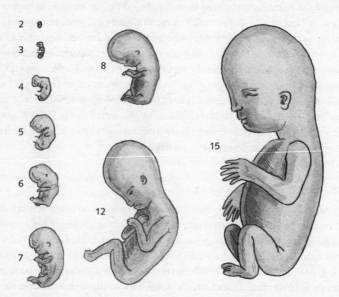

Fig. 7.5 Human embryos from the second to the fifteenth week, natural size. Age in weeks is indicated next to each embryo.

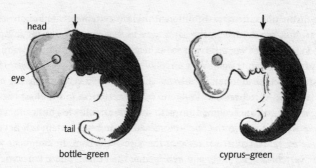

Fig. 7.6 Diagrams representing mouse embryos stained to reveal the expression pattern of a gene needed for bottle-green (*left*) or cyprus-green (*right*). Note that the region of cyprus-green starts further back than bottle-green (*see arrows*).

expression of particular mouse homeobox genes at an early stage of development (a stage corresponding in anatomy to about week 4 in Fig. 7.5). One of the genes codes for an equivalent of the bottle-green protein and the other for cyprus-green protein. You can see that each of these genes is expressed in a discrete region of the embryo, starting at a fixed distance from the head and extended back from this. Most importantly, the starting point for the bottle-green colour is nearer to the head than cyprus-green. That is, the head-to-tail order is the same as in the fly: bottle-green is ahead of cyprus-green. Similar experiments with all the other homeobox genes gave comparable results: the order of expression of the genes in vertebrates was the same as in flies, with apple-green being nearest the head end, followed by bottle-green, then cyprus-green, and eventually ending up with herb-green towards the rear. Together, the various greens gave a series of distinct regions going from head to tail, much as had been observed for fruit flies.

This was a startling result because it showed that there was a common aspect to the organisation of insects and vertebrates: they share an underlying map of colours that follow each other in the same relative order from head to tail. It is reminiscent of Geoffroy's principle of connections, but here it applies to hidden colours rather than bones. Whereas Geoffroy's rule was based on the order of parts of the skeleton, in this case it is the order of hidden colours that is held in common. Vertebrates and insects are united by a set of green hidden colours that are connected or ordered in a similar way from head to tail.

There is still the question, however, of what this map of green hidden colours signifies for vertebrates. In flies, they establish territories in the animal from head to tail, conferring distinctions between the various segments. Take away one or more of the colours, through mutations in the genes, and the identities of some segments are no longer distinguished. Perhaps the hidden colours also provide distinctions between the different parts of vertebrates. Although vertebrates are not segmented in the same way as insects, they are made up of repeated elements of various types from head to tail. This is most obvious in the backbone, which is made of a series of repeating units or *vertebrae* (Fig. 7.7). Going from head to tail in humans there are seven vertebrae in the neck region (cervical vertebrae), followed by twelve vertebrae in the chest region, bearing ribs of various lengths (thoracic vertebrae), followed by another five vertebrae without ribs in the abdominal region (lumbar vertebrae). Beyond this, there are several other vertebrae at the base of the back. In mammals with tails, the vertebrae can continue much further than this. The various types of vertebrae from head to tail are the most obvious place to look for a comparable role of the green hidden colours in vertebrates: perhaps the green colours provide distinctions between vertebrae from head to tail, similar to the way they distinguish between insect segments.

If the hidden green colours are important in providing distinctions between vertebrae, mutants that lack one of the colours might be expected to show alterations in the identity of particular vertebrae. This could be tested in mice because a method had been developed that allowed specific mouse genes to be

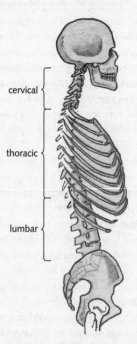

cervical

thoracic

lumbar

Fig. 7.7 Human backbone in side view showing the various vertebrae with and without ribs.

inactivated by mutation. This method was used to inactivate particular homeo-box genes, producing mutant mice that lacked the corresponding hidden colours. Some of these mutants did indeed show striking changes in the identities of specific vertebrae. An example is shown in Fig. 7.8, taken from the work of Hervé Le Mouellic, Yvan Lallemand and Phillipe Brûlet, working in Paris in 1992. As in humans, mice have a series of thoracic vertebrae, each bearing a pair of ribs, followed by a series of lumbar vertebrae that do not have ribs attached to them. The region of transition from thoracic to lumbar vertebrae is illustrated for a normal mouse in the left part of Fig. 7.8.

The diagram on the right of Fig. 7.8 illustrates a mouse which has a mutation in one of its homeobox genes, a gene contributing to the grass-green hidden colour. In this mutant animal, the first lumbar vertebra—normally lacking ribs—has a pair of small ribs attached to it. The lumbar vertebra is displaying a feature, a pair of ribs, normally associated with the thoracic vertebrae that lie ahead of it in the backbone. Indeed, given that thoracic vertebrae are defined as those that bear ribs, we might say that the first lumbar vertebra has been replaced by a thoracic vertebra: its identity has changed from lumbar to thoracic. In other words, as with flies, the hidden colours are needed to provide distinctions along the head–tail axis, in this case between vertebrae. Mutations giving a loss of colour lead to a lack of distinction, such as a lumbar vertebra assuming a similar identity to a thoracic vertebra. You will notice that the transformation is not complete: the extra set of ribs is much smaller than the ribs that come off a normal thoracic vertebra. This is most probably because there are four copies of the homeobox gene cluster in mice, which can act to

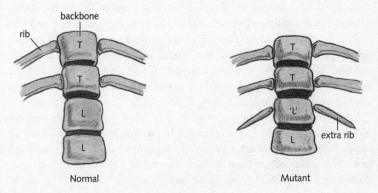

Fig. 7.8 Diagram of part of the skeleton of a normal mouse (*left*) showing thoracic (T) and lumbar (L) vertebrae, compared to a mutant (*right*) which carries a mutation in a homeobox gene contributing to the grass-green colour. Note the extra pair of ribs in the mutant that partially transforms a lumbar ('L') vertebra into a thoracic vertebra.

some extent as backups for each other. Mutation in a gene from one cluster will, therefore, not completely eliminate a hidden colour, such as grass-green, because there are genes in the other clusters that can act to some extent as substitutes.

The analysis of mutations in other homeobox genes gave similar results. Mutant mice that lacked a homeobox gene contributing cyprus-green had alterations in their cervical vertebrae, resulting in the second neck vertebra assuming an identity more like that of the first vertebra.

I have emphasised vertebrae because they are the easiest feature to look at with respect to the head-to-tail axis of vertebrates, but the hidden green colours also affect other distinctions along this axis. In other words, the set of green hidden colours provide distinctions between territories from head to tail that will give rise to various structures during development, of which the vertebrae are the simplest example.

To summarise, as with flies, the green hidden colours of vertebrates are needed to provide distinctions between various regions from head to tail of the animal. Loss or reduction in one of these hidden colours results in a lack of distinction between regions along the head–tail axis, most easily seen as some vertebrae that are normally quite distinct starting to assume similar identities. Thus, the anatomical layout from head to tail in vertebrates depends on an underlying map of green hidden colours similar to that found in insects.

From colour to anatomy

If insects and vertebrates contain a similar set of green hidden colours, how come they end up looking so different from head to tail? Recall that the hidden colours do not correspond to a set of instructions that specify how a structure should be made, like the construction of a particular type of segment: they simply provide a frame of reference that is then interpreted by genes. So although two organisms may have a similar map of green hidden colours, the way this is interpreted and eventually becomes manifest in the anatomy of the animals can be very different.

We are still very far from understanding the precise relationship between hidden colours and the final anatomy of any animal, be it a vertebrate or an insect. Nevertheless, I shall try and outline some of the contributing factors to give a better idea of why these animals can end up looking so distinct, even though they share a similar underlying map.

In Chapter 5, we saw that hidden colours correspond to master proteins that can bind to particular sites in the regulatory regions of interpreting genes. The emerald-green master protein, for example, may recognise one short sequence of DNA, an E-site, whereas the grass-green protein recognises a slightly different sequence, a G-site. By binding to these sites, the master proteins can influence whether an interpreting gene is on or off. In the simplest scenario, an interpreting gene with an E-site in its regulatory region will be switched on wherever the emerald-green protein is to be found in the organism. Similarly, a gene with a G-site will be expressed wherever the grass-green protein is located. Now although vertebrates and insects have a common map of green master proteins along the head–tail axis, many of the interpreting genes that respond to each master protein might be different in each case. We can imagine, for example, that many of the interpreting genes with an E-site in insects may be quite different from those with an E-site in vertebrates. During the long evolutionary time that has separated vertebrates from arthropods, the sites in the regulatory regions of interpreting genes could have changed. This would mean that the interpreting genes expressed in the emerald-green protein territory in vertebrates would differ from those expressed in the emerald-green territory of insects. The map of colours may be the same but their interpretation would differ.

The situation is, of course, much more complicated than my simplified scenario implies. For one thing, interpreting genes have several binding sites in their regulatory regions, allowing them to respond to a combination of hidden colours in the organism. There are many other types of master protein, in addition to the green family, each contributing a hidden colour to give a complex overlapping patchwork. The response of a gene to one or more green colours has to be set in the context of these other hidden colour patterns, some of which might be similar between vertebrates and insects, whereas many others might differ. Furthermore, the proteins encoded by the interpreting genes may themselves have evolved so that their effects on the organism can be different in insects and vertebrates. The way the interpretation of the hidden colours eventually leads to the visible structures of the animal is a very complex affair that can differ between insects and vertebrates in numerous ways.

All of this means that the process by which a set of hidden colours is interpreted and eventually becomes manifest in the final anatomy of the organism is likely to change in many ways during evolution. In insects, the green hidden colours eventually become manifest in the various types of segments, whereas in vertebrates their effects become apparent in other structures, such as the types of vertebrae. So even though the head-to-tail layout of insects and vertebrates depends on a similar map of hidden colours, you might never guess this on the basis of their anatomy alone.

A hidden unity

We can now take another look at the debate between Geoffroy and Cuvier. Cuvier's emphasis on function led him to see insects and vertebrates as being constructed on entirely different principles. There was no meaningful way of comparing these two types of animal. Geoffroy on the other hand was struck by the similar arrangements of parts in groups of animals, such as the mammals, irrespective of what particular function the parts served. Having revealed what he believed to be an underlying law of similarity, the way the bones were connected together, he tried to extend this law to other animals, going as far as insects. To do this he had to establish some sort of correspondence between the parts of an insect and a vertebrate. His solution was to match the outer skeleton of insects with the inner skeleton of vertebrates. Once this was done, detailed correspondences could be drawn, such as between the vertebrate backbone bearing ribs and the outer body of the insect bearing legs. Both systems could be seen to fall under the same umbrella of connections.

With the benefit of hindsight, we can see that there were some merits to Geoffroy's view. There is an underlying set of connections that is similar between insects and vertebrates: the map of hidden colours. Both types of animal have a family of green hidden colours that are arranged or connected in the same order from head to tail, apple-green at the head end, through to herb-green at the tail end. In both cases, the colours provide regional territories, leading to structures with distinct identities developing along the head–tail axis. However, the way that this common map is interpreted and eventually becomes manifest in the visible features of the organism is quite different in each case. In insects, it becomes manifest as a distinction between the various types of segments of the animal whereas in vertebrates it affects different structures, most notably the types of bones arranged along the backbone. Anatomy therefore reveals this underlying map of hidden colours only indirectly: through the way the map is interpreted and eventually becomes manifest in the animal.

It follows that even if animals have a similar set of underlying colours, this unity might only be dimly perceived at the level of their anatomy. That is why many of Geoffroy's detailed correspondences, such as between vertebrate ribs and insect legs, proved to be unconvincing: he was trying to establish a unity of plan based purely on anatomical features. In this sense Cuvier was right to disparage some of Geoffroy's comparisons. Yet although we might fault Geoffroy on some of his particular claims, his overall insight that there was a common underlying map that unified animals as different as insects and vertebrates did prove to be correct.

From an evolutionary point of view, these limitations of anatomical studies can be seen to depend on the degree of relatedness between the animals being compared. There is no difficulty, even from the most superficial inspection, in establishing the correspondence between a human hand and the hand of an ape. It is more difficult to see the relationship between a human arm and a horse's leg or a whale's flipper. In these cases, if we only look at them from the outside, we might conclude that they are entirely different types of appendage. The correspondence is greatly clarified, however, by looking at the arrangement of bones in the skeleton. What looks superficially different is seen to reflect a common arrangement of bones. But when we get to more distantly related animals, such as insects and vertebrates, even anatomical comparisons become of limited use. Based on these, we might have reasonably concluded that the basic layout of these animals had nothing in common (unless, like Geoffroy, we believed in a fundamental unity of plan). The anatomical layouts could have evolved completely independently of each other. To get a deeper insight, we need to look at the map of hidden colours that underlies the anatomy of the organism. Then we see that there is a unity even though it is interpreted and becomes manifest in very different ways in the anatomy of the animal. The map provides a hidden skeleton, a set of underlying connections that does allow meaningful comparisons between very diverse organisms to be made.

I do not mean to imply that the green hidden colours provide an absolute or immutable map. There have been alterations in the homeobox genes needed for the green colours during evolution, such as extra duplications of individual homeobox genes. Even entire clusters of homeobox genes have been duplicated to give the four clusters present in mice and humans of today. These alterations may have allowed the map of green colours to have been modified to some extent. Nevertheless, the basic order of green hidden colours from head to tail, the connections in the map, does not seem to have changed in a fundamental way during the six hundred million years that separate insects and vertebrates from their common ancestor.

What role might the homeobox genes have had in this common ancestor of insects and vertebrates? It seems most likely that, as with organisms alive today, its homeobox genes would also have coded for a set of hidden colours that gave distinctions from head to tail. In this case, however, the various territories of colour may not have been interpreted to give distinctive segments, like those of flies, nor different type of vertebrae, as in mice. Rather, they could have been manifested in a different way again, as I shall now explain.

A good illustration comes from the study of homeobox genes in the nematode *Caenorhabditis elegans*, a tiny worm-shaped animal (about one millimetre long) that belongs to a quite distinct group (phylum) from insects or vertebrates. This worm has neither segments nor vertebrae, yet it still has a cluster of homeobox genes (the worm has only a set of four genes in its cluster as compared to the eight or more observed in insects and vertebrates). As with the other animals I have mentioned, each worm homeobox gene is expressed in a distinct region of the animal, producing territories of green hidden colour from head to tail. The hidden colours are then interpreted by genes so that different parts of the worm assume distinct identities along the head–tail axis.

In this case, though, the identities do not refer to segments or vertebrae, but to groups of cells that are repeated along the worm's length. These cell groups are not surrounded by a hard outer casing, like insect segments, nor do they produce bony regions like vertebrae; they simply form characteristic regions of the worm. In other words, the hidden colours still provide a map or frame of reference from head to tail, but this is interpreted and eventually becomes manifest differently for the worm than for vertebrates or insects.

We can conclude that the common ancestor of insects and vertebrates (and nematode worms) most likely had a cluster of homeobox genes that provided distinctive identities along its head–tail axis. However, the way this hidden map was interpreted and became manifest in this early animal was probably quite different from what we see in humans or flies today. There is a remarkable unity in the map of some hidden colours between animals, which has been preserved for hundreds of millions of years; yet the way this becomes manifest in their anatomy can be very different. We shall return to the question of why such a deep unity can be discerned with hidden colours in some of the chapters towards the end of this book. In the next few chapters I want to deal with a fundamental problem that I have so far skipped over: how the pattern of hidden colours is itself established during development.

CHAPTER 8

The expanding canvas

So far in this book, I have described how the development of organisms depends on a patchwork of hidden colours. This patchwork provides a frame of reference that can be interpreted by numerous genes (interpreting genes), allowing them to be expressed at particular times and places in the organism. This pattern of gene activity, in turn, underlies the complex anatomy of plants and animals. But I have yet to explain how the patchwork of hidden colours itself originates. I have taken it for granted that the patchwork is there without explaining how it came about.

In the next few chapters, I shall try to deal with this problem. I will show that the pattern of hidden colours arises through a chain of events, involving one set of hidden colours building on another set of hidden colours, which in turn depend on another set. In the present chapter I want to give a broad sense of what is being achieved as hidden colours build on each other in this way. To do this, I will draw an analogy between establishing a patchwork of hidden colours and painting a picture. In both cases, a complex arrangement of distinctive regions is produced: in the case of a painting, this is done within the confines of a two-dimensional canvas; whereas in biological development, the territories of hidden colour are distributed within the three-dimensional framework of the organism. In following this analogy, I shall have to introduce some bizarre notions, like canvases that grow and expand or artists that multiply and proliferate. Although these images may seem fanciful at first sight, their purpose is to highlight some of the salient features of development.

Refining a pattern

There are many different ways of painting a picture. Perhaps one of the most systematic approaches is painting by numbers: a picture is carefully drawn out and each region labelled with a number to indicate where each colour should go. All the spatial information is already there in the drawing; the act of painting just transforms numbered regions into coloured ones.

The trouble with painting by numbers as an analogy for the origin of a hidden patchwork is that it does not add any new spatial information. If we were to say that each territory of hidden colour arose by filling in a series of numbered regions, it would simply beg the question of where the numbered regions came from. Indeed, in the sense I am using hidden colours—as a way of distinguishing one territory from another—they are essentially no different from numbered regions (although they are more convenient than numbers for conveying territories). We would be trying to account for one set of territories simply by reference to an identical set of underlying territories that is coded in a different way, hidden numbered regions rather than hidden colours. It takes us no further because the origin of the numbered regions is just as difficult to explain as the origin of the colour patterns.

There is another style of painting which provides a much more useful analogy than painting by numbers. Look at Fig. 8.1, which shows three stages during a self-portrait, taken from a teach-yourself book called *Creating a Self-Portrait*. In the first stage, the artist has broadly defined a few regions of colour, showing approximately where the face, hair, and shirt will go. At this stage, the outlines of each region are somewhat blurred and crudely defined. In the second stage, colours have been laid over the first ones to provide more detail: the position of the eyes, nose, mouth, shirt collar, etc. The first stage has acted as a broad framework that has then been refined during the second. This process of refinement has continued even further in the third stage, which now shows details such as the eyes' pupils, nostrils, chequered markings on the shirt, etc. We can see that each stage is a response to what went before, with colours being added to build up and refine the picture, the brush strokes becoming finer and more detailed at each step. This is how one of the other artists mentioned in the same self-portrait book, Roy Freer, describes his method:

> My approach to watercolour painting is very similar to that with oil or pastel. I start with a broad description of the essential masses and then progress through several stages of refinement until I have achieved the final result.

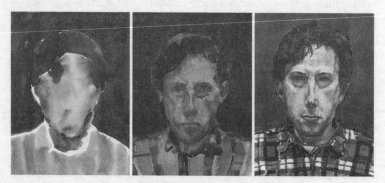

Fig. 8.1 Three stages in a self-portrait by Francis Bowyer (from Coates 1989).

Why doesn't the artist go directly to the detailed last stage, circumventing the other steps? What purpose do the previous stages serve? One problem with going too quickly into painting details is that they can be easily misplaced. If you start by painting the eyes in great detail, you might find later on that they are not in quite the right place in relation to the other features. You would have to go back and start all over again. By first giving the overall layout and gradually narrowing down to the details, the correct relations between features can be established gradually at each scale, from the overall location of the face, to the position of eyes, to the details within each eye. To help them achieve this, some artists screw up their eyes when looking at the subject during the early stages of a painting, blurring their vision on purpose so as to more easily see the broad arrangements and avoid being distracted by details.

Basing one set of colours on a previous set in this way can help to refine the details of a picture. We shall see that this provides a good analogy for how hidden patchworks arise during biological development. The pattern of hidden colours is built up and refined gradually as an organism develops rather than being simply applied in one step to a pre-existing framework, as occurs in painting by numbers. Hidden colours are established through an interactive process in which one set of colours depends and elaborates on what went before.

The growing canvas

There is an important aspect of internal painting that is not captured by the visible painting process as I have described it so far: organisms can *grow* at the same time that their patterns are being refined. It is as if the canvas is not of a fixed size but it is continually enlarging. The extent of growth that can occur during development is quite remarkable. Here is how the eighteenth-century biologist Albrecht von Haller described the situation:

> The growth of the embryo in the uterus of the mother is almost unbelievably rapid. We do not know what its size is at the moment of its formation, but it is certainly so small that it cannot be seen even with the aid of the best microscopes, and it reaches in nine months the weight of ten or twelve pounds. In order to clear up this speculation, let us examine the growth of the chick in the egg. We cannot in this case either measure its size at the moment when the egg is put to incubate but it cannot be more than 4/100 in. long, for if it were, it would be visible, and yet 25 days later it is 4 ins. long. Its relation is therefore as 64 to 64 millions [in weight] or 1 to 1 million.

Since Haller's time, the details of human embryonic growth have been established. At the time of fertilisation, the size of a human egg is about 0.1 mm in diameter and it weighs about one-millionth of a gram. After about seven weeks, it has become about 17 mm long with a weight of one gram: an increase in weight of one million fold. By the time an adult is formed, the size will have increased by a further factor of about one hundred in linear dimensions and one hundred thousand in weight. Overall, the growth from fertilised egg to adult will involve an increase by factors of about ten thousand in linear dimensions and one hundred billion in weight. Similar calculations can be made for plants. A typical plant egg cell is of the order of one-hundredth of a millimetre in diameter, yet it may grow to form a tree more than ten metres tall with roots perhaps extending for a comparable distance underground: an increase of about one million fold in linear dimensions. We therefore need to think of hidden colours being built up within a growing framework rather than one of fixed size.

In terms of the painting analogy, we can imagine the canvas expanding whilst the painting proceeds. I have illustrated this in Fig. 8.2 by showing the same three stages of the portrait in Fig. 8.1, but now with their overall size increasing as they go through the various stages. Between each stage, I have enlarged the portrait by a factor of 4 in linear dimensions, corresponding to a factor of 16 in area, giving a total enlargement over the two stages of 16 times in linear dimensions, or 256 times in area. Of course in this case, growth only occurs in two dimensions, whereas an organism grows in three dimensions.

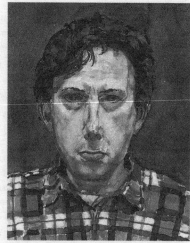

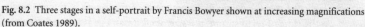

Fig. 8.2 Three stages in a self-portrait by Francis Bowyer shown at increasing magnifications (from Coates 1989).

You will notice that in enlarging the canvas in this way, instead of the artist having to apply finer and finer brush strokes at each stage, he or she can continue to work at a constant scale. This is because it is the scale of the canvas that changes rather than the brush work. Indeed, I have chosen the canvas sizes to make this point, keeping the scale of the newly applied brush strokes about the same at each stage. The overall portrait is being refined as it enlarges, because the quality and detail of the picture increase. However, the scale at which the refinements are being made has not changed at all, because the absolute size of each newly applied brush stroke is unaltered.

In one way, working on a growing canvas would greatly simplify the painting process. We would not need a whole series of brush sizes to elaborate the picture: we could repeatedly use a standard brush of fixed scale, and let the expanding canvas do the rest for us. To define a large area of the picture with a broad sweep of colour, we would not need a large brush, but could simply use the standard brush early on, when the picture is small. To mark some fine detail, we could use the same size brush again but wait until the canvas is relatively large.

A similar consideration applies to the elaboration of hidden colour patterns during the development of organisms. There are several examples I could give to illustrate this point but I shall choose one that happens to be familiar to me.

Painting in bud

In 1987, I was walking together with my colleague Rosemary Carpenter through a field of *Antirrhinum* plants, looking for mutants with altered flowers. We were rather stunned to come across a plant with no flowers at all, amidst a sea of normally flowering plants. Now a plant with no flowers sounds about as useful as a bucket with a hole in it. But because of the reverse logic of genetics, a mutant that lacks flowers most likely has a defect in a gene whose *normal* significance is to promote the formation of flowers. This mutant therefore offered an opportunity to study a gene normally involved in establishing flower development.

Looking more carefully at the mutant, we saw that wherever a flower would normally arise, a leafy shoot was being produced instead. We named the mutant *floricaula*, or *flo* for short, because a stem (Latin *caulis*) was replacing each flower. To clarify what has happened in the *flo* mutant, I will first need to describe two types of bud on a normal *Antirrhinum* plant: shoot-buds and flower-buds.

(1) A *shoot-bud* produces a stem bearing leaves. During the early period of a plant's life, when it is a seedling, there is a single shoot-bud at the tip of the plant which is responsible for most growth above ground level. The shoot-bud

climbs ever upwards as it generates more stem and leaves beneath it, just as a man standing on a brick tower may raise himself higher by adding more bricks to the structure beneath him. If this was all there was to plant growth, the plant would simply form a long stem with leaves around it. However, additional buds are also produced on the side of the plant, located in the angle between the leaves and the stem. These additional buds may themselves be shoot-buds; that is, they may also produce stem and leaves to form a shoot or branch coming out from the side of the plant. In the early phases of plant growth, all the extra buds will be of this type: all will form shoots. But later on, buds of a different type are produced: flower-buds.

(2) *Flower-buds* grow to form flowers comprising four whorls of organs: sepals, petals, stamens and carpels. Unlike shoot-buds, which can continue growing almost indefinitely, the growth of a flower-bud is strictly limited to making only the four whorls, after which it stops making any more organs. The flower-buds are produced at a relatively late stage of plant life, and are borne from the upper part of each stem, in the so-called inflorescence or flowering region. This region is shown schematically in the left part of Fig. 8.3 (the leaves in this region are smaller than those lower down on the plant and are sometimes referred to as bracts).

So much for the normal *Antirrhinum* plant. If we now turn to the *flo* mutant for comparison, we note that all the flower-buds have been replaced by shoot-

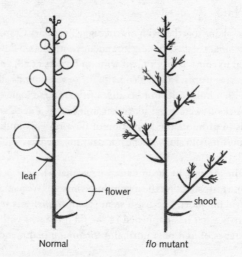

Fig. 8.3 Schematic illustration of flowering region (inflorescence) of a normal *Antirrhinum* plant compared to the corresponding region in a *flo* mutant. Note that in the mutant, each flower is replaced by a reiterating shoot.

Fig. 8.4 *Circle Limit 1* (1958), M. C. Escher.

buds (see right part of Fig. 8.3). Wherever a flower would normally have formed, side-shoots are now produced. These side-shoots can in turn produce further shoot-buds in the angles of their leaves, which will grow to form even more side-shoots. The process can continue in this way, reiterating shoots upon shoots, reminiscent of a picture by M. C. Escher (Fig. 8.4), where the same image is endlessly repeated (such patterns, in which the same level of detail is repeated at an ever smaller scale, are known as fractals). It is as if there is a failure in *flo* mutants to distinguish between two different types of bud: flower and shoot. Without this distinction, shoot-buds are continually reiterated by default.

As with some of the other mutants I have described, we can think of the *flo* mutant as having lost a hidden colour. For convenience, I shall name this hidden colour *floral-pink*. The floral-pink colour normally gives a distinctive identity to flower-buds. If a bud has this hidden colour, it will go on to form a flower; if not, it will form a shoot. In normal plants, floral-pink will only be present in the buds of the upper part of the stem. Being coloured floral-pink, these buds will go on to form flowers rather than shoots. In *flo* mutants, floral-pink is entirely missing in the plant so the only type of bud that can now be produced is a shoot-bud. Thus, in the absence of floral-pink, all the buds that would have made flowers now form shoots instead.

Putting this in terms of genes, we can say that the gene affected by the *flo* mutation is normally needed to produce the floral-pink hidden colour. By convention, genes are named after the mutations that affect them, in this case *flo*. We may summarise by saying that plants that have a mutation in the *flo* gene are unable to make floral-pink, and therefore produce shoot-buds in place of flower-buds.

We can compare floral-pink to some of the other hidden colours I previously mentioned: those that distinguish between the whorls of organs in a flower (sepals, petals, stamens and carpels). The distinction between these organ types depends on three hidden colours, symbolised by amarone-red, burgundy-red and claret-red (Chapter 6). The combinations of these reds establish four concentric regions of colour, providing distinctions between the whorls of organs. Floral-pink plays a comparable role to these various red colours, except that it affects the identity of the *whole* bud rather than specific whorls within the bud. In other words, the effect of floral-pink is more comprehensive than any of the reds, influencing the identity of the entire bud rather than a subregion within it.

I previously described how the distribution of the various red colours could be directly seen in the bud, by staining with probes that specifically detected their corresponding RNA or protein molecules (Chapter 5). How does their distribution compare to that of floral-pink? Fortunately, our research group was eventually able to isolate *flo*, the gene needed to produce floral-pink. We could then make a probe to look at where the *flo* gene was expressed in the normal plant (i.e. where *flo* was switched on to make RNA and protein). Effectively this allowed us to determine directly when and where the floral-pink colour was being produced. The results showed that *flo* was expressed only in flower-buds, at a very early stage in their development. At this stage, each bud is just a tiny group of cells, about one-thirtieth of a millimetre across. This is before the time that the various hidden reds appear: they are typically seen a few days later when the bud is about one-fifth of a millimetre across.

We can conclude that floral-pink appears in the bud *before* the various reds, when the size of the bud is *smaller*. I have shown this in Fig. 8.5, where the flower-bud is schematically shown as a two-dimensional disc. At the early stages, when the bud is very small, floral-pink is produced and this establishes a distinctive hidden colour for the bud as a whole, ensuring that it will form a flower rather than a shoot. Later on, when the bud has grown a bit bigger, the

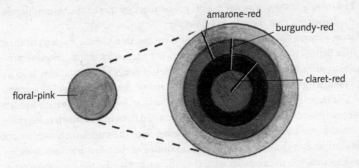

Fig. 8.5 Flower-bud represented as a disc at two stages of development. In the earlier stage (*left*) the entire bud has the floral-pink hidden colour, distinguishing it from a shoot-bud. Later on (*right*), when the bud has enlarged, concentric patterns of various red hidden colours arise, resulting in distinct identities for the four whorls of organs.

concentric pattern of reds appears, defining the territories that result in distinct identities for the four whorls of organs. The scale of colouring is about the same at each stage, but the size of the framework is different. More comprehensive territories, such as those spanning the entire flower-bud, are established earlier on, when the bud is very small. More detailed territorial distinctions, such as those between whorls, are depicted later on, when the bud is larger. As with painting a canvas, the broad areas are defined before the narrower ones, but because the framework is growing, the scale of colouring remains about the same.

There is a further parallel with painting: the later colours *depend* on the earlier ones. If floral-pink is missing from the early buds, as is the case in *flo* mutants, the concentric pattern of red territories will not appear later on. One set of hidden colours depends on those that went before, just as when painting a canvas each colouring step depends on the colours that were previously applied. We shall encounter this notion again in later chapters when I try to explain in more detail how hidden colours can elaborate on each other.

Multiplying artists

So far I have emphasised growth in terms of the overall size of the developing organism, but there is another important aspect of growth that also needs to be considered. In addition to growing in size, multicellular organisms also grow in terms of cell numbers. As an individual develops, the cells divide to give more cells, which can in turn divide again and again. In many cases, the increase in cell number and the growth in overall size go together: the organism gets bigger at the same time that the cells are dividing.

Now in terms of our painting analogy, we can think of the nucleus of each cell as being equivalent to an artist. This is because the application or production of hidden colours depends on a set of genes in the nucleus, such as the identity genes needed for the green or red families of colour. The expression of these genes, whether they are on or off, will determine which hidden colours (master proteins) appear, just as an artist determines which colours are applied to the canvas. So if the cells, together with the nuclei within them, are dividing and multiplying while the overall size of the organism increases, we have to imagine the number of artists increasing whilst the canvas enlarges as well.

To give a sense of what this means, imagine a square canvas laid on the ground. It is initially 1 metre on each side, but gradually enlarges until it gets to a final size that is 10000 times greater in linear dimensions, making a square that measures 10 kilometres along each side. This would correspond to the relative growth in linear dimensions of an egg developing into a human adult. Initially, there is just one artist standing on the canvas, applying, say, an overall background colour. A bit later, the canvas has enlarged and there are now a few more artists, each standing on a section of canvas with their own brushes, roughly painting in the basic layout of the picture on top of the earlier colour. As the canvas enlarges further, the artists also multiply, each now working on and elaborating one of the rough areas marked out earlier on. A further enlargement leads to yet more artists joining in, each using the colours previously placed on the canvas as a framework for further elaboration. As far as the artists are concerned, they are always working and making patterns at about the same scale, but the canvas as a whole is getting more and more intricate with a finer pattern. Each artist adds more refinement to his or her part of the picture, the overall elaboration coming from successive refinements of the same type.

All the artists are identical (each nucleus has the same set of genes) but they end up doing different things because they are working on different parts of the canvas. Each artist *reacts* to what is already present on his or her part of the canvas as well as adding colours, elaborating the pattern in some way. All of this is carried out at the local scale of each artist, so an artist working on one part may have no idea what the overall picture looks like. In fact, only the first few artists will have been aware of the whole canvas at a very early stage, when the picture was in a very crude form.

This notion of a growing canvas, with the artists multiplying at the same time, captures many of the key features of the process by which patterns of hidden colour are established within organisms. The process is not, however, always as regular as this. For example, during an early stage of development in many animals, the fertilised egg goes through a series of divisions without any overall growth in size, so the egg becomes divided up into smaller cells, a process called *cleavage*. This would correspond to a phase of painting when the canvas is relatively large and the artists multiply up whilst the same overall size of the canvas is maintained.

Back to the beginning

So far I have considered development from the time of fertilisation to the appearance of a mature organism. I now want to complete the life cycle and look at how the fertilised egg, the starting canvas in the analogy, itself originates. The DNA in the nucleus of the fertilised egg comes from both the mother and father: the egg and sperm cells each contribute a half-share of DNA (Chapter 2). Now although both sexes make an equal donation to the nucleus, their contribution to the rest of the cell—the cytoplasm with its surrounding membrane—is far from equal. The bulk of this is derived from the *mother*. The egg is much larger than the sperm and its contribution of cytoplasm is therefore predominant. The female's egg can be thought of as both the product of one generation, the mother, and also as providing the starting material for a new generation.

In terms of the painting analogy, we need to imagine that the material in the initial canvas is largely derived from a piece of mother canvas. I have illustrated this in Fig. 8.6, using Leonardo's *Mona Lisa* as the mother figure. We normally think of this picture as the end product of a painting process, but I now want to consider it as also providing the source of a fresh painting. Imagine taking a tiny portion of this picture to provide a starting canvas for a new painting. Two such fragments of the *Mona Lisa* are shown magnified in Fig. 8.6, to illustrate what a starting canvas derived in this way might look like. The original *Mona Lisa* is about 50 × 80 cm, and these fragments correspond to a region of 2 × 2 cm, so they each represent about 1/1000 of the total area of the painting. They have a relatively simple overall colour, which may be entirely uniform (upper fragment) or still have some variation in colour (bottom fragment). In either case they clearly lack the complexity of the whole *Mona Lisa*.

We can now think of the life cycle of a multicellular organism in terms of a cycle of paintings. Starting with a relatively small uniform canvas, colours are gradually built up and elaborated by an ever growing number of identical artists as the canvas enlarges. Eventually a large complex painting, such as the *Mona Lisa*, is produced. A small piece of canvas is then removed from the painting, which acts as the starting material for a new painting. This starting canvas does not depict the whole painting that will later emerge, it just provides a relatively uniform framework upon which a new painting can be built. The new painting

Fig. 8.6 The *Mona Lisa* (1503–6), Leonardo da Vinci, (Louvre, Paris), with two tiny parts of the canvas shown magnified.

will not be identical to the *Mona Lisa* because the artists working on it will derive half of their character from the father. So although the mother provides the initial canvas, the way it is further elaborated will depend on contributions from both mother and father. (In the case of female aphids that can reproduce without sex, the situation is simpler and would be equivalent to continually generating the same picture: lots of identical *Mona Lisas*).

This painting cycle gives an overall sense of what is being achieved when a pattern of hidden colours is being elaborated as an organism develops. The egg starts off with a relatively simple distribution of hidden colours, largely derived from the mother. After fertilisation by a sperm cell, the pattern of hidden colours starts to become more elaborate as the egg divides and grows, giving more and more cells. We might call this whole process of hidden colour elaboration and refinement *internal painting*. The information needed for this process is contributed equally from the sperm and egg, through their DNA, but it is the egg that provides the bulk of the starting material, including the initial hidden colours. Eventually a mature organism develops with a complex internal patchwork of hidden colours. A tiny portion of this organism, its egg or sperm cells, then contributes to the next generation. The egg contributes the bulk of the starting cell material to the offspring, together with a half-share of nuclear DNA. The sperm's main contribution is a half-share of DNA, with relatively little else. With this overall view in mind, I now want to deal with the problem of how the patterns of hidden colour can be elaborated during development: how the process of internal painting proceeds from one stage to the next.

CHAPTER 9

Refining a pattern

Several years ago I went to an evening class on learning how to paint. I sat down at an unoccupied easel, while the teacher carefully arranged various vegetables and an empty bottle of wine on a table-cloth. 'Today we're going to do a still life', he announced. The other members of the class, who were mostly regulars, then set about busily sketching and painting the scene, while the teacher went around commenting on their various efforts: 'Very good, John, but what about some darker tones over here' or 'I like the green, Anne, but I wonder if your bottle is a bit on the small side.'

Finally he got round to me. He looked at my blank piece of canvas board. 'You don't seem to have got very far', he said. 'I've no idea where to start', I explained. He then suggested I begin by depicting the various objects on the table with some broad areas of overall colour. I followed his advice and found that setting out some coloured regions in this way was indeed a very effective way of getting started.

In this chapter I shall describe some first steps in internal painting—the process whereby hidden colours are elaborated during development. The starting point will be a fertilised egg, primed with some basic hidden colours from the mother. This will provide the initial canvas. We will then see how more hidden colours are added to this framework, giving the early embryo a pattern of distinctly coloured regions, much as an artist might lay out some broad areas of colour early on. Unlike an artist painting a picture, however, the internal painting will all be done with no hands: it will involve only a series of molecules interacting with each other.

To describe this in a satisfactory way, I will need to go into some of the detailed mechanisms involved. These may seem rather complicated at first, but having gone through them, a relatively simple overall message will start to emerge. The reader might wonder whether it is possible to avoid going through all the details and go straight to the major conclusions. My answer would be that it is only by getting an appreciation of the mechanisms involved that we can begin to understand how a pattern can be elaborated purely in molecular terms, without the help of an external guiding hand. Once this has been funda-mentally understood, we will be able to look again at the question of how biological development compares to a creative process.

I have chosen fruit flies as my example of internal painting. This is because research on flies has made such remarkable progress in the last two decades that flies now provide the best understood case of early development. Before describing how the system works, I want to explain the experimental approach that played a key part in unravelling the story.

Screening the dead

The reverse logic of genetics says that if you want to understand a process, you begin by looking for mutants in which the process is defective or altered in some way. To study early fly development, you might therefore look for mutants in which fertilised eggs do not develop properly. Perhaps you could screen lots of animals, searching for rare mutants in which the embryos do not form normally. By studying the genes that have been altered to give these defective embryos, you might then be able to build up a picture of how development normally occurs. Although this might sound fine in principle, there are several problems with this strategy.

First of all, any fundamental defect in early development is likely to give rise to an embryo that dies long before adulthood. In the case of humans, ten to fifteen percent of all pregnancies terminate in spontaneous abortion within twenty weeks of fertilisation, often because of major abnormalities in the developing embryo. In flies, the remains of the dead embryos are to be found in eggs that fail to hatch. Looking inside these eggs and trying to see what has happened to the embryos can be a very time-consuming business, particularly if you have to look at many dead embryos in this way.

Another problem with screening for mutant embryos is that it is more difficult to study mutants when they leave no direct offspring. If you are trying to analyse a defective gene in detail, you have to be able to keep it going in some way. This is no problem if your mutant survives to adulthood because its defective gene gets passed on, together with its other genes, to the next generation of offspring. A mutant that dies as an embryo, however, leaves no descendants. (There is a way round this but it is rather laborious: the defective gene has to be propagated indirectly, by breeding from parents or siblings of the dead embryo. This is possible because every gene is present in two copies, one coming from each parent. Typically both of these copies need to be defective to have a significant effect on an organism. So although both copies of a gene will be defective in a dead embryo, the defective gene might still be present in a single dose in the normal-looking parents or siblings.)

Apart from these practical considerations, there is a more theoretical objection to screening embryos for mutants. Early development may be so complicated and finely balanced that any mutation with a significant impact on it might completely mess the whole thing up. The mutants obtained would be so deranged that it would be impossible to make head or tail of them: working out what had happened would be like trying to decipher scrambled eggs. So even if embryo mutants could be obtained, they would not tell you very much because of their being so badly messed up.

For all these reasons (amongst others), the traditional wisdom of fruit fly genetics was to concentrate on mutations affecting the *adult* rather than the *embryo*. By analysing how mutations affected the characteristics of adult flies, you could then try to work out how the fly embryo developed in the first place. Looking directly for mutant embryos was not considered to be the most fruitful avenue of research. It was not until the late 1970s that all this changed, with a series of screens for embryo mutants pioneered by two scientists: Christiane Nüsslein-Volhard and Eric Wieschaus, working in Heidelberg.

Nüsslein-Volhard realised that if they were going to have any success at all in screening for these sorts of mutants, they would need a quick and easy method for looking at embryos. To do this she used a special solution that made most of the egg contents transparent and allowed the *cuticle* or hardened skin of the embryo to be seen very easily. You may recall that Ed Lewis developed a similar procedure to sort out mutants of the Bithorax Complex that died early (Chapter 4). But whereas Lewis was using this method to study mutations he had already produced, Nüsslein-Volhard wanted to use it for a primary screen, to identify the mutants in the first place.

Nüsslein-Volhard and Wieschaus then tried out the method on some existing stocks of flies that had been reported to lay some defective eggs (i.e. eggs that never hatched). When they looked carefully at these eggs, they saw that some of them contained mutant embryos with very specific defects in their cuticle pattern. The embryos were obviously not all scrambled up, as conventional wisdom might have predicted, but had very precise alterations, like particular regions of the body being missing. This was very important because it meant that embryo mutants might reveal genes with specific roles in early development. The mutations were not perturbing development in a non-specific way, they were affecting genes with particular roles that might be deciphered. Encouraged by this, Nüsslein-Volhard and Wieschaus decided to go ahead and screen a large population of flies for more embryo mutants. Because of the complication that their mutants would die, this was quite a challenge in logistics, involving lots of flies being kept and recorded systematically. Nüsslein-Volhard, who had a bent for strategic planning, recalls how they set about it:

> I organised it more or less because I was the most practical person; I mean I am very untidy, but when I want to do an experiment, I try to be very well organised. Otherwise we did everything together. We had four people working with us: two technicians and two animal caretakers. One collected virgins (unfertilised female

flies) all the time and the other one made cuticle preparations and the other technicians scored plates and fixed embryos. When you start to get going the flies come, the flies come!

In the evenings Eric and I looked at the prepared embryos together. It was very necessary that this was done by two people. We had this microscope with a bridge where two people could look at the same preparation at the same time, very very important. And we had our discussions.

Through their observations and discussions, they classified and sorted out many different types of mutants, each with a specific defect in the embryo. This early screen was followed by several others, establishing a systematic collection of mutants with altered embryo patterns. Once these mutants had been identified, it was then possible to isolate some of the genes involved and study them in detail. They were mining a seam of gold, providing an invaluable collection that was to allow the early events in fly development to be unravelled for the first time.

Bicoid makes black

I now want to describe one of the mutants that emerged from these sorts of embryo screening experiments: a mutant called *bicoid*. Normal flies can be divided into three main regions: head, thorax and abdomen. The *bicoid* mutant embryos had a rather fundamental defect: they seemed to have almost no head or thorax! The appearance of the *bicoid* mutant embryo compared to a normal one is shown in Fig. 9.1.

To get an idea of what an equivalent mutation might look like in humans, look at the painting in Fig. 9.2 by René Magritte—although in this case, it is an adult that lacks the head and chest, rather than an embryo.

Because the head and thorax are missing in the *bicoid* mutant, the *normal* significance of the *bicoid* gene (i.e. the gene altered by the *bicoid* mutation) is to *promote* the formation of a head and thorax. The *bicoid* gene normally plays a role in establishing the head and thorax regions of the body, so without it, these regions fail to develop. We are talking about a gene that has a very

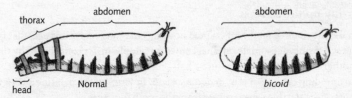

Fig. 9.1 Normal fruit fly embryo compared to *bicoid* mutant (side view).

Fig. 9.2 *The Symmetrical Trick* (1928), René Magritte. Private collection.

fundamental role in development: it is needed for a major part of the animal to develop, and a pretty important part at that. Without a head or thorax, a fly would not be much to speak of.

The *bicoid* gene codes for a master protein—a protein that can bind to genes and switch them *on* or *off*. Master proteins are equivalent to hidden colours (Chapter 5), so we can think of *bicoid* as producing a particular type of hidden colour. For convenience, I am going to call this hidden colour, and its corresponding master protein, *black*. It will help if you can remember that *bicoid makes black*. The black hidden colour is needed for the region including the head and thorax to form. In *bicoid* mutants, no black is produced and so this region of the body does not develop.

This black colour will provide a useful starting point for our exercise in internal painting. Before proceeding any further, however, I will need to give a few more details about the starting canvas.

Sizing up the canvas

In the previous chapter I mentioned how the bulk of the fertilised egg, that is, the cell membrane and its internal contents (cytoplasm), is contributed by the mother. Both parents contribute a half-share of DNA to the nucleus but the surrounding cytoplasm is of maternal origin. In terms of the painting analogy, most of the starting canvas comes from the mother picture.

Now, in many organisms the egg cell can grow to be quite large within the mother before it is fertilised. It is as if the new canvas gets a head start, expanding inside the mother to form quite a large expanse. A human egg cell, for example, grows to be about one-tenth of a millimetre in diameter before it is fertilised. This is about five to ten times larger in linear dimensions than a typical human cell. In the case of fruit flies, this expansion of the egg cell is even more dramatic: it ends up being about one-half of a millimetre long. This is an enormous cell, being about one-sixth of the entire length of the female fly. If you were to magnify the fly to human size, the egg cell would be about the size of a rugby ball (or an American football). A female fly may produce more than two hundred of these rugby balls during her life: no mean feat by any standards.

After a large egg cell has been fertilised, it typically starts to divide without much growth in its overall dimensions. The very large cell effectively gets divided up or cleaved into smaller cell portions. Although the total number of cells increases, their average size gets progressively smaller at each division, eventually approaching a size that is more typical for a cell. For example, four days after fertilisation, a human egg will have gone through several rounds of division to give a clump of about one hundred cells, but the overall size of this clump will be roughly the same as that of the initial egg: about one-tenth of a millimetre across. There are more cells but they are each much smaller than the fertilised egg cell. Eventually, this process of cleavage comes to an end and the embryo starts to grow bigger in overall dimensions as its cells grow and divide.

Fruit fly eggs also go through a phase of cleavage without overall growth, but there is an additional complication. After fertilisation, the nucleus in the egg cell divides to give two nuclei, but the membrane around the cell does not divide. In other words, you end up with two identical nuclei in the same cytoplasm. Each nucleus contains the full complement of DNA (derived from both father and mother), but it shares its cytoplasm with the other nucleus. This process repeats itself several times until many nuclei are formed, all immersed in the same cell fluid (Fig. 9.3). At this point the embryo is one large cell with lots of nuclei inside it. The nuclei then migrate to the outer part of the

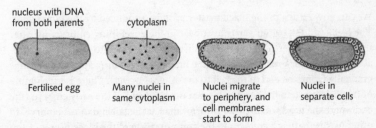

nucleus with DNA from both parents   cytoplasm

Fertilised egg | Many nuclei in same cytoplasm | Nuclei migrate to periphery, and cell membranes start to form | Nuclei in separate cells

Fig. 9.3 Early development of the fertilised egg of the fruit fly.

large cell, where a membrane starts to form around each nucleus. The free nuclei have now become enclosed by membranes to form separate cells, giving a ball with an outer layer of about six thousand cells (Fig. 9.3). Throughout this process there is no overall growth, so the final ball of six thousand cells is about the same size as the initial fertilised egg (one-half of a millimetre long).

You can think of these cells of the embryo as having arisen in two phases. In the first phase, nuclei divide freely in the same cell fluid, with no cell membrane separating them. In the second phase, the nuclei acquire their own bit of membrane, collectively forming a ball of six thousand cells. All of this happens very rapidly: it takes about three hours from having a single nucleus to becoming a ball of six thousand cells. The various structures of the fly larva, such as the head and body segments, then develop from this ball: the head forming from one end and the tail at the other.

Putting this in terms of painting, the new canvas starts off very large, with one artist (the nucleus) in the middle. The artist then starts to divide to give two artists, which in turn multiply, so that eventually the large initial canvas is populated with thousands of artists. During this process there are no barriers (cell membranes) that clearly demarcate the bit of canvas each artist is painting; the artists just work on the region around them. Later on, some lines of demarcation are drawn, so that the activity of each of the six thousand artists becomes more confined. Throughout this process, the overall size of the canvas does not change.

Whilst this example of development may be slightly atypical, it does simplify our initial painting exercise in two respects. First of all, because there is no overall growth, we do not need to worry about the canvas expanding at the same time as it is being painted. Secondly, the lack of barriers between nuclei early on means that there is relatively free communication between different zones of the developing embryo. This will greatly simplify matters when we have to consider how the hidden colours in nearby regions are coordinated.

A gradient of colour

We can now return to the black master protein, produced by the *bicoid* gene. The black hidden colour starts to appear in the developing embryo just after the egg has been fertilised. If the fertilised egg is probed to reveal the whereabouts of the black master protein, the intensity of staining appears to change gradually from one end of the cell to the other (Fig. 9.4). There is much more of the black protein in the cytoplasm at the head end of the early embryo (the end where the head will eventually form) than at the tail end of the embryo. In other words there is a *gradient* in the concentration of the black protein from the head to the tail end of the cell (by concentration I mean the number of black

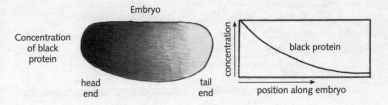

Fig. 9.4 Gradient of black protein in the early fruit fly embryo.

master protein molecules per unit volume, equivalent to the *intensity* of hidden colour). The concentration gradient is shown graphically next to the embryo in Fig. 9.4. You can think of the head end as being black and the colour gradually diminishing through a series of lighter and lighter greys as you go towards the tail end.

The gradient arises because the black protein is only made in the cytoplasm at the head end of the cell: it is only at this end that the black protein is produced by translation from RNA. The black protein then diffuses through the rest of the cytoplasm, gradually diminishing in concentration towards the tail end. The head end acts as a source of black hidden colour which gets gradually more dilute as you get further and further away. This means that the fertilised egg already has a basic pattern of hidden colour from one end to the other, a gradient of black protein in its cytoplasm. It would be as if our initial canvas is primed with a graded wash of colour, with say black on the left and gradually merging to white on the right.

You may wonder why the black protein should only be made at one end of the embryo. I shall return to this important question later on in the chapter. For now, I want to look at how this gradient of colour can be interpreted and built upon to establish a more detailed picture.

Interpreting the grey scale

The role of hidden colours (master proteins) is to provide a frame of reference that can be responded to by interpreting genes (Chapter 5). Recall that each interpreting gene is divided into two regions: a coding region that determines the type of protein made by the gene; and a regulatory region that influences when and where the gene is expressed. The regulatory region contains a set of binding sites—short stretches of DNA (locks) recognised by master proteins (keys). In the case of the black master protein, we can call the binding site, or short sequence of DNA it recognises, a B-site. In the simplest scenario, if an interpreting gene has a B-site in its regulatory region, it would be switched on

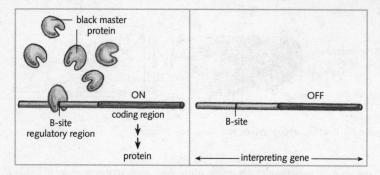

Fig. 9.5 Activity of an interpreting gene with a B-site in its regulatory region in the presence (*left*) or absence (*right*) of the black master protein.

when the black master protein is present (Fig. 9.5, left). On the other hand, if no black protein is present, the interpreting gene will be off (Fig. 9.5, right).

I have previously described how an interpreting gene can respond to the presence or absence of hidden colours in particular regions of an organism (Chapter 5). I now want to look at how an interpreting gene might respond to a *gradient* of hidden colour. Responding to a gradient is more complicated because we now have to deal with the *concentration* of master protein, not just its presence or absence.

To see how the concentration of black protein might affect matters, you need to appreciate that the binding of black protein to a B-site is a *reversible* process. That is, when a black protein molecule encounters a B-site it may bind for a period of time and then come away. As long as there are many black protein molecules around, as soon as one comes off the B-site, another will be there to bind in its place, so effectively the B-site will always be occupied.

We can now consider how such an interpreting gene might respond to the concentration of black protein at various positions in the early fly embryo, going from the head end towards the tail. As we have seen, shortly after fertilisation, the embryo comprises a large cell with lots of nuclei dotted all around the cytoplasm (Fig. 9.3). Each nucleus will contain our interpreting gene, so we can look at how the gene will respond in nuclei at various positions in the cell. Starting with nuclei at the head end, where the concentration of black protein is high, the interpreting gene will be on because the B-site is almost always occupied. This might continue to be true as we move towards the tail end, through various dark shades of grey. But there will come a point, say halfway along the egg, at which the concentration of black is no longer enough to ensure the B-site is always occupied. In other words, the concentration of black protein has started to become so low that there is insufficient of it to

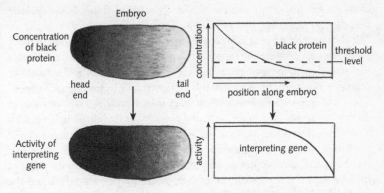

Fig. 9.6 Response of an interpreting gene to the gradient of black protein.

guarantee that there will always be some bound at the B-site. At this point, the gene might sometimes be on (when black protein happens to be bound), and other times off (when no black protein is bound), giving a reduced overall *average activity* for the interpreting gene (Fig. 9.6). In other words, rather than producing its protein at a maximum rate, the interpreting gene will be working at a slightly lower level on average. As we go further tailward, the average activity of the gene drops further still as the concentration of black goes down, until eventually the activity may drop down to zero and the interpreting gene will be off at the tail end (Fig. 9.6). The gene has responded to the gradient by being on in the head half, but gradually diminishing in activity in the tail half.

According to this scenario, the point at which the interpreting gene starts to decline in activity depends on where the concentration of black protein falls below a certain level. Above this *threshold level* the gene is pretty much fully on, whereas below it, gene activity starts to decline. In the example I gave, the threshold level of black protein was reached about halfway along the cell. In principle, the position in the cell where this happens could be shifted by changing the black concentration gradient. The black protein is only made at the head end of the cell, so if more black protein was produced from this source, more would diffuse through the cytoplasm, increasing the concentration to some extent at each point in the cell. This would mean that the point at which the concentration of black falls below the threshold level would no longer be reached halfway along the cell but further towards the rear. In other words, the expression of the interpreting gene would extend further back, say two-thirds of the way along the egg rather than halfway. Conversely, if the source of black protein was reduced at the head end, the concentration would fall below the threshold level sooner, nearer to the head. This would mean that the expression of the interpreting gene was more restricted, say to one-third of the way along the cell.

If the interpreting gene is indeed responding to the concentration of black protein, we would therefore expect that its activity would vary in a predictable way according to the gradient of black. Christiane Nüsslein-Volhard and her colleague Wolfgang Driever tested this experimentally by looking at the development of eggs with various levels of black master protein. They were able to do this by varying the number of *bicoid* genes in the fly (*bicoid* makes black). Typically, a fly has two copies of the *bicoid* gene, one from its father and one from its mother. By various genetic tricks they produced flies with other copy numbers, such as no copies, or one, three, or four copies. The more copies of the *bicoid* gene in the fly, the more black master protein will be made at the head end of the egg cell (actually it is the number of *bicoid* copies in the female laying the egg that counts, for reasons that will become clear later on), and the further towards the rear the activity of an interpreting gene should extend. As shown in Fig. 9.7, this is exactly what Nüsslein-Volhard and Driever observed. With no copies of the *bicoid* gene (no black protein) the interpreting gene did not come on anywhere. With one copy of *bicoid*, the interpreting gene did come on, but only near the head end. As the number of *bicoid* copies was increased further, so the expression of the interpreting gene extended further and further to the rear. These results confirmed that the interpreting gene was indeed responding to the grey scale, only starting to diminish in activity when the intensity of black fell below a critical level.

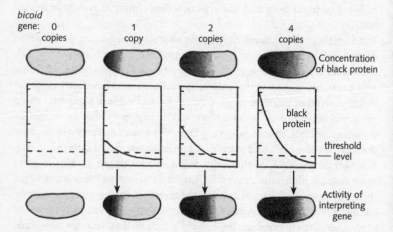

Fig. 9.7 Extra copies of the *bicoid* gene lead to more black protein being made at the head end of the fruit fly embryo, which in turn leads to a steeper gradient. In response, the activity of the interpreting gene extends further towards the tail end.

The idea that gradients might be involved in development was not new; it goes back to the late nineteenth century. At that time Thomas Hunt Morgan was studying the ability of worms with their heads chopped off to regenerate new heads (this was before he started working on fruit flies). He found that the ability to regenerate a head decreased as the cut was made further and further back. It was as if the regenerating powers were gradually diminishing along the worm. As he summarised: 'Perhaps for want of a better expression, we might speak of the cells of the worm as containing a sort of stuff that is more or less abundant in different parts of the body. The head stuff would gradually diminish as we pass posteriorly.' The trouble with this proposed gradient of head 'stuff' was that it was very hypothetical. As Morgan himself noted: 'I do not pretend that this explains anything at all, but the statement covers the results as they stand.'

The importance of the work on *bicoid* is that for the first time it was possible to identify a particular molecule, a master protein, that varied in concentration from one end to the other and was responded to in a particular way. Instead of a mysterious 'stuff', here was a defined molecule that played a key role in early development whose effects could be studied in detail.

Let me summarise the main points so far. By screening embryos for various types of mutant, a key mutant, *bicoid*, was found that lacked both head and thorax. This mutant has a defect in a gene (also called *bicoid*) that normally produces a particular type of master protein, symbolised by a black hidden colour. In normal embryos, the black master protein is produced only at the head end, from where it diffuses to form a gradient of decreasing concentration towards the tail end. This gradient can be responded to by an interpreting gene with a B-site (lock) in its regulatory region, recognised by the black master protein (key). The response may lead to the interpreting gene being very active (switched on for most of the time) at the head end but dropping in activity further towards the rear. The point in the embryo at which this drop happens depends on where the concentration of black master protein falls below a critical threshold level.

Varying the response

I have described how an interpreting gene can make one sort of response to a black gradient. I now want to look at other sorts of response: other ways of interpreting the gradient. Rather than varying the gradient, as I did in the previous section, we shall keep the gradient fixed and vary the response. What I mean by this will soon become clear.

For any given gradient, the activity of an interpreting gene will start to decline at a particular distance from the head. Now the point at which the drop starts to occur depends on how well the black protein sticks to the B-site: how much *affinity* the black master protein has for the sequence of DNA it is binding to. If the black protein sticks very well, having a *high affinity* for the B-site, then even a low concentration of black protein might be enough to keep the interpreting gene on for quite a while. In this situation, activity of the interpreting gene might only start to drop significantly near the tail end of the embryo, where the level of black becomes very low (Fig. 9.8, left).

On the other hand, if the black protein does not stick well, having a *low affinity* for the B-site, it would take a relatively high concentration of black to keep the interpreting gene on: the black protein would always be coming off the B-site, so many black protein molecules would be needed to ensure that the B-site was occupied for a significant period. This would mean that even in the dark grey areas, there may not be enough black protein to keep the gene mostly on. In other words, the activity of the interpreting gene would start to decline quite near to the head end of the embryo (Fig. 9.8, right). So a low affinity would lead to a narrow range of activity towards the head end, whereas a high affinity would lead to activity over a more extended length of the embryo. This was confirmed experimentally by Driever and Nüsslein-Volhard, who showed that lowering the affinity of binding sites in an interpreting gene did indeed lead to expression in a more restricted region towards the head.

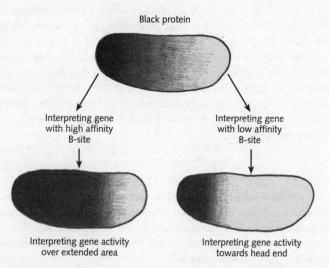

Black protein

Interpreting gene with high affinity B-site

Interpreting gene with low affinity B-site

Interpreting gene activity over extended area

Interpreting gene activity towards head end

Fig. 9.8 Fruit fly embryos showing two different responses to the gradient of black by interpreting genes with high or low affinity B-sites in their regulatory regions.

In molecular terms, the affinity between the black master protein and the B-site depends on how well their shapes match each other. If the master protein fits very precisely into a particular B-site, like a key fitting perfectly into a lock, the affinity will be high. If the fit is not so good, say because the DNA sequence in a particular B-site is not a perfect match for the black master protein, then the affinity will be low. This means that a slight modification of the DNA sequence of the B-site, varying its capacity to match the black master protein, can change its affinity for black.

The importance of all this is that it allows the gradient of black to be interpreted in *several different ways*, according to the particular DNA sequence of the B-site in the interpreting gene. An interpreting gene with a high affinity B-site in its regulatory region will respond differently from one with a low affinity B-site. In other words there are various ways in which the same gradient can be responded to. The way any particular interpreting gene responds will depend on the affinity of the B-site in its regulatory region. (I have over-simplified things. The response is more complicated than this because an interpreting gene would typically have several B-sites, not just one, and there can be other hidden colours in the background that may also influence how black binds and influences gene expression. These additional factors mean that the activity of the interpreting gene tends to change more rapidly near the threshold level than I have indicated.)

A rainbow of colours

We have seen that a gradient of black master protein in the developing egg can potentially be responded to in several ways by different interpreting genes. This means that any particular interpreting gene may be very active (mainly on) in some regions of the embryo, and inactive (mainly off) in others, according to the way it responds to the black protein through its regulatory region. Now the consequence of an interpreting gene being switched on is that it will produce its own type of protein from its coding region (i.e. the coding region will be transcribed to give RNA which will in turn be translated into protein). The key to the next step in the painting exercise is to look at what the protein produced by the interpreting gene might itself do. In other words, we need to consider the role of the proteins produced by the interpreting genes responding to black. We shall see that the products of many of these interpreting genes are themselves master proteins: these genes produce more hidden colours!

Now I am in danger of contradicting myself here. I previously classified genes into two types: those that produce hidden colours (such as identity genes) and those that respond to them (interpreting genes). I did this in order to avoid confusion between the production and interpretation of hidden colours. This classification of genes has acted as a ladder, helping us to get to this level of understanding. Having got to this level, we can now dispense with the ladder. *I am now saying that genes which produce hidden colours are also interpreting genes.* This is because the genes for hidden colours also have their own regulatory regions, their own molecular antennas. Like all interpreting genes, each gene for a hidden colour has two regions: a regulatory region and a coding region. But in this case, the type of protein produced from the coding region is itself a master protein, a hidden colour. Whilst it is still true that only a subset of genes produce hidden colours (master proteins), essentially *all* genes are able to interpret them, including the genes for the hidden colours themselves. This may sound like a circular argument at first, with genes for hidden colours being able to interpret themselves, but its meaning should become clearer as we encounter some specific examples.

To show how this allows us to elaborate patterns, I am going to give a simplified story. I will begin by introducing *seven* different interpreting genes that respond to the black gradient through their regulatory regions. As well as responding to a hidden colour, each of these genes also produces its own hidden colour, a particular type of master protein, from its coding region. We can name the master proteins produced by the seven genes after the colours of the rainbow: *red, orange, yellow, green, blue, indigo* and *violet*. Each of the interpreting genes therefore has a coding region that encodes a particular master protein (red or orange or yellow, etc.); and a regulatory region that acts as its own individual molecular antenna, ensuring that it is expressed at particular times and places.

We will assume that the regulatory region for each gene contains a B-site with a particular affinity for the black master protein, allowing the gene to respond in a particular way to the black gradient. As we saw in the previous section, a low affinity B-site will lead to a narrow range of activity towards the head end, whereas a high affinity B-site will lead to activity over a more extended length of the embryo. We will assume that the gene for red has the lowest affinity B-site, so it is expressed only near the head end of the developing embryo. The gene for orange has a slightly higher affinity B-site, so that its region of expression overlaps with red but extends a bit further tailwards. The gene for yellow has a B-site with a slightly higher affinity again, so it extends further back than orange; and so on up to the gene for violet, which has a very high affinity B-site and is expressed all the way from head to tail. The result is a series of overlapping colours, with all seven colours being made at the head end and gradually fewer and fewer being produced towards the rear, ending with mainly violet at the tail end. I now want to show how this *overlapping* pattern can get converted into a rainbow of *separate* colours. Before doing this it will help to give an analogy.

Bands around a mountain

Mountains are diverse habitats, rich in many different plant species. The vegetation near ground level is similar to that of the surrounding countryside but as you climb higher up, the temperature gradually drops and the predominant plant species progressively change to those that are better adapted to live in cooler conditions. The gradient in temperature results in bands of particular species around the mountain, with warm-loving species towards the base and cool-loving species towards the top (for our purposes we can ignore other environmental factors that change up the mountain, such as rainfall or soil quality).

In addition to the physical environment, there is another constraint on the distribution of the plant species: they *compete* with each other. To see this more clearly, we will start with a mountain having no plants growing on it, and then sprinkle some seed of various plant species all over it in the spring and follow their growth over several years. All the plant species will be annuals, living for one year and then setting seed for the next. In addition, we shall assume that the seed they produce is unable to travel very far from the parent plants.

In the simplest case, we could sprinkle seed for one species, let's call it violet, at low density all over the mountain. We will assume that the violet species is able to survive in almost all temperature conditions to some degree. On its own, it would therefore grow to cover the entire mountain, setting seed at the end of the year to give more violet plants next year. Eventually the whole mountain would be covered with a dense population of violet (Fig. 9.9, left).

Things get a bit more interesting if we start by sprinkling some seed for two species which compete with each other, violet and indigo. Perhaps indigo does not grow very well on the highest regions, so violet tends to win there. But below this, we will assume that indigo is a more effective competitor and tends to win over violet. At the end of the first year of growth, there is more violet at the top and more indigo lower down. This means that seed for the following year will not be evenly distributed over the mountain: more violet seed will have been

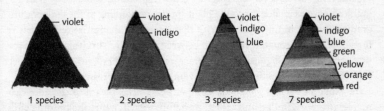

Fig. 9.9 Distribution of various plant species on a mountain following competition between them.

shed in upper regions and more indigo seed in lower regions. The advantage to violet in upper regions now becomes twofold: its seed is more abundant, and it grows better than indigo. Similarly, in lower regions indigo seed outnumbers violet and it competes better in growth. You would therefore expect that in the second year of growth, violet predominates even more over indigo in upper regions, and indigo outdoes violet even more in lower ones. Repeating this over several years you would tend to end up with two distinct zones: violet at the top and indigo occupying the remaining mountain (Fig. 9.9, middle left).

Things would have been slightly more complicated if we had started out by sprinkling seed for three species over the bare mountain: blue, indigo and violet. We will assume that blue is an even better competitor than indigo but doesn't like growing quite as high up. Just as violet and indigo competed in the upper regions of the mountain, so indigo and blue will compete further down. Indigo will tend to do better than blue at higher altitudes but blue will win out lower down. Over several generations of competition we might now end up with violet at the top, then a band of indigo, and finally the rest of the mountain below covered in blue (Fig. 9.9, middle right).

We could keep going in this way, sprinkling more and more species on the initial mountain (green, yellow, orange and red), each new species being slightly more sensitive to the cold but being a better competitor on the lower slopes, until we end up with a rainbow pattern, ranging from a band of red at the bottom through to violet at the top (Fig. 9.9, right).

The important message is that the final pattern on the mountain does not just depend on how the species respond to the temperature gradient but also on how they respond to each other over several generations. On its own, each species might occupy a large territory extending from the base of the mountain upwards to various extents. But put together, the species compete and restrict each other to narrower zones.

A conflict of colour

We can think of the seven interpreting genes responding to the black gradient in the embryo as analogous to the plant species responding to the temperature gradient on the mountainside. With the mountain, the initial response was further refined through competition between species for survival. We shall see that something analogous occurs in the case of the embryo, involving competition between interpreting genes over which one gets expressed.

To explain how this molecular competition works, I want to consider two master proteins: indigo and violet. As before we will name the binding sites recognised by each master protein after its colour: violet binds to a V-site and indigo binds to an I-site. Now each of these master proteins is encoded by a

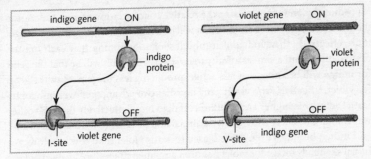

Fig. 9.10 Interactions between indigo and violet genes. On the left, the indigo gene is on and switches the violet gene off. On the right, the violet gene is on and switches indigo off.

gene with its own regulatory region attached. Suppose the gene for violet has an I-site in its regulatory region, as shown on the left part of Fig. 9.10. This would mean that the indigo protein could bind to the violet gene and influence its activity. We will assume that when the indigo master protein binds to the violet gene, it switches the gene off (i.e. the binding of the indigo protein interferes with or blocks the production of RNA from the violet gene). This means that wherever indigo hidden colour is produced in large quantities, the gene for violet will be off. In other words, if the indigo gene is on, producing lots of its protein, the violet gene will be off.

The right part of Fig. 9.10 shows the converse situation. In this case, the gene for indigo has a V-site in its regulatory region, allowing the violet protein to bind. Whenever violet binds there, indigo is switched off. This means that if the violet gene happens to be on, producing a lot of its protein, the indigo gene will be off.

We can now imagine a situation in which there are two competing outcomes. If we combine both the left and right parts of Fig. 9.10, such that the violet gene carries an I-site, and the indigo gene carries a V-site, there will be two conflicting interactions: indigo tends to switch the violet gene off, whilst violet tends to switch the indigo gene off. Indigo and violet master proteins tend to inhibit each other's expression, much as two different species might compete for a place on the mountainside. Which colour wins? Will indigo eliminate violet or will violet supersede indigo? We shall see that the outcome depends on two factors: the *initial activity* of each gene and the *affinity* of the master proteins for their binding sites.

Let us first take the effect of *initial activity* (i.e. the degree to which the genes are on or off to begin with). Suppose that we start with the gene for violet having a slightly higher activity than the gene for indigo (i.e. the violet gene is on more often than the indigo gene). This will mean that more violet protein is produced than indigo, raising the relative concentration of violet. Because the violet protein is more abundant, it will be able to inhibit the gene for indigo more effectively than indigo can inhibit violet (assuming that each master protein binds with a similar affinity to its sites). The result will be that the gene for indigo will become even less active, producing less protein, than the gene for violet. Effectively the violet gene now has two advantages over indigo: its initial activity is higher and it inhibits indigo more effectively than the other way round. The effect will be that the initial difference in activity between violet and indigo becomes even more biased towards violet than it was before. An initial slight advantage for violet has been amplified to produce an even greater advantage. The process can repeat itself again and again until, eventually, violet comes to predominate, with indigo making only a minor contribution.

In this example, the bias towards one colour came from an initial difference in activity, but we could also create a bias by changing the *affinity* of the binding sites in the regulatory regions. For instance, suppose that the I-site has a higher affinity for the indigo master protein than the V-site has for the violet protein. This would mean that at equal concentrations, indigo will switch off violet more effectively than the other way round. So even if we start with each gene at the same level of activity, producing each protein in equal abundance, violet will be inhibited more than indigo. This will in turn affect the overall concentration of the proteins, leading to a higher level of indigo than violet. Now indigo has two advantages over violet: a higher concentration as well as being a more effective inhibitor. The result will be that even more indigo is made than violet, further increasing indigo's preponderance. Eventually, indigo will take over and violet will be a very minor colour.

To summarise, the final outcome of this molecular competition between indigo and violet will depend on both the initial activity of the genes and the affinity of the master proteins for their binding sites. The colour that wins will be the one with the higher initial activity or affinity.

To avoid any possible misunderstandings, let me be quite clear that the sort of competition I am talking about is purely at the molecular level. It is not about one gene trying to outdo another in terms of survival. Both violet and indigo genes are in the DNA of the same organism, and I am using competition simply to refer to the sort of molecular interaction that has evolved between them. In terms of the organism's survival, the genes are actually cooperating, helping to refine each other's expression pattern, as we shall now see.

Painting a rainbow

With this notion of molecular competition in mind, we are going to look at how the pattern of rainbow colours may be refined in the early fly embryo.

Recall that I mentioned seven interpreting genes, each producing a master protein named after a colour of the rainbow: red, orange, yellow, etc. We shall be dealing with these genes at the stage when the embryo is a large cell containing many nuclei dotted around in the cytoplasm. All seven interpreting genes will be present in each of these nuclei, but to make things easier I will treat the genes one by one, just as more species were gradually added to the mountain.

Let us first consider the gene for violet. Initially, the activity of this gene depends on how it responds to the gradient of black protein in the embryo. Earlier in this chapter, I proposed that the violet gene has a high affinity B-site in its regulatory region, allowing it to be switched on even where the concentration of black is very low. The violet gene therefore starts off by being active in nuclei throughout the embryo. This means that each nucleus will be producing RNA from this gene which will go into the local cytoplasm around it to make violet protein. Every nucleus is effectively acting as a local source of violet, so together all the nuclei will ensure that the violet protein is present throughout the cell cytoplasm. There are now two hidden colours in the cell, a gradient of black and a ubiquitous violet.

Enter a second gene, the one for indigo. This gene also responds to the gradient of black through the B-site in its regulatory region. But because the affinity of its B-site is slightly lower, the activity of the indigo gene does not extend quite as far to the rear as violet, declining somewhat towards the tail end. Most of the nuclei in the embryo will therefore act as sources for indigo, except those at the rear which will not produce very much.

I now want to look at how the genes for indigo and violet respond to this situation. As before, we will assume that the genes for violet and indigo compete with each other over which one gets expressed (that is, each has a site in its regulatory region recognised by the other master protein, such that the gene will be switched off when it is occupied). We will also assume that the indigo protein has a higher affinity for its binding site, so that other things being equal, indigo will tend to outcompete violet. Now let the competition begin!

The higher affinity of indigo means that for most of the cell, where the initial activity of the two colours is about the same, violet will be eliminated whilst indigo asserts itself. As we get near the tail end, however, the initial activity of violet becomes higher than that of indigo (because of the response to the black gradient). There will now be two conflicting biases: violet is better off in terms of initial activity, but indigo has the higher affinity. The numerical advantage of violet gets better as we go further tailward, so we might expect that there would be a critical point at which violet gets the upper hand and starts to beat indigo. This will in turn mean that violet inhibits indigo even more effectively, and so violet becomes more predominant. The final result will be a zone of

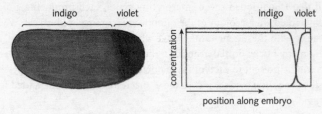

Fig. 9.11 Molecular competition leading to indigo in most of the cell and violet just at the tail end.

violet at the tail end of the embryo. So instead of the violet and indigo colours largely overlapping, violet will now tend to be restricted to just the tail end, and the rest of the cell will be occupied by indigo (Fig. 9.11).

There will still be a narrow region of overlap between violet and indigo where their zones meet. Because there are no cell membranes between the nuclei, the colours can slowly diffuse through the cytoplasm. This means that the colour from one zone will leak or diffuse into the other, giving a fuzzy zone in between (Fig. 9.11). The process of diffusion tends to even things out, ensuring that hidden colours change in a gradual and coherent way from one part of the cell to another.

The overall result is that starting with a gradient of black, we have refined the pattern to produce two zones: indigo and violet. I say *refined*, because the distinction between the zones is *sharper* than it was when only black was around. In other words, the gradient that separates violet from indigo is steeper than the original gradient of black. This is because of the way the genes for violet and indigo have competed with each other, so that they have become more tightly restricted than at the start. The refinement has arisen because the genes for violet and indigo are *both producing and responding to hidden colours*.

We can now apply the same principles to genes for the other rainbow colours. For example, what happens when we bring in a third gene, the one for blue? The expression of this gene starts to drop off a bit sooner than indigo (the blue gene has B-sites with slightly less affinity for black than the indigo gene). Blue and indigo now mostly overlap except for a region towards the tail end, where blue starts to decline in intensity. Now suppose that the indigo and blue compete with each other (they each have a site in their regulatory region recognised by the other colour, such that the gene will be switched off when it is occupied). We will also assume that blue binds with a slightly higher affinity to its site than indigo. As with the competition between indigo and violet, blue will predominate over indigo for the most part because of its higher binding affinity (I am assuming that blue can also outcompete violet in this region). But towards the tail end, where the concentration of blue starts to decline,

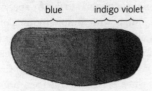 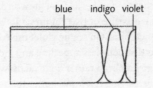

Fig. 9.12 Result of molecular competition between three hidden colours.

indigo will get the upper hand. Eventually we will end up with three zones: blue occupying most of the length of the cell, then a region of indigo, and finally a zone of violet at the tail end (Fig. 9.12).

We can carry on in this vein, considering more and more genes until we get a full rainbow of hidden colours, red through violet, from head to tail. The original gradient of black has been interpreted and refined to give seven regions of colour (Fig. 9.13).

You can get a feel for what is going on by thinking in terms of artists painting a canvas. Imagine you are one of many artists (nuclei) working on part of a large canvas. Your part of the canvas is not clearly demarcated from that of your neighbours, but overlaps with theirs (there are no cell membranes between nuclei). The canvas has been primed with a gradient of colour ranging from black at one end (head), through various shades of greys, to almost white at the other end (tail). Suppose you are at the tail end and you respond to the very light grey around you by applying a dilute mixture containing mainly violet together with a little bit of indigo (and even less of the other rainbow colours). Being struck by the mainly violet colour now on the canvas, you start to prefer it over the other colours and decide to increase the proportion of violet in your next mix. After painting on this new mix, you find violet even more appealing relative to the other colours, and eventually end up putting only violet on the canvas. Your neighbours do the same, so this whole region becomes painted violet.

Now imagine you are slightly further away from the tail end. Here the canvas

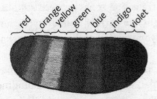 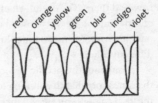

Fig. 9.13 Final pattern of rainbow colours in the embryo.

is a slightly darker shade of grey, and you initially respond to this by applying a dilute mixture of two colours, violet and indigo, with minor amounts of the other colours. On seeing violet mixed with similar amounts of indigo, the violet looks distinctly less appealing and indigo now stands out as preferable. Whereas indigo had been rather unattractive when it was only a minor part of the initial mix with violet, it looks better than violet when present in equal amounts. So you respond by applying more indigo on the canvas in preference to violet with your next mix. The process repeats itself so that eventually you and your neighbours paint the whole region indigo.

If you were to start further up the canvas again, at an even darker shade of grey, you might begin by placing a dilute mixture of three colours on the canvas, blue, indigo and violet. But after you see them, the blue stands out as preferable, and from then on you and your neighbours start to increase the proportion of blue. In a similar way, green would be made further along, and so on up to red. The final pattern arises because each time you add new colours, you respond to what you see.

Now let us look at what happens in regions of the canvas where different zones of colour meet, say around the junction between indigo and violet. In this area you might be in two minds about which colour to favour. Subtle differences in the initial shades of grey might lead some artists to prefer violet, others indigo. If each artist was working alone on his or her own region of canvas, you could end up with a higgledy-piggledy pattern of violet and indigo patches. But remember that the area you are painting is not clearly demarcated from that of your neighbours: it is shared with them to some extent. This means that you will be influenced by what they are doing. Say you are in the indigo region, near to where it meets violet, and you happen to deviate slightly from what your neighbours are up to, by starting to apply a bit more violet. Your neighbours, who are applying indigo, would soon smother your bit of canvas as well, raising the amount of indigo to a similar level to theirs. You would be outnumbered by your colleagues and would tend to conform to whatever they were doing. Of course, there will come a point towards the tail end where the artists will gradually switch to applying violet, but they will tend to do this in a collective way, rather than as individuals going off and painting a bit of violet here and there. The overall result is that the transition zones between regions of different colour will tend to be quite straight and narrow, with everyone on one side conforming to one colour, and everyone on the other side applying a different colour.

Throughout this painting process you need have no idea of where you are on the canvas or what the overall picture looks like. You are simply responding to the local colours you see and then modifying them in some way by adding more colour. To someone high above it may look *as if* all the artists know exactly where they are because each seems to be applying a colour that fits in with the overall pattern. But the artists themselves are oblivious to this; they simply see the canvas immediately around them and respond accordingly. They are no more aware of their position than a plant knows where it is on a mountainside.

Let me summarise some of the essential ingredients that have allowed the gradient of black to be interpreted and refined. First of all, there is a large population of nuclei (multiple artists) distributed throughout the embryo that can respond to the initial gradient. This allows different regions of gene activity to be established (if there was only one nucleus, the system would not work because there could only be one response at any time). Secondly, there are no barriers between the nuclei, so that neighbours tend to coordinate their activities. This makes sure the response is coherent rather than too much of a hotchpotch. Finally, the genes within each nucleus are able to both produce and respond to hidden colours. Each of the seven genes interprets the local colour whilst at the same time being able to modify it by adding a certain amount of its own colour. Taken together, this allows an initial pattern to be interpreted and refined in an interactive manner, with colours continually elaborating on what went before.

Filling in some gaps

I have described the whole process of producing a rainbow as a painting exercise, giving a general sense of how a gradient of hidden colour might be built upon and refined. The significance of this for the organism is not, however, to depict some pretty patterns but to establish a basic frame of reference for development. To see this, we need to look more closely at the biological effects of these hidden colours.

So far I have only mentioned the effects of one of these genes for hidden colour: the *bicoid* gene that makes black. Recall that mutants without a *bicoid* gene produce embryos without a head or thorax. In other words, the black colour is needed for the development of about a half of the embryo, towards the head end.

What happens if a mutation inactivates a gene for one of the rainbow colours? It turns out that these mutations give embryos with several adjacent segments missing (normally there are 14 main segments in the fly, Chapter 4). If, say, green is absent from the rainbow, the middle of the embryo (i.e. the region that is normally green) will not develop, giving an embryo with some middle segments missing. Similarly, if violet is absent, an embryo will develop with several segments at the tail end missing. In other words, each of the rainbow colours is needed for the corresponding part of the embryo to develop fully. In fact this is how the genes for the rainbow colours were identified in the first place: they were noted by Nüsslein-Volhard and Wieschaus in their screens for embryo mutants, as a class of abnormal embryos with chunks of their body missing. Because of this deficit or gap in the embryo, the genes affected by these mutations were christened *gap* genes. The gap genes are what I have been calling the genes for rainbow colours. Each gap gene codes for a master protein that is expressed in a particular region or band of the embryo, giving it a hidden colour. Together, they divide the early embryo into distinct territories from head to tail, like the rainbow pattern I described, but not quite so simple (there are actually about nine gap genes, rather than seven as given in my account).

According to my simplified scheme, all the gap genes needed for the rainbow colours are activated by responding to the gradient of black. This would imply that a mutant without black should also lack all the rainbow colours. You might expect such a mutant to develop as an embryo that was all gaps, perhaps forming a tiny blob with no segments at all. This is *not* what is seen in *bicoid* mutants. As we have seen, the *bicoid* mutant lacks a head and thorax but it still produces an abdomen. The reason is that the egg is primed with a few other colours (not previously mentioned for the sake of simplicity) in addition to black. These other colours can to some extent compensate for the lack of black in *bicoid* mutants, particularly in the tail half of the embryo. So rather than the *bicoid* mutant lacking all the rainbow colours, it still has those in the tail half, say green, blue, indigo and violet, allowing the abdomen to develop. If all the initial or priming colours needed to set up the rainbow are missing (as in mutants that have several genes inactivated in addition to *bicoid*), the embryo does indeed form a tiny blob without a head, thorax or abdomen.

So in practice, the expression of the gap genes is not established by interpreting just one initial colour, black, but several others as well. By studying how the gap genes respond to these initial colours and to each other, researchers have been able to start working out how the final pattern of rainbow colours is actually produced. The story that has emerged is quite complicated but the general principles are not too different from my simplified scheme of rainbow painting: gap genes respond to their local hidden colours as well as adding to them, eventually establishing territories of gene expression through molecular interactions and competition.

Beyond the rainbow

I have described the painting of a rainbow in detail so as to show the basic principles by which one set of hidden colours can be built upon and refined through interactions between molecules. We can imagine similar principles being applied again and again to elaborate the pattern further. Just as the gradient of black acted as a frame of reference over a large region of the embryo, each rainbow colour provides a steeper gradient that can be further interpreted and elaborated by other genes. The process can continue in this way, gradually refining the colours more and more. Other considerations will of course come into play in these further stages, such as cell membranes forming between nuclei, and the organism growing larger. We shall deal with some of these issues in later chapters. The point I want to make here is that making a rainbow is just one step in an extensive process of internal painting. The rainbow colours are just a prelude to many more hidden colours that will be added, further refining and elaborating this pattern.

One set of genes that respond to the rainbow are those involved in providing distinctions in segment identity. Recall that there are eight of these identity genes, producing apple-green to herb-green in various territories from head to tail (Chapter 6). The green colours appear after the rainbow and have more restrictive effects. Whereas the rainbow colours are needed for the *formation* of each type of segment, the various greens only affect the *identity* of each segment: remove the rainbow and no segments develop; remove only the various greens and the segments develop but they all have the same identity.

Now the head-to-tail layout of the green territories itself depends on the previously established rainbow colours: without the rainbow, the various zones of green would not be established. We can think of the whole process as a series of painting steps: starting with a broad gradient of black, then a rainbow with fuzzy boundaries, then the series of greens with sharper boundaries, as schematised in Fig. 9.14. The internal painting process does not stop here: it continues, based on the greens together with other hidden colours, to produce an even

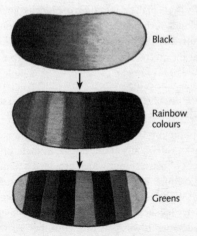

Black

Rainbow colours

Greens

Fig. 9.14 Schematic illustration of some hidden colours at various stages of development.

more elaborate and refined picture. There is a changing internal landscape of colour that becomes more detailed as the organism develops.

This means that there is a specific sequence of colours in *time* as well as *space*. Black comes before the rainbow colours, which in turn precede the greens. I have described the sequence of a few hidden colours, but the same principles apply to numerous other colours, so that a complex three-dimensional patchwork is gradually elaborated as the fly develops. The shifting internal landscape of colour provides a four-dimensional frame of reference for genes to interpret (i.e. the three dimensions of space plus the one dimension of time), allowing each gene to be expressed at particular times and places.

The sequence of colour patterns can be interpreted by essentially all the genes in the organism, each responding according to the binding sites in its regulatory region. Whereas only a subset of genes are involved in producing the hidden colours, all genes can respond to them. As we have seen in this chapter, some of these interpreting genes themselves code for hidden colours. But the majority produce proteins with various other roles in the cell, such as modifying cell structure, or influencing how a cell divides, grows, moves or even dies (we shall deal with some of these aspects later on in this book). It is by affecting the expression of these sorts of genes that the hidden frame of reference eventually becomes manifest in the shape and anatomy of the organism.

I have described the internal painting process as if each colour persists once it appears on the scene. However, this is somewhat misleading. In practice, master proteins have a limited lifespan and get broken down after a while. Like plants on a mountainside, there is a continual turnover of master proteins, with new ones being produced by genes while some existing ones get broken down. We should really think of the elaboration of hidden colour in this more dynamic sense. Some colours, such as the black and rainbow colours, are present for only a relatively short period of early development and then disappear. Other colours, such as the greens, may persist for much longer as they continually get replaced by more of the same.

Back to mother

The starting point for this exercise in painting was a large fertilised egg producing black protein only at one end. This then diffused through the cell, giving a gradient of black colour which provided a framework that was elaborated by other colours. But where does the initial asymmetry in the fertilised egg come from? Why does the egg only make black protein at one end?

The asymmetry in black traces back to the development of the egg cell *before* it was fertilised, when it was part of the mother. As we saw in the previous chapter, the bulk of material in the fertilised egg comes from the mother.

Specific cells in the mother's ovary (called *nurse cells*), lying outside the egg, are connected to what will be the head end of the egg. These nurse cells transcribe the *bicoid* gene to make RNA copies. The RNA molecules then enter the egg and are deposited in the cytoplasm of the egg cell while it is still growing in the mother (*bicoid* RNA is therefore not produced from the *bicoid* gene inside the nucleus of the egg cell, but is imported from maternal cells at one end). Once the RNA enters the egg cell, it becomes anchored at the head end, and is then translated to make the black protein after the egg has been fertilised. The black protein is only made at the head end because previous development in the mother has ensured that *bicoid* RNA is only present at that end.

To show that this localisation of *bicoid* RNA was indeed responsible for setting up the gradient of black protein, Wolfgang Driever, Christiane Nüsslein-Volhard and colleagues injected purified RNA transcribed from the *bicoid* gene into the other end of the egg cell, where the rear end would normally form (Fig. 9.15). They found that black protein was now made at both ends of the fertilised egg, as predicted. Furthermore, a symmetrical embryo developed with a head and thorax at both ends! A good approximation to this in humans is shown on the Kings, Queens and Knaves of most packs of playing cards. Here was a dramatic demonstration that the initial asymmetry in the location of *bicoid* RNA was indeed setting the orientation of the head and thorax pattern. By making black protein at both ends, two gradients of black are produced, which in turn are interpreted to give a pair of inverted half-rainbows (red, orange, yellow: yellow, orange, red), eventually leading to the double-headed embryo.

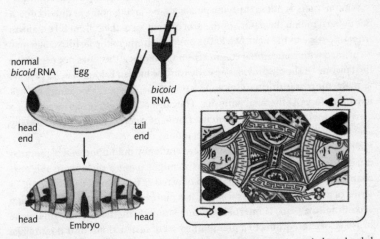

Fig. 9.15 If *bicoid* RNA is injected into the tail end of a fruit fly egg, a symmetrical two-headed embryo develops, similar to what is seen on a playing card.

The key point is that in tracing hidden colours, such as black, further back in development, we do not come across an absolute starting point for the whole process—we simply get deeper into the development of the mother. If, for example, we were to ask why only the cells at one end of the egg transcribe *bicoid* to make RNA, we would have to find out what makes these cells different from others in the ovary. This would in turn bring us to the problem of how territories of hidden colour are elaborated in the developing ovary. Tracing the problem further back still, we would get into earlier and earlier stages of the mother's own development. It is in a sense arbitrary whether we see the growth of the egg as a late step in maternal development or a very early step in the next generation of development. We can equally regard it as a fragment of the maturing mother canvas or the starting canvas of a new picture.

We could of course choose the time of fertilisation as a convenient starting point for development. This after all is when the DNA from the two parents comes together for the first time. But in doing this, we must be clear that we are not starting with a *tabula rasa*. The fertilised egg is primed with hidden colours that mainly come from the mother. And as we have seen, in the case of the fruit fly some of these colours owe their asymmetry to events that occurred before fertilisation. The production of asymmetry before fertilisation should not be taken as an essential requirement for development. In the seaweed *Fucus*, for example, asymmetry is established in the egg after it has been released from the mother plant and fertilised. Even in *Fucus*, though, the fertilised egg will start off with some hidden colours, in the same way that the starting canvas of every painting has to have a colour to build on, even if it is only a uniform white.

What matters is not identifying an absolute starting point in development but understanding the nature of the process that takes the cells of an organism from one stage to the next. We have seen that although the fertilised egg may be primed with some initial colours from the mother, this does not depict the final picture. It simply provides an early wash of hidden colour that is a prelude to many more transformations and elaborations. These early colours are not like miniatures of the final painting. They just act as very broad sweeps on a canvas, providing a very elementary frame of reference which is gradually elaborated through a series of interpretations.

I have shown how hidden colours are gradually built upon and elaborated during development, much as an artist might paint a picture. In the next chapter I want to look at the question of whether this should be thought of in terms of a creative process, or whether it is more like manufacturing something according to a plan. Is internal painting simply a matter of executing some brush strokes according to a preordained set of instructions, or is it more like the process of creating an original picture?

CHAPTER 10

Creative reproduction

On the face of it, the reproducibility of organisms, their similarity from one generation to the next, would seem to imply that some sort of plan or set of instructions is being followed as they form. Internal painting, the elaboration of hidden colours, occurs in pretty much the same way every time a fruit fly develops, suggesting that it is much more like slavish manufacture than a creative process. To pursue this issue any further we will need to take a closer look at what underlies reproducibility. What determines the consistent development of organisms every generation? Once this has been explored, we will be able to make a more informed comparison with art and see whether there is a fundamental contradiction between creativity and reproducibility.

Although biological reproduction tends to produce similar individuals each generation, they are not all exactly alike. We are most aware of this in our own species, as variation in the facial features and physique from one person to the next. But we might detect a similar level of variation in other species if we were only as sensitive to their individual characteristics. The ornithologist Peter Scott, for example, learned to recognise individual swans by noting the colour markings on their bills. It is just that normally we are less aware of variation in other species because it does not seem terribly important to us. Distinguishing between people is an essential requirement of everyday life but we can get by without being able to recognise individual daisies or fruit flies.

Much of the variation between individuals in a species comes from genetic differences—slight variations in their sequence of DNA. If the DNA of two individuals is exactly the same, as in identical twins, they resemble each other much more closely than individuals with distinguishable DNA sequences. The extent of biological reproducibility therefore depends on the similarity in genetic make-up. For maximum consistency in development, the DNA sequence in the developing egg should always be the same: the fertilised egg should always begin with exactly the same genes. When we say that development is a completely reproducible process, it is in the context of a very similar, preferably identical, genetic background.

In addition to genetic make-up, the environment also influences the way an organism develops. A seed sown in one type of soil and climate may produce a very different looking plant from a similar seed grown under other conditions.

In some cases, you might even take two plants as belonging to different species simply because they have grown under different conditions. Similarly, environmental conditions have important effects on the way animals develop. If fruit fly larvae are exposed to a relatively high temperature for short times, they will often develop into adults showing major abnormalities in their body structure; defects that are just as severe as those produced by mutations in genes. Another striking effect of the environment is illustrated by certain species of butterfly that develop in different colour forms from one season to the next, say one form appearing in spring, the other in summer (a phenomenon called polyphenism). The two colour forms may look as distinct as two different species, yet they only differ because of the environmental circumstances that prevailed at the time of their development (the relevant environmental factors can include temperature and daylength). In the case of mammals, the environment of the embryo is carefully controlled in the mother's womb and buffered from the conditions outside the mother. But if the embryo's environment is altered in some way, it may lead to alterations or abnormalities in the way the embryo develops. For example, if a mother has German measles (rubella) during early pregnancy, the resulting child may be born with severe abnormalities. The rubella virus is able to cross the placental barrier between mother and child, and infect the embryo, modifying its development.

When we say that an organism develops in a consistent and reproducible manner, it is always in the context of a specified environment and genetic make-up. Both of these factors must be defined from the outset if development is to proceed in a predictable way. To make a fair comparison between development and artistic endeavour, we should therefore standardise the initial conditions for painting a picture in the same way that we have done for a developing organism.

We can think of the surroundings of an artist as equivalent to the environment of a developing organism. In the case of Leonardo painting the *Mona Lisa*, for example, there is the subject herself and the background around her. The general lighting conditions are also a very important ingredient, as Leonardo himself was at pains to point out:

> when you wish to paint a portrait, paint it in bad weather, at the fall of the evening, placing the sitter with his back to one of the walls of the courtyard. Notice in the streets at the fall of the evening when it is bad weather the faces of the men and women—what grace and softness they display! Therefore, O painter, you should have a courtyard fitted up with the walls tinted in black and with the roof projecting forward a little beyond the wall; and the width of it should be ten braccia, and the length twenty braccia, and the height ten braccia; and you should cover it over with the awning when the sun is on it, or else you should make your portrait at the hour of the fall of evening when it is cloudy or misty, for the light then is perfect.

As part of the general environment we might also include the musicians and jesters that, according to the biographer Vasari, were employed by Leonardo so as to keep the subject of the *Mona Lisa* smiling while he painted her.

To standardise the environment of a painting, we should therefore only look at works that have been produced under the same conditions, keeping the subject, light and all the other elements in the surroundings fixed.

These conditions alone do not, however, determine the type of painting that is produced. Much depends on the particular artist involved. Compare the *Mona Lisa* to the portrait of an Italian woman, *L'Italiana* by Vincent van Gogh (Fig. 10.1). Clearly they differ greatly in style even though the subjects are comparable. If you were to imagine transporting exactly the same model that Leonardo was painting, together with her surrounding courtyard and entertainers, to the late nineteenth century for van Gogh to paint, there would be little difficulty in distinguishing the result from the *Mona Lisa*, even though the immediate environment would have been the same in each case. This is because each painter works in a very different way. It is not simply that Leonardo's painting is more realistic than van Gogh's. Leonardo used various stylistic devices to portray the subject in a particular way. The shadows, particularly around the eyes and mouth of the *Mona Lisa*, are left purposely diffuse and blurred to create an ambiguous and slightly mysterious effect (a technique invented by Leonardo, called *sfumato*). Also the background that Leonardo paints is an imaginary landscape, not the one that would have been behind her. Conversely, although van Gogh's style with its strong brush strokes may seem less realistic, the painting may convey certain features of the sitter's personality and appearance more effectively. Leonardo and van Gogh simply painted in different styles, each of which can be seen to have its own merits.

We might compare the paintings of Leonardo and van Gogh to the development of two different species. Even if the immediate environment were to be kept similar, each would produce a very different result. In the case of the painters, this is because of their distinct cultural and biological heritage. Leonardo was an illegitimate child of the fifteenth century, brought up during the height of the Italian Renaissance. Vincent van Gogh had a different biological heritage and was born four hundred years later when Impressionism was coming to its peak. This means that when each one starts a painting, they come to it with a very different set of circumstances behind them. Their brains operate and react in different ways because of their distinct biological and cultural heritage.

Similarly, two species can develop in very different ways because they have a distinct evolutionary history. In this case, the legacy of the past is held in the genetic make-up and properties of the fertilised egg. If these differ substantially,

Fig. 10.1 *Mona Lisa* by Leonardo da Vinci, above (Louvre, Paris), compared to *L'Italiana* by Vincent van Gogh, below (Musée d'Orsay, Paris).

so will the appearance of the organism that develops, even if the environment is similar.

The reproducibility of organisms with a given genetic make-up and environment should therefore be compared to a series of paintings in which both the artist and the surroundings are standardised. We might want to look at a series of paintings by Leonardo, with him always in the same surroundings. Strictly speaking, we really need to standardise not just the artist but his *state of mind* at the beginning of each painting. This would be the equivalent of identical twins starting with exactly the same genetic make-up. But now we come to a problem. After Leonardo has painted the *Mona Lisa* once, his state of mind will not be the same as when he started. The very act of painting the picture will have changed his outlook. For one thing, he will now have the painting of the *Mona Lisa* before him whereas this was not available when he first started out. If he were to paint the same subject a second time he would know of the previous result. Perhaps he would avoid the second painting being too much like the first, exploring novel features that were not covered so well in the first painting. Or he may no longer be interested in the subject so much after having already spent a lot of time on it. Whatever the case, he will not start the second painting with the same mental state as he did the first time round.

To get around this problem, we might imagine that Leonardo has a bout of amnesia after he finishes each painting. Suppose that every time he completes the *Mona Lisa*, he forgets everything he did previously during the painting. He wakes up in the same mental state after finishing it off as when he started to paint the picture the first time. Of course there also has to be a conspiracy amongst his friends so that someone hides the finished painting from him each time and no one tells him that he has done it all before.

Given these admittedly peculiar circumstances, Leonardo might start to produce a whole series of versions of the *Mona Lisa* over time. We now have a situation that is equivalent to the development of an organism with a fixed genetic make-up in a standardised environment. Of course the numerous *Mona Lisa* paintings might not look exactly the same because of chance variations that are out of our control, but then again identical twins do not look exactly the same.

The point I want to make is not so much to do with whether the paintings are precisely the same each time, but with the nature of the painting process. An outsider who is unaware of the conspiracy, and simply sees all the versions of the *Mona Lisa* being churned out, might conclude that Leonardo is simply manufacturing lots of slightly different copies of the same painting. Based on the consistency of the output, it may seem they are being produced by copying from an original or by always following the same routine set of instructions.

But this is not at all the way that Leonardo would see it. To him each painting is a genuine and novel creation. He is not copying because he does not know what the finished picture will look like. Nor is he manufacturing them according to an independent plan or set of instructions. He is creating the pictures through a highly interactive process in which there need be no clear separation between plan and execution. Each time he might start out with some idea of the sort of picture he wants to paint and apply a few brush strokes. He may react to these in one way or another, adding some here, modifying some there. The picture emerges through a creative process rather than one of fabrication or copying. Although all the versions may look similar, they have been created as originals, without any advance knowledge of the final outcome. The external observer who judges that Leonardo is copying or manufacturing based on the consistency of the output would have completely misunderstood what Leonardo was actually up to.

At this point the reader might object that although Leornardo need not have had a detailed plan in mind while creating the *Mona Lisa*, he nevertheless might have had one. Perhaps Leonardo did have a very precise idea of how he was going to paint the picture: a clear mental vision of how he was going to proceed before he even put his brush to the canvas. Wouldn't this mean that he was being creative while at the same time following a plan? Now I do not believe that in practice most artists do anticipate the outcome of their work in detail, because this would be to deny the importance of how they interact with the medium. Nevertheless, let us ignore this and suppose, for the sake of argument, that Leonardo did have a precise plan for all the brush strokes needed to paint the *Mona Lisa*, and a clear vision in mind of what the final picture would look like before he painted it. The process of painting would then involve a clear separation between plan and execution.

But this would raise the question of where the plan in Leonardo's mind itself came from. How did Leonardo come up with this detailed vision? It is no good saying that the vision was itself derived from an earlier mental plan, because this would simply beg the question of where this earlier plan came from, leading to an infinite regress. The answer would surely be that the vision was a *creative* product of Leonardo's mind: the plan was itself the outcome of a highly interactive mental process. This process may have occurred entirely within Leonardo's brain, or perhaps it might have also involved some preliminary sketches, carried out by Leonardo to help him formulate the plan more clearly. Either way, the plan would have arisen through an interactive process that could not itself have been a matter of following a plan (otherwise we are back to an infinite regress). Now the key point here is that if Leonardo had set about painting the *Mona Lisa* in this way, the most creative part from his point of view would have been coming up with the plan or vision, rather than the act of painting. Once he had arrived at a detailed mental conception of how to paint the picture, executing the plan by applying paint to the canvas would have been a purely mechanical exercise, equivalent to the process of writing down a poem once the lines have been thought up. To put it another way: if he had a perfect vision in mind of what the final picture would look like, the act of painting would simply correspond to *copying* this vision. So even in this case, the creative process would not involve the following of a plan. Rather, it would be concerned with the events before the painting had begun, when Leonardo was coming up with his detailed vision.

If the *Mona Lisa* had arisen in this way, we could distinguish between two phases in its production: a primary creative process of arriving at a plan, and a secondary technical process of executing the plan. So long as Leonardo wakes up from his bouts of amnesia in a mental state that precedes both of these phases, he would still feel he was being genuinely creative each time, although in this case the creativity would occur in the phase of coming up with a plan rather than during the application of the paint. He would still wake up with no clear idea of what the final painting would look like, having to arrive at his vision each time through a creative process. As before, the external observer seeing similar paintings being churned out might misunderstand what was going on. It might look like a case of routine manufacture; whereas it actually involves a highly creative process that is not primarily derived from following a plan.

I have spent some time going through this painting analogy in order to show that creativity and reproducibility are not necessarily incompatible. We can imagine a situation in which a process is both genuinely creative and yet has a more or less reproducible outcome. Of course, in practice artists do not have bouts of amnesia every time a painting is finished, so it is a reasonable assumption that a series of very similar pictures are more likely to have been copied or fabricated in some way than created.

Now when it comes to development, we are struck by the reproducibility of organisms and are therefore drawn to analogies with a process of manufacture according to a plan. We naturally equate development with someone consistently following instructions. But in my view this is misleading. We are making the same mistake as the external observer seeing the output of Leonardo the amnesiac. The process of development is much more like creating an original than manufacturing copies, in the sense that the final product, the adult, is not there from the beginning but gradually emerges through a highly interactive process, in which each step builds on and reacts to what went before in a historically informed manner. Like the amnesiac Leonardo, the fertilised egg has no detailed sense of what the final picture will look like. It is not so much that the egg forgets how the previous picture looked; it never got to see the picture in the first place. It is simply a tiny part of the earlier canvas, one cell amongst many others that nevertheless carries all the information needed to produce another canvas, given certain environmental conditions.

The notion of each step being historically informed is very important here, and distinguishes development and human creativity from many other sorts of interactive process. There are many processes in which the various steps are not historically informed, even though the steps build on each other in complex ways. Take the formation of a snowflake. This depends on a complex sequence of events in which water solidifies at the surface of a growing ice crystal. This process is highly interactive: the shape of the crystal influences how it will grow, which in turn will influence the subsequent shape of the crystal. Each step in the growth of a snowflake therefore depends on what went before and influences what is to come (slight variations in the growth conditions may therefore affect the final form in quite complex ways, accounting for the many different shapes of snowflakes). However, none of these steps is historically informed: the way a crystal grows is not influenced by the experience of crystals that have grown in the past. It simply depends on the prevailing physical conditions. By contrast, each step in biological development is informed by the evolutionary history of the organism, involving countless previous generations of natural selection. The creativity of human beings also depends on their biological inheritance, as well as on their cultural heritage and experiences. In this respect, the development of a fertilised egg and a creative process are much more similar to each other than to processes in which there is no historical input of this kind.

Please do not misunderstand. I am not trying to suggest that the fertilised egg is driven by some mysterious creative force. Nor do I wish to imply that the egg is actually thinking about what it is doing as it develops. My point is that human creativity can give us a better overall sense of the nature of developmental processes than comparisons with fabrication according to a plan. If we think of development in terms of manufacture, we will start to look for a set of instructions as distinct to whatever it is that follows them, or we will search for an independent plan that is then executed, or try to identify software as distinct from hardware. It is the notion of reproducible manufacture that impels us to look for these distinctions. But as we have already seen with the exercise of internal painting in the previous chapter, and as will become more apparent during this book, development is an interactive process in which each step builds upon and reacts to what went before in a historically informed manner. We will try in vain to make a separation between plan and execution because the two are deeply interwoven. And to my mind this is similar to what happens in creative processes. Being external observers, the consistent output of development leads us to equate it with manufacture. But once we delve into the internal mechanisms, we see they have more features in common with creative activities, even though they give reproducible results.

CHAPTER 11

Scents and sensitivities

In his essay *Possible Worlds*, the biologist J. B. S. Haldane considers how the world would appear to a being with different senses and instincts from our own, such as a dog. Haldane's dog is much more intellectual than any that exists, a dog with a brain as complex as our own, though organised along rather different lines. This is because unlike for us, a dog's life is dominated by its nose: it investigates its surroundings by smelling everything. For us, seeing is believing, and smell is more subjective and less reliable. We have little problem in listing the objects we see in a room, but find it much harder to describe the various fragrances given off by a glass of wine. This is not because fragrances are less objective, but simply because our sense of smell is poorly developed relative to our sense of vision. How would the dog's life, dominated by smell, compare to ours?

Haldane's dog orients itself and finds its way round by smell. Instead of visual signs, the dog decorates its environment with odours, marking territories and pathways with scent. It finds its way round by following sequences and trails of smells rather than by reading a map. Our dog's house might even be decorated with smells, perhaps with a different odour on the walls of each room. A patent for odorous wallpaper was actually filed in 1943, proposing that the wallpaper be impregnated with smelly compounds which would be slowly released into the air. Humans don't seem to have taken to the idea, but perhaps our dog would. The dog also recognises its friends by their smell. When meeting a fellow dog in the street it sniffs it first rather than recognising it by sight.

Because the most important quality of an object is its odour, Haldane's dog classifies things according to their type and intensity of smell rather than their appearance. Our dog is a very good chemist: many pure chemicals look white and indistinguishable but they can be differentiated by smell, and its close relative, taste (taste and smell are sometimes called the *chemical senses* because they seem to detect directly the properties of chemicals). Our dog is also an excellent cook, conjuring up dishes with the most elaborate scents. Indeed, cooking is one of the highest art forms in our dog's world, with the greatest canine chefs achieving a level of subtlety and sophistication in terms of smells that matches painting or sculpture (perfumery is also high on the list of arts).

Our dog's idea of a 'thing', which we normally think of as being of fixed dimensions and occupying a particular part of space, is very different from ours. For the dog, things are defined by their smells and have a much more change-able existence, often transforming from one type into another. When a cake is baked it ends up giving off different smells than when it started. When milk turns sour it becomes something new. In each case, one thing is turning into something quite different because of a chemical change. Chemical transfor-mations play a role in our dog's notion of how things are, that is perhaps equiva-lent to the role of motion in space for us. To our dog it might seem as if a cake in the oven is continually 'moving' from one state to another in the world of smells, much as we think of an object moving from one position to another in space.

Mathematically, our dog has no problem with counting and arriving at the concept of number, as these depend on being able to sequence events in time, not space. It can count smells just as well as we count bricks. The ideas of geometry, though, are far less accessible to it. The notion of a three-dimensional space in which every point has a different set of coordinates does not appeal to our dog. A geometrical transformation, such as reflecting an object in a mirror, seems like a mystical notion that is hard to grasp in comparison to a change in smell brought about by a chemical reaction. Compared to us, the dog is weak in geometry but strong in chemistry.

Now when it comes to understanding development, our dog may have a few advantages. It seems to me that many of the difficulties we have in thinking about development stem from our over-reliance on external geometry. To us, the most striking feature of an egg developing into an adult is the *increase* in geometrical complexity: an egg is more homogeneous in terms of spatial organisation than the adult that eventually develops from it. Yet from a purely geometrical perspective this seems unintelligible, because the basic transfor-mations in geometry applied to an object do not increase complexity but *preserve complexity*. Take a simple geometrical transformation, such as moving an object in a straight line from one place to another, a process mathematically called translation. Although this changes the object's position, it does not change its complexity in any way. Or consider other geometrical changes, like rotating an object, or reflecting it in a mirror, or magnifying or reducing its size. Although all of these operations change the object's appearance in some way, they do not change its quality or complexity. We simply end up with the same object presented from a different viewpoint.

In my view it is this geometrical perspective that draws us towards the idea that there is a preformed version of the adult in the egg (Chapter 1). Prefor-mation is an attractive way of explaining development because it accounts for the whole process as a series of purely geometrical transformations: enlargements of a miniature that is just as complex as the fully grown adult. It is easy to comprehend because it appeals to our visually dominated mind.

However, in trying to understand development we need to get away from this purely geometrical viewpoint, and think of the transformations more in terms of their underlying chemistry. We need to think more like our dog: through our nose rather than our eyes.

Now I seem to have boxed myself in here. I am suggesting that we should try and think more like a dog; yet we are fundamentally visual animals. We cannot simply change the way we perceive the world by starting to think in terms of odours rather than images. We comprehend things much better through getting a mental picture of what is going on than by forming some sort of internal smellogram. That is why we commonly acknowledge that we understand something by saying 'I see what you mean' rather than 'I smell what you mean'.

I shall try to deal with this problem in very much the same way that a chemist might, by still using the concepts of geometry but applying them at the molecular level rather than at the level of what is externally visible. Much of our understanding of chemistry is based on the twin notions of molecular geometry and energy. We think of a chemical reaction in terms of a redistribution of energy as the atoms in molecules are rearranged in space. Rotten eggs give off a horrible smell because of a chemical reaction in which hydrogen and sulphur atoms combine in a particular spatial arrangement to form a pungent gas called hydrogen sulphide. What is perceived at one level as a change in the properties of a chemical, such as its smell, can be thought of at the submicroscopic level as a rearrangement of atoms, a change in molecular shape (recall that I am using shape to refer to the distribution of electric charges as well as the overall three-dimensional arrangement of atoms). A change in smell can give us an overall intuition that a reaction is going on, but to get a more exact idea of what is happening we have to turn to molecular geometry.

In this chapter, I shall use the notion of smell to help explain a key aspect of development: how cells communicate. Just as I used colours to give an intuition of how territories in an organism are defined, so smells will now help to take the process further by conveying how cells interact with each other. As with the colour metaphor, I shall use smell to give an overall idea of the process, whilst at the same time providing a more precise explanation of what is happening in terms of molecular shapes and interactions. We will get a more dog-like overview (or oversmell) but it will be based on a foundation of molecular geometry.

Although I shall not talk about the detailed energy changes involved in these molecular events, you should bear in mind that all of the processes require energy, stored in the form of chemical bonds, to drive them along. As described in Chapter 2, this energy ultimately comes from the sun. It is through channelling this energy that more geometrical complexity can eventually emerge as development proceeds.

Receptor proteins

A good point to begin our exploration of smells will be to take a closer look at our nose. The nose is not confined to the visible protuberance on our face, but penetrates inwards for several centimetres. A group of sensory cells lying deep in this cavity detect smells and send the information they receive to the brain. The sensory cells can recognise molecules in the air because of particular types of protein in their membranes, called *receptor proteins*. It is the shapes of these proteins that allow us to distinguish between different chemicals entering the nose. Each receptor protein sits in the membrane of a sensory cell and points into the nasal cavity, where it may encounter molecules that are inhaled along with air (Fig. 11.1). There are hundreds of different types of these receptor proteins in the nose, each with a shape that matches a particular range of molecules. If a molecule matches or binds to a receptor protein, it will influence the shape of the protein. I have indicated this in the figure by showing a bump forming at one end of the protein. Recall that the shape of a protein determines what it does—the sorts of reactions it encourages to happen. So when the receptor protein changes its shape, it facilitates a new chemical reaction in the cell. This in turn sets off a chain of other reactions, eventually leading to an electrical nerve impulse being sent from the sensory cell to the brain.

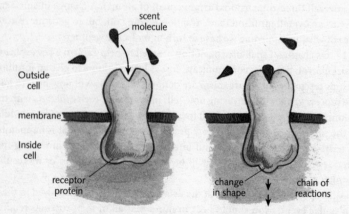

Fig. 11.1 Receptor proteins in the sensory cells of the nose are triggered when molecules of scent bind to them, setting off a chain of reactions.

How does the brain know which particular receptor proteins have been triggered when we smell something? Although many different types of receptor protein are produced by the sensory cells of the nose, each individual sensory cell is thought to make only one or very few types of receptor protein. So when we sniff cinnamon, the inhaled molecules will only trigger a particular set of cells, those that have receptor proteins matching cinnamon. These then send a set of nerve impulses to the brain. It is this combination of impulses, coming from a particular set of sensory cells in the nose, that gives us the sensation of having smelled cinnamon. Sniffing almonds will stimulate a different combination of sensory cells, which we register as the smell of almonds. Even humans, with their relatively undeveloped sense of smell, are thought to be able to distinguish between about ten thousand different odours in this way. We can distinguish more smells (thousands) that the number of different types of receptor protein (hundreds). This is possible because the inhaled molecules do not trigger just one type of receptor protein but several different types. It is the *combination* of receptor proteins triggered by a molecule that provides the molecule with its distinctive chemical signature.

Not everyone is able to smell things to the same extent. After eating asparagus, some people will describe a very strong smell in their urine, whereas others will not detect anything remarkable. In a series of tests, in which people sniffed urine collected from a man after he had eaten 450 grams of canned asparagus, it was shown that some people were very sensitive to the smell whereas others were not. It seems that everyone produces pungent compounds in their urine after eating asparagus but some people can't smell them very well. Similarly, some people are insensitive to the smell of freesia flowers. The insensitivity to particular smells is called *specific anosmia*, the olfactory equivalent to being colour-blind.

The cause of specific anosmias is not yet known but in several cases they have been shown to be inherited. That is to say, they are passed on by genes. A good guess is that some of them result from mutations in genes for receptor proteins. Like all other proteins, each receptor protein in the nose is coded for by a gene. If someone has a mutation in one of these genes, a change in its DNA sequence, their corresponding receptor protein may be defective. This would in turn lead to a reduced ability to detect particular molecules, such as those in the scent of freesias. Thus our *sensitivity* to smells depends on our repertoire of genes that code for receptor proteins.

Our sense of smell gives us direct information about molecules in the air because it employs receptor proteins that can discriminate between their shapes. When a molecule binds to a receptor protein in the nose, it sets off a chain of chemical events in the cell that leads to a particular response, the sending of a nerve impulse. Our ability to smell can be understood in terms of geometry: the geometry of how atoms and charges are arranged to give molecules their shape. That is how smell can monitor the molecules that are wafted into the air as a cake is baked, as milk turns sour or as an egg turns rotten.

Taste is thought to operate in a similar way to smell, though in this case there are far fewer types of receptor involved. We can only discriminate between four types of taste (sweet, salty, sour and bitter) through sensory cells in our tongue. Much of what we commonly call taste is actually due to us smelling the food while we eat it (hence our inability to taste much when we have a blocked nose).

Smell thy neighbour

As well as allowing organisms to smell or taste their external environment, receptor proteins also play a fundamental part in communication between cells within the body. In the case of our nose, the molecules that are detected by receptor proteins come from the surrounding air. Most cells in our body, though, are not surrounded by air but by other cells. Even the sensory cells of the nose have air only on one side of them and are surrounded on other sides by cells. We shall see that receptor proteins provide a very important mechanism that allows cells to respond to their neighbours.

In Chapter 9, we saw that the refinement of patterns in the early fly embryo depended on communication between neighbouring nuclei. Without some sort of communication, each nucleus would go off and do its own thing irrespective of its neighbour, giving a hotchpotch rather than a coordinated elaboration of a pattern. In this case there was no problem with communication because there were no cell membranes that separated the nuclei. The nuclei were immersed in a common cytoplasm so hidden colours (master proteins) could diffuse from one region to another.

This, however, is a rather exceptional situation. For the most part, development occurs while nuclei are located in separate cells. Master proteins cannot diffuse freely from one cell to another because of the cell membranes that are interposed between them. This has two important consequences. On the one hand, it means that diffusion is not continually mixing up hidden colours from neighbouring regions, so that the patterns can be more stable and distinct: the hidden colour in one cell will not be continually leaking into its adjacent cells, so it is easier to keep territories of hidden colour separate. On the other hand, there is no longer a straightforward means of communication between neighbours: nuclei are no longer in free contact with each other so they cannot coordinate their activities through diffusion. But without some sort of contact, the system would tend towards cellular anarchy, with each cell doing its own thing.

This is where receptor proteins come in: they allow cells in the developing organism to communicate with each other even though they are separated by membranes. Patterns can be refined in a coordinated way because cells can send molecular signals to each other via receptor proteins. We shall see that unlike our true sense of smell, triggering receptor proteins during development does not lead to nerve impulses being sent to the brain, but to a different sort of response—typically a change in hidden colour. It is as if cells are continually sniffing each other, modifying their hidden colours according to the scents they detect.

Induction in newts

One of the first pieces of evidence that chemical communication between cells might be important for development goes back to the 1920s, to Hans Spemann and his school in Germany, working on newts. Spemann was taking small pieces from one newt embryo and implanting them in another to see how the implanted tissue might develop. In most cases, he found that if the transplantation was carried out at an early stage of development, there was little disturbance in development: the implant joined in the development of its host. However, he noticed that when a particular region of the early embryo was transplanted, there was a dramatic effect. The implant did not simply join in the development of its new surroundings; on the contrary, a second embryo formed at the site of implantation. It seemed that the transplantation of this particular region could result in the formation of an almost entire second embryo.

At first Spemann thought this second embryo was derived simply by the growth and development of the implant. However, a few years later he changed his mind, when his collaborator Hilde Mangold repeated the experiments using two newt species with embryos that could be easily distinguished by virtue of their different pigmentation (Fig. 11.2). By taking a piece of embryo from a lightly pigmented species and transplanting it into a darkly pigmented host, the implant could be permanently recognised by its lighter colour. Spemann and Mangold then saw that when a secondary embryo formed, it was not just made from the light-coloured implanted tissue but also contained large amounts of darkly pigmented host tissue. It was as if the implant had somehow organised or *induced* its nearby cells from the host to participate in producing a new embryo.

A possible explanation was that the implant was sending some sort of chemical signal that induced the nearby cells to behave in this way. If the chemical could be identified, the nature of how cells coordinate their activities

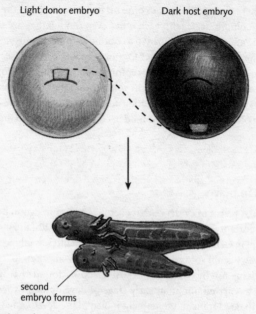

Light donor embryo   Dark host embryo

second
embryo forms

Fig. 11.2 Experiment by Spemann and Mangold in which part of an early newt embryo from a lightly pigmented donor was implanted in a darkly pigmented host. This induced the formation of a second embryo containing large amounts of host tissue.

during development might be revealed. A crusade followed to try and identify the chemical, but unfortunately this met with more frustration than success. Joseph Needham summed up the situation in 1939:

> In conclusion, it may be said that although the progress made in the last ten years in these fields has been very great, we can nevertheless see now that owing to the special difficulties of the subject . . . it may be more like fifty years before we can expect to have certain knowledge concerning the chemical nature of the naturally-occurring substances involved in embryonic induction. Like so many other biological problems, this has turned out to be more complex than the first explorers thought.

Needham's estimate of fifty years was not too far out. It was only in the late 1980s and early 1990s that the nature of these sorts of chemical signal involved in development started to be unequivocally identified. An important turning point came from molecular studies on specific genes, as I shall now illustrate with an example taken from fruit flies.

A molecular marriage

During the 1960s, Seymour Benzer's laboratory at Caltech was screening fruit flies for mutants that did not respond much to light. If you shine light on normal flies, they tend to move towards the light source. Benzer was looking for exceptional individuals that did not behave normally and stayed put instead of wandering over to the light. One of the mutants he obtained was later shown to have a specific defect in its eyes, effectively making it partially blind to the light.

To understand the defect in Benzer's partially blind flies, you need to know some background about insect eyes. The eye of a fruit fly is made up of about 750 identical facets or *ommatidia* (Fig. 11.3). Each ommatidium contains 20 cells of various types, each cell type being specialised in a different way: one cell type secretes the main lens, another produces pigment, and yet other cell types are responsible for detecting the light. These 20 cells are arranged in a very precise and stereotyped way, with the same pattern of cells being repeated in every ommatidium. I am going to concentrate on the interaction between just two of these cells, called R7 and R8.

One of Benzer's mutants was partially blind because it lacked a particular type of cell from every ommatidium: the R7 cell. Without the R7 cell in its ommatidia, the fly could not see properly and therefore did not respond to light in the same way as normal flies. Because of this deficiency in R7, the mutant was named *sevenless*. The striking thing about the *sevenless* mutant is that apart from lacking R7, the ommatidia otherwise seem to be normal. To get a graphic illustration of what this means, look at the painting by Magritte shown in Fig. 11.4. It shows one room with a man reading a paper, and three other rooms that are identical except that they are unoccupied. We can think of the eye of a normal fly as equivalent to 750 identical rooms (ommatidia), each with a person in it (the R7 cell). The eye of the *sevenless* mutant would then be like

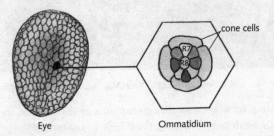

Eye   Ommatidium

Fig. 11.3 The eye of a fruit fly is made up of many ommatidia. Some of the cells inside one ommatidium are shown so as to reveal the relative positions of R7, R8 and cone cells.

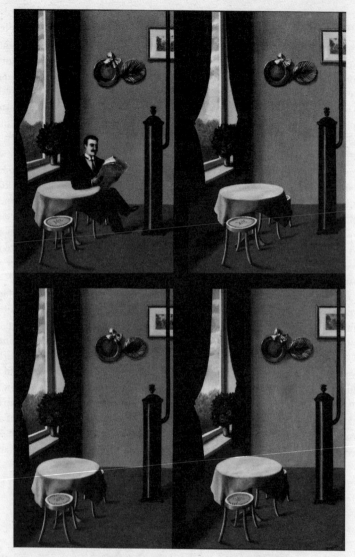

Fig. 11.4 *Man with a Newspaper* (1928), René Magritte. Tate Gallery, London.

an identical set of rooms with the man missing in each case. In the *sevenless* mutant, one element is lacking, the R7 cell, but otherwise the pattern of cells in each ommatidium is unchanged; just as in the empty rooms of the painting, only the man and his paper are missing while the rest remains unchanged. It is a very specific change in the pattern, affecting one element independently of the others. Strictly speaking, the cell that would have become R7 is not really missing in *sevenless* mutants, but has become like one of the other types of cell normally found in the ommatidium, called a *cone cell*. Everywhere an R7 cell should have been made, a cone cell develops by default. In the same way, in Magritte's painting, the area of the canvas occupied by the man has not been cut out in the empty rooms but has been replaced by elements from the background, such as wall and floor.

The *sevenless* mutant flies carry a defect in one particular gene. By convention, this gene is given the same name as the mutant—*sevenless*. But to avoid confusion between the mutant and the gene, from now on I shall refer to the gene by the abbreviation *sev*, only using the full name *sevenless* when referring to the mutant fly. Thus, a normal fly has a functional *sev* gene allowing the development of an R7 cell in each ommatidium. In *sevenless* mutant flies, the *sev* gene is defective and as a consequence of this, the R7 cell does not form (a cone cell forms instead). That is to say, the *normal* significance of the *sev* gene is to *promote* the formation of an R7 cell; so that when *sev* is defective, R7 is missing. You should be clear that although the *sev* gene only influences the R7 cell in each ommatidium, the gene is also present in the DNA of every other cell in the body. It is just that the only significant effect of the *sev* gene is on the development of one particular type of cell, R7.

I now want to look at how the *sev* gene exerts its effect on the development of the eye. Although the *sevenless* mutant highlights the normal significance of the *sev* gene—its requirement for the development of R7—it does not tell us how the *sev* gene actually works. How does a functional *sev* gene normally lead to the development of an R7 cell? Like other genes, the *sev* gene codes for its own particular type of protein, and to appreciate how it works, we need to know what this protein does. An important breakthrough in addressing this issue came in 1987, when Ernst Hafen, Gerry Rubin and colleagues, working at Berkeley, were able to isolate the *sev* gene. They could then show that the protein encoded by *sev* had a structure that was typical of *receptor proteins*. Remember that receptor proteins are able to recognise and bind to specific molecules coming from outside the cell. Once a receptor protein has been triggered by binding to a molecule, it changes in shape, setting off a chain of chemical events within the cell. The *sev* gene codes for a particular protein of this type, that I shall from now on refer to as the *sev-receptor protein*. Finding that the *sev* gene coded for a receptor protein implied that some sort of molecule existed, equivalent to a scent, that was triggering the receptor. However, unlike what happens when we smell something, the response to the scent was not the sending of a nerve impulse to the brain, but a developmental event: the formation of an R7 cell in each ommatidium. In flies without the sev-receptor protein (*sevenless* mutants) there would be no such response, so the R7 cell would fail to develop.

The question now became: what was the nature of the scent, the molecule that the sev-receptor protein was detecting? An important advance in answering this came in the late 1980s, with some studies by Lawrence Zipursky and colleagues at the University of California, Los Angeles. These researchers found another type of mutant fly that lacked R7 cells, which they called *bride-of-sevenless* (the reason for this curious choice of name will soon become apparent). In this case, the defect was not in the *sev* gene but in a different gene, which I shall refer to as *boss* (the abbreviation of *bride-of-sevenless*). Like *sev*, the *boss* gene is needed for the R7 cell to develop: in *bride-of-sevenless* mutants, the *boss* gene is defective, so R7 is missing. There are therefore two different types of mutant which lack R7—*sevenless* and *bride-of-sevenless*—each due to a defect in a particular gene, *sev* or *boss*. But whereas the *sev* gene codes for a receptor protein, it turns out that the *boss* gene codes for a protein of a com-plementary type, called a *signalling protein*. Signalling proteins are a general class of molecule that match the shape of receptors, much as scents match the shape of receptor proteins in the nose. The difference is that whereas scents come from the outside, signalling proteins involved in development are pro-duced by cells within the organism. Nevertheless, it will be helpful to use the notion of *scents* as a metaphor for such signalling proteins, so long as we remember that they are derived internally. This is similar to the way I have been using hidden colours as a metaphor for master proteins. We can summarise by saying that the *boss* gene produces a particular scent, the boss-signalling protein; whereas the *sev* gene produces a particular receptor, the sev-receptor protein.

Now there is a partnership between *boss* and *sev*, because the boss-signalling protein *matches* the sev-receptor protein. In other words, the boss-signalling protein is the molecule, or internal scent, that the sev-receptor protein is sensitive to.

To see how this partnership between *boss* and *sev* influences eye development, look at Fig. 11.5, which shows two cells in the ommatidium of a normal fly: one that will become R7, and a neighbour that will become another type of cell, called R8 (their relative positions in the eye are shown in Fig. 11.3). First look at what happens in the cell on the left that will form R8. In this cell, the *boss* gene is switched on, and it therefore produces the boss-signalling protein (scent). The signalling protein sits in the cell membrane, with part of it exposed to the outside of the cell. Now look at the cell on the right. In this cell, the *sev* gene is switched on, so the sev-receptor protein gets made. This protein moves to the cell membrane, where it then makes contact with the boss-signalling protein in the adjacent cell. The shape of the boss-signalling protein matches the sev-receptor and this sets off a chain of events in the responding cell, leading

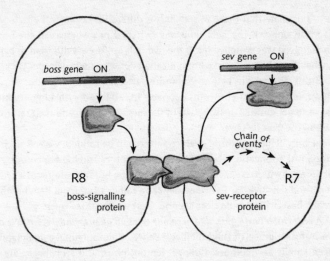

Fig. 11.5 Relationship between scents, sensitivities and cell fate in a fruit fly ommatidium. The cell on the left (destined to become R8) produces a signalling protein from the *boss* gene, which triggers a receptor protein (produced by the *sev* gene) in the adjacent cell. This sets off a chain of events which eventually lead to the cell acquiring an R7 identity.

to its developing as R7. It is as if the responding cell is sniffing its neighbour, and when it detects a particular scent—the boss-signalling protein—it responds by becoming an R7.

The partnership between *sev* and *boss* is a dog-like marriage, based on a chemical match between two partners (hence the name *bride-of-sevenless*). In mutants with a defective *sev* gene, the marriage breaks down because no sev-receptor protein is produced, rendering the responding cell blind to the scent. In mutants defective for *boss*, no signalling protein is made, so there is no scent to stimulate the receptor.

In many ways, the conclusions from these studies are similar to those of Spemann and Mangold with newt transplantations. In both cases, cells communicate with each other, influencing the way they develop. The difference is that in the case of the eye, it was possible to identify the signalling molecules and the way they were responded to (the signalling molecules thought to underlie Spemann and Mangold's experiment have now been identified as well).

Oranges and lemons

I have given an example of how cells within the same organism can influence each other through molecular signals and receptors. But in describing this case,

I had to postulate that there was already a difference between the two cells involved: in one cell the *boss* gene was switched on, whereas in the other cell the *sev* gene was switched on. If the two cells are already different to begin with, what does this chemical communication between them achieve? We shall see that this process can help to elaborate patterns of hidden colour during development. To explain how this happens, I shall begin with a hypothetical case so as to get some of the key parts of the message across, and then return to the case of the fruit fly's eye.

To see how cellular sniffing enables patterns to be refined, we will need to provide some connections between scents, receptors and hidden colours. To get us going, I will make three rules that relate the hidden colours of a cell to its scents and sensitivities. The rules may seem rather arbitrary at first, but their molecular basis will become clear later on. Here are the three rules:

(1) *A cell with a particular hidden colour gives off an associated scent.* We can get an intuitive grasp of this rule because we commonly make associations between smells and colours. Orange, lemon, plum and strawberry are all colours, but they are also associated with particular scents: the smells of oranges, lemons, plums and strawberries. It is not that the colour pigments in the fruits are directly responsible for the smell, but the two are associated with each other. An orange fruit is both coloured orange, due to its pigment molecules, and gives off a particular scent, because it produces other molecules that can enter the air and be detected by our nose. In the same way, each hidden colour is not directly responsible for the scent given off by a cell, but is associated with it. Each hidden colour is a master protein of a particular type, whereas each scent is a signalling protein of a particular type.

(2) *Each hidden colour is associated with a sensitivity to a particular scent.* This is slightly more tricky to grasp than the previous rule. You have to imagine that if a cell has a hidden colour it will also respond to a particular type of scent. For example, a cell with a lemon hidden colour might respond to the scent of oranges. That is to say, it produces a receptor protein that matches the scent given off by an orange cell. Similarly, the orange cell might produce a receptor that is triggered by another scent, say the odour of plums. In this way each hidden colour is associated with a specific receptivity.

(3) *When a cell detects a scent, it responds by changing its hidden colour.* If a lemon cell should detect the scent of oranges, it will respond by changing to a new hidden colour, say from lemon to plum. Similarly, an orange cell that detects the smell of plums will react by changing to a different colour, say from orange to strawberry. It immediately follows from the first two rules that the new hidden colour will lead to a new scent (rule 1) and a new sensitivity (rule 2). The particular colour a cell changes to may vary according to the hidden colour it started off with and the type of scent it is exposed to. Hidden colours

can change in response to the local scents, much as a chameleon might change its visible colour in response to its environment.

I now want to use these three rules to elaborate the pattern of hidden colours in a row of four cells. We shall start with two of the cells having one hidden colour, *orange*, and the other pair having a different hidden colour, *lemon* (Fig. 11.6). This initial arrangement will be built upon using our three rules to make a more elaborate pattern.

Each cell will both be giving off a scent and be receptive to a scent. Let us suppose that the lemon cells are sensitive to the scent of oranges (i.e. they produce a receptor that matches the scent given off by orange cells). Now assuming that the scent does not travel very far, only the lemon cell adjacent to an orange cell will be bathed in the orange scent, so only this lemon cell will respond. The response will be a change in hidden colour, say turning from lemon to *plum* (Fig. 11.6, step 1). Thus our first step has been to switch the colour of one of the four cells.

There is now a cell with a new hidden colour, plum, that has its own

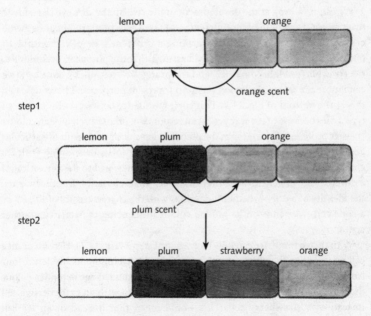

Fig. 11.6 Elaboration of patterns through scents and sensitivities. In step 1, a lemon cell responds to the scent of oranges by turning plum. In step 2, an orange cell responds to plum scent by turning strawberry.

associated scent and sensitivity. Suppose that orange cells are sensitive to the scent of plums (i.e. the plum scent matches the receptor proteins on orange cells). The result will be that the orange cell adjacent to the plum cell—the one bathed in plum scent—will respond by changing colour, say from orange to *strawberry* (Fig. 11.6, step 2). Note that the other orange cell does not change colour, even though it is receptive to the scent, because we are assuming that the scent does not travel more than one cell along. Also the lemon cell next to plum does not change colour because it does not have the receptor that recognises the plum scent. The system has allowed us to elaborate the pattern to give a predictable sequence of four different colours—lemon, plum, strawberry and orange—having started with only two.

This process of elaboration is possible because the hidden colour in one cell is able to both influence and respond to the hidden colours of its neighbours via scents and sensitivities. For example, in the first step, the orange cell influenced its neighbour to turn plum-coloured. But then the plum cell in turn influenced the orange cell to turn strawberry-coloured. Cells are continually modifying and reacting to the hidden colours of their neighbours through mutual sniffing, allowing patterns to be refined in an interactive manner.

We can now look at the development of the cells in the fly's eye in terms of this process. Each of the four differently coloured cells produced in our scheme can be thought of as eventually forming a different type of cell within an ommatidium of a normal fly. We will assume that the plum cell becomes R8, the strawberry cell becomes R7, and the orange cell becomes a cone cell (we can ignore the fate of the lemon cell for our present purposes). I have indicated this at the bottom of Fig. 11.7. This correspondence between colour and cell type arises because of the way genes can respond to, or interpret, hidden colours (master proteins). For example, the plum hidden colour may result in particular genes being switched on or off in a cell, leading to its developing with R8 characteristics. Similarly, the strawberry colour may lead to the activation of a different set of genes, those that confer R7 characteristics. In other words, the identity of a cell—whether it develops with the characteristics of R8, R7 or a cone cell—depends on its hidden colour influencing the activity of other genes.

With this in mind, we can now look again at step 2 (Fig. 11.7). Here an orange cell responds to the plum scent of its neighbour, changing its hidden colour from orange to strawberry. Now if for some reason this change in hidden colour did not occur (i.e. step 2 failed), you would end up with an extra orange cell instead of a strawberry cell. This would mean that instead of an R7 cell developing (strawberry = R7) you would end up with an extra cone cell (orange = cone cell). This is precisely what happens in *sevenless* or *bride-of-sevenless* mutants: the R7 cell is missing and a cone cell develops instead.

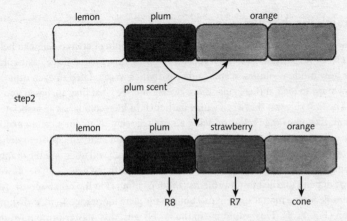

Fig. 11.7 Relationship between hidden colours and cell types in the ommatidium of a fruit fly. Step 2 is shown, in which an orange cell responds to plum by turning strawberry.

The failure to switch from orange to strawberry could come about in two ways. In one case, the receptor in the orange cell might be missing, rendering the cell insensitive to the scent of plums. The orange cell would fail to turn strawberry because it lacks the receptor protein that detects plum scent. This is what happens in *sevenless* mutants: they lack the receptor protein. Another way in which the switch might fail is if the plum cell does not produce its scent. Without the plum scent the adjacent orange cell will not turn strawberry-coloured. This is the situation in *bride-of-sevenless* mutants; they are unable to make the signalling protein. In both types of mutant, the R7 cell fails to form but it is for a different reason: either because of a lack of sensitivity (*sevenless*) or a lack of scent (*bride-of-sevenless*).

Notice that both orange cells in our scheme are sensitive to the plum scent, but only one responds because it is much nearer to the source of the scent. In the same way, the sev-receptor protein is actually present in several cells of the developing ommatidium, including those that will form R7 and cone cells. But it is only the cell next to R8 that is exposed to the boss-signalling protein and therefore responds by becoming R7.

We can see from this that the main purpose of the molecular courtship between *sev* and *boss* is to help refine the pattern of cell types in a coordinated way: it results in an R7 cell developing in a particular position relative to its neighbours in each ommatidium of a normal fly. Without this signalling system, the pattern would be less refined because the R7 cell type would be missing from the eye, and an extra cone cell would be there in its place.

Examining the rules

I have tried to give the gist of how molecular signalling between cells can help to elaborate patterns of hidden colour. In doing this, I listed three basic rules for how hidden colours, scents and sensitivities were related to each other. I now want to look at these rules more closely and see what they are based upon.

The first rule was that a cell with a particular hidden colour has an associated scent, like an orange cell producing an orange scent. Now for a cell to exude the orange scent, the gene that codes for the relevant signalling protein needs to be switched on. Unless this gene is expressed the cell will not make the orange scent. But as we have seen in previous chapters, whether a gene is on or off will itself depend on the master proteins (hidden colours) in the cell. Suppose, for example, that a master protein can bind to the gene for orange scent, switching it on (Fig. 11.8). This would mean that a cell with this master protein would automatically produce the orange scent. We have established a link between a cell's hidden colour (master protein) and the scent (signalling protein) it produces. In my hypothetical scheme, the hidden colour and the scent it switches on were simply given the same prefix, *orange*. It is the *orange* hidden colour that binds to and switches on the gene for the *orange* scent. It is important not to confuse the orange scent, which is a signalling protein, and the orange hidden colour, which is a master protein.

Similarly, the lemon master protein binds to the gene for lemon scent, switching it on. The association between hidden colour and scent arises because

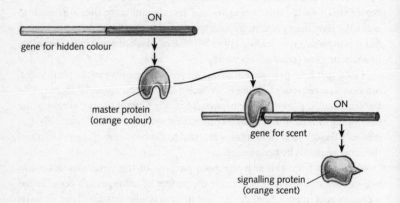

Fig. 11.8 Molecular explanation for the association between hidden colours and scents. In this case a gene for the orange hidden colour produces a master protein which then binds to a gene for orange scent, switching it on.

master proteins are able to influence which of the genes for signalling proteins are switched on or off (in some cases, the proteins made by these genes may not be the scents themselves, but may act more indirectly by affecting the synthesis of a scent).

A similar explanation lies behind the second rule, that each hidden colour is associated with a sensitivity to a particular scent. The ability to detect a scent depends on a receptor protein, which is also encoded by a gene. There will be one gene for the receptor protein that matches orange scent, another gene for the receptor that matches lemon scent and so on. As before, whether a cell expresses one of these genes will depend on its master proteins. For example, the lemon master protein might switch on the gene producing receptor protein for orange scent (Fig. 11.9). This would mean that lemon cells would automatically be sensitive to orange scent because they express the gene for the matching receptor protein. Similarly, the orange master protein may switch on the gene producing the receptor protein for plum scent, ensuring that orange cells are sensitive to the aroma of plums. As with scents, the association between hidden colours and sensitivities arises because the master proteins are able to influence which of the relevant genes are on or off.

To explain the third rule, that a cell changes its hidden colour when it responds to a scent, I shall need to introduce another feature of proteins. Recall that what a protein does depends on its shape. This in turn depends on the sequence of amino acids in the protein, specified by the gene that codes for it (Chapter 2). Now although the amino acid sequence largely determines the

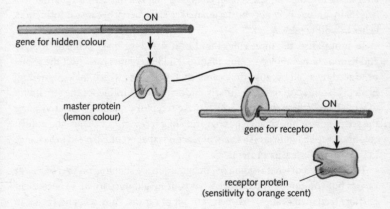

Fig. 11.9 Molecular explanation for the association between hidden colours and sensitivities. In this case a gene for the lemon hidden colour produces a master protein which then binds to a gene for the receptor sensitive to orange, switching it on.

shape of the protein, there are other factors that can influence matters. In some cases, the shape of a protein may be affected by molecules that bind to it. We have already seen that in the case of a receptor protein, its shape changes when the matching molecule docks into it (page 184). In other cases, the shape may be altered by a slight chemical modification of the protein, typically the removal or addition of a phosphate molecule.

The importance of these alterations is that in changing a protein's shape, they will also affect what it does. A protein with an extra molecule bound or attached to it may promote a different chemical process in the cell from what it might otherwise have done. By influencing the shape of a protein, a modification can affect what happens in a cell.

This provides the basis for how receptor proteins work. When a scent molecule binds to its matching receptor, the shape of the receptor protein changes. As a result of this, a new chemical reaction is promoted in the cell. In the case of receptor proteins in the nose, the reaction leads to a nerve impulse being sent to the brain. However, with many of the receptor proteins involved in development, the reaction sets off a chain of events that eventually leads to a *modification in the shape of a master protein*. This may involve a new molecule binding or being attached to the master protein. Now because the master protein has an altered shape, it will bind to genes and affect them in a different way. Some new genes might be switched on, others might go off. In terms of hidden colours, we might summarise this by saying that the colour of the master protein, which reflects what it does in the cell, has changed. The master protein has changed its colour because it now binds to genes differently, giving it an altered significance for the cell. In changing shape, a master protein that was originally orange may now be designated as strawberry because it influences genes in a different way.

To summarise, the three rules that I postulated for my scheme of oranges and lemons have a molecular foundation. Hidden colours can affect the scents produced by a cell because they can switch the genes which make signalling proteins (scents) on or off. Similarly, hidden colours influence the sensitivity of a cell by switching the genes which make receptor proteins (which recognise the scents) on or off. Finally, the triggering of a receptor can change the hidden colour of a cell by setting off a chain of reactions that eventually lead to a change in the shape of a master protein.

To get an idea of what is going on, imagine you are standing on part of a large canvas but you cannot directly see what your neighbours are up to because of barriers (cell membranes) between your bit of canvas, your cell, and theirs. In addition to being able to apply colours to the canvas, you and your neighbours are also able to produce scents and receptors. The scents given off by your neighbours trigger reactions in your surrounding membranes that can result in the colour of your canvas undergoing a change. You may have applied a nice lemon colour to the canvas, only to find that it has turned plum-coloured because of an incoming scent of oranges. On seeing the new plum colour, you react in various ways. You start to produce a plummy scent and also a new receptor for inclusion in the membrane. The plum scent may in turn influence your neighbours, changing the colour of their piece of canvas, say from orange to strawberry, and the new receptor may influence the way your cell membrane responds to further scents. You and your neighbours may carry on in this way, influencing and reacting to what the others are doing.

In my hypothetical scheme, I considered each cell to produce only one type of master protein (hidden colour), one signalling protein (scent) and one receptor protein (sensitivity). In practice, things are a lot more complicated than this. Every cell will typically produce multiple master proteins, receptors and signalling molecules, giving it an overall combination of hidden colours, scents and sensitivities. This allows the influence and response to scents to be more elaborate than I have implied.

When a cell is triggered by a scent, the particular colour change will depend not only on the scent, but also on what master proteins (hidden colours) are already there. There is a comparable situation with our own responses to scents. Some time ago, a Chinese student came to our lab to study for his PhD; soon after he had arrived from China, I invited him round for a meal. When we got to the cheese course, he started to look rather worried. I was surprised to learn that many Chinese do not make or eat cheese; it is not part of their diet. To the Chinese student, who had never seen or tasted cheese before, giving him cheese was like waving smelly socks under his nose and expecting him to enjoy the experience (he is now used to cheese and quite likes it, although he still has problems with the smelliest varieties). Our response to a smell not only depends on the chemical nature of the molecule being sniffed, it is also affected by the general state of the sniffer.

In the same way, the response of a cell to a signalling molecule depends on the state of the cell, its hidden colour. A scent may trigger a change in hidden colour, but precisely what that change involves will depend on what the colour was to begin with. This means that cells in one part of the body may respond to a scent quite differently from those in another part of the body, because the scent is being detected in a different context of hidden colours. Indeed, some scents are used repeatedly during development, having a variety of effects according to the context in which they are sensed. Because of this, the number of types of scent does not need to be as great as the number of types of cell, as my simplified scheme of oranges and lemons had indicated.

Extending the range

So far I have assumed that the molecular signals or scents produced by cells do not travel very far. In the story of oranges and lemons, the scent only travelled as far as the neighbouring cell and no further. This is valid for some cases, like in the example from the eye, where the signalling protein appears to remain attached to the membrane of the cell that produces it. In other cases, however, signalling molecules may be released from the cell into the surroundings, allowing them to diffuse further away. A scent may be able to travel a distance involving several cells in this way. How does this affect the process of refinement?

We can answer this by using the same system of orange and lemon cells but this time we will assume that each scent can diffuse several cells in distance instead of only moving one cell along. We will also start with three times as many cells: twelve instead of four. As before, half of these cells will be orange and half lemon (Fig. 11.10). Now because the scent diffuses away this time, we would expect that the *concentration* of the orange scent will gradually decline as you get further away from orange cells. There will be a relatively high concentration near to the orange cells, a strong scent of oranges, but this will drop as we go further into the lemon region. There may come a critical point, say four cells away, where the orange scent is no longer intense enough to trigger a lemon cell. Effectively, the orange scent only has a range of three cells; beyond this the concentration will be too low to stimulate a response. This means that three of the lemon cells will be bathed in sufficient scent to make them turn plum-coloured. We now end up with three lemon cells, three plum cells and six orange cells (Fig. 11.10, step 1).

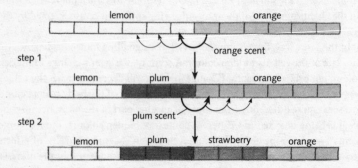

Fig. 11.10 Elaboration of patterns in which scents can diffuse away from their source. In step 1, three lemon cells are exposed to sufficient levels of orange scent to turn them plum. In step 2, three orange cells are exposed to sufficient plum scent to turn them strawberry.

For the next step, we can essentially repeat the process. The plum cells will act as a source of scent that diffuses away, so there will be a critical point, say four cells away, where the scent of plums is too faint to trigger the orange cells. Effectively the plum scent will only be able to influence three of the nearby orange cells, switching them to strawberry. The net result is three lemon cells, three plum, three strawberry and three orange cells (Fig. 11.10, step 2). The final pattern is the same as we obtained when the scent could travel just one cell along; the only difference is that each coloured region now comprises three cells instead of one.

The overall effect of increasing the range of the scent is that the process of the refinement occurs at a slightly different scale in cellular terms, in this case three cells rather than one. If the scent could be detected five or ten cells away from the source, the scale of refinement would be even coarser.

There is, however, one further feature that comes into play when the intensity of the scent varies in this manner. It becomes possible for cells to respond in *different ways* according to the concentration of the scent. Instead of only one response to the scent, a range of responses is possible. The way this can happen is formally very similar to the process described in Chapter 9, in which a gradient in concentration of black master protein was responded to in various ways. For example, suppose the triggering of a lemon cell by the scent of oranges results in *two* different master proteins undergoing a change in colour, with one becoming plum-coloured and the other becoming avocado-coloured. We will also assume that these colours compete with each other (Chapter 9), in such a way that plum tends to switch off avocado, and avocado tends to switch off plum. This means that if both of these colours are produced, as will happen when there is a high concentration of orange scent, only one colour will eventually predominate: the best *competitor*, say plum. In other words, initially the cell will start to turn both plum and avocado, but eventually plum will take over completely. But there is also another consideration that needs to be borne in mind: the degree of *sensitivity* to the scent. Suppose that the change to avocado is more sensitive to the scent than the change to plum. That is to say, the change to avocado can occur even when the concentration of scent is quite low; whereas the change to plum requires a higher concentration of scent (this could be because the chain of reactions triggered by the scent are more efficient in modifying one master protein than the other). This would mean that at low concentrations of scent, there will initially be more avocado than plum. This numerical advantage to avocado might compensate for its lesser ability to compete, so that under these conditions avocado is eventually victorious over plum. Thus, instead of three lemon cells switching to plum as in our original scheme, perhaps only the one nearest to orange would turn plum (where the concentration of scent is high), and the two further away would turn avocado (where the concentration of scent is lower). This provides another variation on the ways of refining a pattern, with cells near to a source of scent reacting by turning one colour and cells further away reacting by turning another.

Now although the distance over which scents act during development may vary, they typically do not extend more than of the order of ten to one hundred cells (we shall come across some exceptions to this in the next chapter). That is to say, the scale at which patterns are refined is pretty much confined within a relatively small range. This seems to be true whether we are talking about refinements that occur early on in development or at later stages. Nevertheless, the pattern of hidden colours in the organism does get progressively refined to a greater and greater degree as it develops. This is because the number of cells increases as the patterns are elaborated. At an early stage of development, a distance of ten cell diameters might cover a significant fraction of the organism's length, whereas at a later stage it might correspond to only a tiny proportion.

In terms of the painting metaphor, the number of artists is increasing as the canvas develops. Each artist is working on a bit of enclosed canvas and both influencing and reacting to what the neighbours are up to through scents and sensitivities. The range of influence of each artist does not change very much as the canvas develops: each may only be able to detect scents within a radius of about ten neighbours. At an early stage, when there are few artists, a distance of ten neighbours accounts for quite a large proportion of the canvas length. Each artist therefore gets a sense of what is going on over a relatively large part of the picture and has a significant overall influence. But as the artists multiply, each one individually influences and responds to an ever shrinking fraction of the painting. To someone overhead, the pattern is becoming ever more elaborate and refined, but as far as the artists are concerned, they are always operating at a similar scale of very local interactions.

Keeping on track

A good illustration of how cells progressively lose their sense of the overall canvas during development comes from some experiments carried out by Hans Spemann in 1918 (we already came across some of Spemann's work on newts earlier in this chapter, on page 187). Spemann was exchanging small pieces between pigmented and unpigmented newt embryos. For example, he took a piece that would form brain in one embryo and exchanged it with a piece that would form skin from the other. As long as he did the exchange at a very early stage, the transplanted pieces developed according to their new location, irrespective of where they had come from: a piece that would normally form brain tissue would switch to develop into skin if it was moved into the skin-forming region. If the same experiment was done a bit later, however, the transplanted pieces ignored their new general surroundings and went on to develop according to how they would have done in their original location. It was as if at later stages the cells had determined whether they would form skin or brain and it didn't matter where you put them: they would carry on doggedly, acting on what their previous history told them.

This is what you might expect of a refining process in which each step is concerned with a progressively smaller part of the picture. At an early stage, molecular signals are involved in making distinctions between cells that will give rise to broadly different regions, like brain and skin. If transplantations occur during this time, cells will respond appropriately to their new surroundings and scents. At a later stage, the cells have moved on to the next level in refinement, such as distinguishing between particular bits of brain or between certain regions of skin. The cells have already determined whether they will form brain or skin and are now using scents and sensitivities to elaborate a narrower set of territories (narrower in terms of the overall picture). Transplantations carried out at this stage do not result in brain switching to skin because those distinctions were already established at an earlier stage, when there were fewer cells in the picture. Because patterns are elaborated through building on what went before, the cells become concerned with a more restricted range of decisions and possibilities as the organism develops (I shall deal with some apparent exceptions to this in the next chapter).

This raises an important problem: for the system to work, there has to be continuity that links early and late events. In a process in which each step elaborates and builds on what went before, it is essential that cells keep track of where they are in the sequence of events. A group of cells that temporarily loses its place in the sequence of hidden colours may find it impossible to re-establish it again, because the relevant scents from its neighbours are no longer there.

It is not yet fully understood how cells manage to keep their place in the sequence, but there are several likely mechanisms. One way of keeping continuity is by individual cells passing their hidden colours on to their progeny cells every time they divide. It is continuity by descent, much as tradition or wealth might be faithfully passed on from generation to generation. Every time a cell within an organism divides, its contents get divided up between the two daughter cells so that each inherits a copy of DNA together with a share of surrounding cytoplasm. The spoils can include master proteins (hidden colours), which may be located in the cytoplasm or bound to the DNA. By inheriting these hidden colours, the daughter cells may be able to re-establish the same combination of hidden colours as their parents. In some cases, one or more hidden colours can be maintained for many generations of cell division in this way. For example, the green hidden colours involved in fly segment identity are thought to be perpetuated for most of development by this sort of mechanism (strictly speaking it is not the green hidden colours that get passed on but other hidden colours that switch the genes for green on or off).

In other cases, it seems that continuity is not maintained in such a rigidly aristocratic fashion. If there are several artists working nearby on a piece of canvas, it may not matter too much if one of them has an occasional memory lapse, so long as his or her neighbours are there to help reinforce the right colours, quickly putting the artist back on track. If there are enough neighbours, the chance of them all losing their place at the same time is very small, so continuity is always maintained. In a similar way, continuity can be maintained in a community of cells through signalling between cells which reinforces hidden colours, rather than purely on the basis of individual cell ancestry.

To summarise, cells communicate with each other by signals and receptors, allowing patterns of hidden colour to be continually refined as the organism grows and develops. To the onlooker, the organism acquires an ever more elaborate geometry, and it looks as if every cell must somehow know exactly where it is in order to behave in the right way. But cells are not geometers, calculating their positions in space on a grid. Like dogs, they recognise molecules in their local environment and modify their behaviour accordingly. They do not sense the whole picture, except at the earliest stages when it is in a very crude form. They are only aware of their immediate surroundings, and by repeated local transformations the overall pattern becomes more elaborate as the picture gets larger. As cells build on their past, they become progressively determined to follow narrower and narrower paths; and for this system to work, there has to be a continuity that links steps.

CHAPTER 12

Responding to the environment

The German illustrator Ludwig Richter tells in his autobiography how as a young art student in the 1820s, he and three friends visited a famous beauty spot in Tivoli where they sat down to draw the landscape. All four of them were determined to draw as precisely as possible what was in front of them, not deviating in the slightest detail from nature: 'We fell in love with every blade of grass, every tiny twig, and refused to let anything escape us. Every one tried to render the motif as objectively as possible.' However, when they compared their efforts later on in the evening, they were surprised to see how different each picture was. Although the subject was the same, the pictures were as different as the personalities of the four artists.

The story shows that no matter how hard someone might try to imitate nature, the result is always an individual *response* of the artist rather than a carbon copy. The environment never determines or dictates what the picture will look like because so much depends on the way the individual artist reacts. Painting a landscape or portrait is an active process in which the artist responds to the surroundings rather than passively imitating them. Although the final picture may remind us of a scene, it is always through the eyes and reactions of the individual artist rather than an objective rendition of nature.

Now in describing the relationship between painting and the environment, we need to be careful to distinguish between two very different sorts of environment. First of all, there are the general surroundings of the artist, such as the lighting conditions, the landscape that is being painted, or the person who is sitting for a portrait. All of these conditions are what we commonly mean by the artist's external environment. A second aspect of the environment is the canvas or support on which the artist is working. This is a very particular part of the artist's surroundings, the part that he or she is actively working on.

The relationship of the artist to these two aspects of the environment—the canvas and the general surroundings—is quite different. The canvas is continually changing as a direct result of the artist's actions. Each time a brush stroke is applied, the artist modifies what is on the canvas, defining a new region of colour. The artist then reacts to this. Perhaps the colour is too strong, or it clashes with another colour on the canvas, or it fits in perfectly. Whatever the reaction, it will influence the next colours that are applied. These will in turn modify the canvas further, to produce yet more reactions in the artist. The artist is continually modifying as well as responding to the canvas, so the patterns change in a highly interactive and dynamic way. Although the canvas is physically external to the artist, outside his or her body, it is very much an integral part of the creative process, an inseparable part of the act of painting.

By contrast, the general surroundings of the artist, such as the landscape, model or lighting, are not modified to any great extent whilst the painting proceeds. The landscape may of course change as the sun sets or the clouds move, but these are not as a direct consequence of what the artist is doing. There may be the occasional bit of interference by the artist, such as asking a sitter for a portrait to turn slightly more this way or that, but this is relatively minor compared to the extensive interactions between the artist and the canvas. A movement or change in the subject may even be a source of irritation for an artist who might want everything to stay fixed; whereas changes in the canvas are an essential part of the painting process. So although both the general surroundings and the canvas are physically external to the artist, their relationship to the painting process is quite different. We might say that the overall surroundings provide an *external* environment whereas the canvas is an *internal* environment, an integral part of the creative activity itself.

We can get some curious effects when we confound these two aspects of the artist's environment. Look at Magritte's painting, called *The Human Condition*, shown in Fig. 12.1. It is a painting within a painting: a picture on an easel framed within a room and landscape. As the painting on the easel is exactly contiguous with the landscape behind, it seems to be both a part of the landscape and a representation of it. Magritte said of this picture: 'In front of a window, as seen from the interior of a room, I placed a picture that represented precisely the portion of landscape blotted out by the picture. For instance the tree represented in the picture displaced the tree situated behind it, outside the room. For the spectator it was simultaneously inside the room, in the picture; and outside, in the real landscape, in thought.' Magritte's picture works by confounding the canvas with its general surroundings. However, the environment he shows, the room and window looking onto a landscape, is not truly external; it has been created by Magritte as part of the overall canvas. Although the surroundings of the easel may seem to be part of the external environment, they have actually been put there by the artist. The paradox arises because we confuse the external environment, the one that the artist does not greatly influence, with what is on the canvas itself. But no matter how realistic the painter's style, the general surroundings never become incorporated into the

Fig. 12.1 *The Human Condition* (1933), René Magritte. Private collection.

canvas. The canvas is not part of the external environment, it is internal to the act of painting, something that has been created through mutual interaction with the artist.

Developmental responses

When we look at how an organism's development is related to its environment, we encounter a comparable situation to that in art. As we have seen, the growth and development of plants and animals are highly interactive processes in which each step is an interpretation of what went before. This is equivalent to the artist's internal dialogue with the canvas. Furthermore, as with painting, development does not occur in isolation but is always in the context of a particular set of environmental conditions. Factors such as temperature, humidity and light may all influence the way an organism develops. We shall see that it is not that these factors dictate how development will proceed, but rather that development *responds* in one way or another to them, just as the activity of an artist may respond to the external environment.

There are many different styles in which developing organisms respond to their environment, just as there are a variety of approaches to art. They range from cases in which the environment provides a rather uniform context, much as the general lighting conditions and ambience set the stage for an abstract painting, to cases in which variation in the environment plays an important role, nearer to what happens in representational art, such as landscape or portrait painting. Before getting into the mechanisms of how the internal systems of development are able to respond to the external environment, let me give a few examples of the various styles.

In many cases, the external environment is held relatively constant while the organism develops. The embryo is buffered from the outside world by, say, being protected within a mother's womb, or by being incubated in a nest by a parent. This is perhaps the type of development we are most familiar with, in which the environment provides a relatively uniform context that allows the process to occur.

In other cases, development is more geared to heterogeneity in the environment. We already encountered an example of this in Chapter 10, where I mentioned that some species of butterfly can develop in two different colour forms, say a spring and a summer variety, according to the environmental conditions at their time of development. This developmental response allows their colour to be adapted to different seasonal conditions, just as we vary our dress according to season.

Another example of how development may respond to the environment involves sex determination in some reptiles. Unlike ourselves, where sex depends on which chromosomes we inherit (females have two X chromosomes, males have one X and one Y), the sex of all crocodilians, most turtles and some lizards depends on the temperature at which they develop. Eggs incubated at low temperatures produce one sex, say a male, whereas those incubated at a higher temperature produce the other, a female. The survival of these species depends on variation in the environment: if the environmental temperature was always low or always high, all the offspring would be of the same sex, signalling the end of the species. In these cases, modifying development in response to heterogeneity in the environment is an essential feature of the system.

An even more intriguing form of environmental sex determination is in the sea worm *Bonellia*. The female of this species is about the size of a plum and has a proboscis about one metre in length. The male is a tiny worm-like animal a few millimetres long, and lives as a parasite within the uterus of the female. When the eggs fertilised by the male are liberated, they develop into free-swimming larvae. Any of the larvae that happen to settle on the proboscis of an adult female will develop into males, whereas those that remain free-swimming and gradually sink to the bottom become females. In other words, whether the larva develops into a large female or a tiny male depends on its external environment, whether it happens to land on an adult proboscis or not.

It is in plant development, however, that we see the most extensive modifications in response to environmental variation. Because they are rooted to the ground, plants continually have to modify their patterns of development according to local circumstance. A plant's direction of growth, branching pattern, leaf shape and flower production can all depend on the external environment. Because of this flexibility, plant development is often said to be plastic. In my view, this is misleading because plasticity conjures up the idea of a malleable substance whose shape and form are imposed from without. Plants, however, are not moulded by their external surroundings; they respond to them. For example, most shoots tend to grow towards light. This is not because the light draws or pulls the shoots towards it, but because plant shoots respond in a particular way to light, modifying their pattern of growth such that the tips bend towards the source of illumination. Roots will typically react in a different way, tending to grow away from light.

Similarly, plants can respond in various ways to the direction of gravity. Even when grown in complete darkness, shoots tend to grow upwards, in the opposite direction to the force of gravity, whereas roots respond in the opposite way by growing downwards. It is not that the roots are pulled down by gravity; they change their orientation of growth in response to it.

One of the most striking environmental responses is displayed by some amphibious plants which grow partly submerged in water. The leaves of these plants can develop in different ways, depending on whether they are surrounded by water or air. Leaves produced lower down on the stem, in the aquatic environment, tend to be finely dissected and feather-like, whereas those borne in the air are more compact (Fig. 12.2). The plant effectively displays the height of the water in its own structure. This reflects the way the developing leaves respond to their environment. If a young plant growing completely under water is moved to a dry environment, it will switch from producing feathery leaves to aerial leaves. Conversely, a plant that has been raised entirely in the air and then submerged in water will start to produce leaves with a feathery shape. In each case, the previously fully grown leaves do not change; it is the newly arising leaf-buds at the tip of the plant that develop according to their new surroundings. This response allows the shape of the developing leaves to be adjusted according to their environment: large feathery leaves are

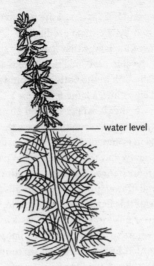

Fig. 12.2 An amphibious plant (*Myriophyllum*) showing differences between leaves grown in water and air.

appropriate for waving about in water whereas compact leaves are more suited to aerial surroundings.

The development of organisms may also respond in various ways to physical damage inflicted by their environment. There are two broad types of response. To avoid confusion, I shall refer to one of these as *restoration*, and the other as *regeneration* (sometimes both of these processes are put together under the general heading of regeneration).

Restoration

The propensity for restoration following damage is particularly well developed in plants. When you mow the lawn or prune a bush, the plants usually manage to recover. To see how they do this, I need to explain more about how plants grow.

Unlike most animals, which have a more defined endpoint in development, plants continue to grow and add new parts to themselves throughout their lives. This is achieved by groups of cells, called *meristems*, that continually divide and replenish themselves whilst at the same time adding more tissue to the plant. Meristems can be thought of as being perpetually *embryonic cells*, which, like Peter Pan, can continually maintain an early condition rather than maturing. For example, there is a *primary meristem* at the main growing tip of a shoot,

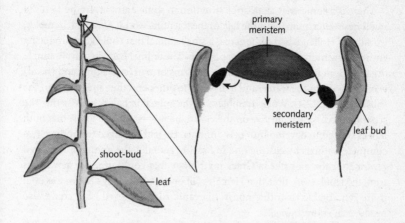

Fig. 12.3 Growing tip of a plant, showing the primary meristem, together with the leaf-buds and secondary meristems derived from it.

which replenishes itself whilst also adding cells to the stem below and to leaf-buds on the side (Fig. 12.3). The primary meristem continues to generate leaf-buds as it climbs upwards, so you end up with a series of leaves of different ages, with the youngest at the top and progressively more mature stages as you go down the plant. The primary meristem also produces additional meristems, called *secondary meristems*, on its periphery. Typically, a secondary meristem is to be found in the angle between each leaf-bud and the stem (Fig. 12.3). Each secondary meristem can be thought of as a very young shoot-bud. Once it starts to grow, the secondary meristem will itself produce stem, leaf-buds and yet more meristems (flower-buds may also arise from meristems but their growth is more limited). Potentially the process could carry on indefinitely, with the plant becoming more and more branched. In practice, however, many of the additional meristems and buds tend to remain dormant for long periods, so that growth is channelled in a limited number of directions.

When plants are cut or pruned, they typically respond by activating growth from meristems and buds that were previously dormant or slow-growing. The system ensures that when a major growing point is lost, others are activated and take over. Similarly, when you mow the lawn, it does not grow back from the cut surfaces of the leaves, but from meristems and buds that were already present in the grass plants, below the level of the cutting blade. That is why you should not set your blades too close to the ground, as you will then remove the buds and meristems. These are all examples of response to environmental damage through the development of meristems and buds that were already present.

There is a somewhat analogous situation in some animals. In the 1740s, a small freshwater animal was the talk of the scientific world. Abraham Trembley, working in Holland, had come across a small animal that could keep producing entirely new individuals after it was cut up. The animal had a relatively simple organisation: it comprised a trunk with arms or tentacles at one end (head) waving about in the water, and a disc at the other end that attached it to the substrate (Fig. 12.4). When Trembley cut the animal in half to see whether the separate pieces could survive on their own, he was astounded to see that both halves started to develop into new animals. He had managed to produce two complete animals by cutting one in half! Eventually, the animal was named *Hydra* after the monster in Greek mythology that could produce new heads from the trunk every time that Hercules cut one off. Trembley's *Hydra* was even more remarkable than the monster because each removed head could also produce an entire animal.

The ability of *Hydra* to behave like this reflects the way it normally grows and propagates itself. A well-fed *Hydra* will form a bud about two-thirds of the way down the trunk. The bud then elongates and sprouts arms at the end. In two or three days, the bud looks like a little *Hydra* and can then pinch off at its base to become an independent individual. Like plants, *Hydra* can grow from buds;

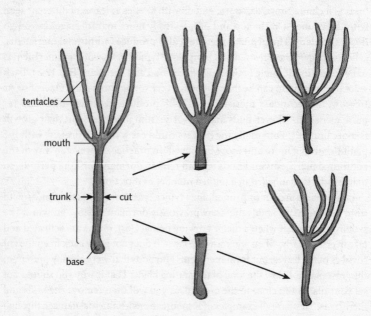

Fig. 12.4 Restoration of two animals of *Hydra* after one has been cut in half.

but whereas plants produce their buds from meristems at the growing tips, in the case of *Hydra* the buds are produced directly from the trunk. The trunk can be considered as equivalent to a plant meristem, the cells maintaining themselves in a somewhat embryonic condition while at the same time producing buds to the side, arms and mouth at the head end, and a basal disc at the other end. Seen in this light, slicing up *Hydra* is equivalent to pruning a plant. Cutting the animal stimulates the production of new structures from the embryonic cells in the trunk, much as pruning a plant can stimulate the development of pre-existing meristems.

Regeneration

Another type of response to damage is regeneration. This involves the production of new structures from *mature* tissue, unlike restoration, which comes from pre-existing cells of a more embryonic nature. For example, if you take a leaf from an African violet or begonia plant and put it onto some soil, new plantlets will form which can eventually grow into full-sized plants. If you cut the leg off a newt, a new one will grow in its place. In all these cases, it is the mature tissue of the organism that is responding to the environmental change by developing new structures. The cells of a leaf or leg are made up of *differentiated* cells: cells that have distinctive features, quite unlike those of the earlier embryonic cells from which they were originally derived. A leg, for example, contains many different types of specialised cells, such as skin cells, nerve cells and bone-producing cells. It is the cells in a developed leaf or leg, rather than groups of pre-existing embryonic cells, that are responsible for regeneration.

At first sight, this phenomenon seems to fly in the face of what I said in the previous chapter. There I described how patterns are elaborated during development through cells building on what went before, with the range of possibilities becoming narrower as the organism grows. The cells of a mature leaf or a leg have gone through an extensive series of restricting steps and elaborations. Yet in the case of regeneration, whole plants or entire limbs can develop from the cells of these mature tissues. How are we to explain this, if development involves a narrowing down of options?

The key is that the cells of the mature tissues do not regenerate structures by continuing with the process of elaboration; they first have to go back to a more embryonic condition. Because patterns are refined through building on previous steps, the only way to get regeneration from cells in mature tissue is for them to return to an earlier state of development, and start to rebuild the picture all over again. This is precisely what seems to occur. When a begonia leaf is cut off a plant and placed on some soil, its cells do not directly switch to become root or shoot cells. They first become *de*differentiated, going back to

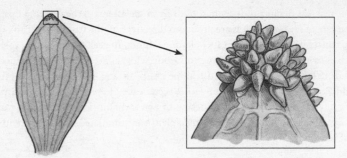

Fig. 12.5 Leaf of *Malaxis paludosa* showing the development of a cluster of embryos at the extreme tip.

the cell types typical of embryos or meristems. Only after this do they start to grow and divide again to form a new plant. They have to go back before they go forward again, as might be expected from a system which builds on a sequence of steps.

Some plants exploit this as a normal way of propagating themselves. In mature leaves of the bog orchid *Malaxis paludosa*, the cells at the tip dedifferentiate—they return to a more embryonic condition—allowing them to develop into tiny egg-shaped embryos (Fig. 12.5). When these embryos become detached from the leaf, they develop into new plants, just like the embryos produced through seed (although unlike plants derived from seed, the new plants are genetically identical to the parent plant because no sex is involved). In all these cases of regeneration from mature tissues, the cells must first return to a more embryonic condition before they can start building up their internal patterns again.

The same applies to regeneration in animals. When a newt's limb is severed, the stump heals over and cells near the tip start to lose their characteristic features. Bone cells, cartilage cells, nerve cells and so on lose their particularities and become a mass of dedifferentiated cells. These cells then start to grow and develop into the missing part of the limb (Fig. 12.6). The new limb does not come from cells of the stump directly changing to those of a new limb. The cells go back to an earlier, more embryonic state, and the limb then regrows just as it did when it first developed. As with regeneration in plants, the cells have to go back before they can go forward.

A sense of the surroundings

I have given these examples to show how the development of organisms can respond in a variety of ways to external influences such as temperature, light,

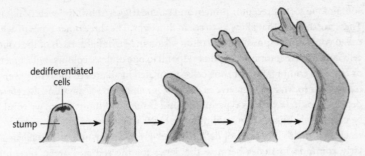

Fig. 12.6 Stages during regeneration of a newt limb.

gravity, water and physical damage. I now want to look at some of the under-lying mechanisms that allow organisms to respond to what is around them. As a first step it will help to look at how our own senses allow us to respond to our surroundings.

We saw in the previous chapter how our senses of smell and taste depend on receptor proteins that are triggered by molecules in the environment. The receptors allow our internal chemistry, the reactions within our cells, to respond to chemical information from outside. Our other senses, those of vision, hearing and touch, also depend on receptor proteins, but instead of responding to molecules in the environment, they are triggered by other factors. The receptors convert stimuli such as light, sound or pressure into the chemical language of cells. The net result is the same as for smell and taste—the internal reactions of cells are altered—but the stimulus is different.

Our sense of vision provides a very good example of how this works. There are a particular group of receptor proteins, called opsins, in the cells of the retina at the back of the eye. Each of these receptor proteins changes shape when it is struck by a particle or photon of light (to be precise, it is a small molecule attached to the receptor protein that changes shape when it absorbs the light, and this then leads to an alteration in the shape of the receptor protein). The altered shape of the receptor protein then sets off a chain of reactions in the cell. The receptor protein effectively converts a physical signal from the outside, light, into the language of chemical reactions within a cell. These chemical reactions in the cell in turn lead to charged atoms (ions) moving across the cell membrane, resulting in an electrical signal, a nerve impulse, being sent to the brain. We can summarise by saying that a light stimulus is converted into a chemical reaction which then leads to an electrical impulse.

We have four types of light-sensitive cell in the retina of the eye, each con-taining a characteristic type of receptor protein. Cells of one type, called rods, contain one sort of receptor protein and can be triggered by low levels of light. These are the cells that allow us to see in dim light. The three other types of cell, called cones, are responsible for colour vision in bright light. Each of the cone cells contains receptor proteins that respond to one of three colours: blue, green or red. The combination of cone cells stimulated in an area of the retina eventually leads to what we perceive as colour. Our ability to see anything therefore depends on these various types of receptor protein. Without them, we would be completely blind. Colour-blindness results when one of the genes for the receptor proteins is missing or defective (red–green colour-blindness is particularly common in males because the genes for the red and green receptor proteins are on the X chromosome. Males have only a single X chromosome in each cell, so if this chromosome carries a mutation in one of these genes, the individual will be colour-blind. Females have two X chromosomes, so even if one of these carries a defective gene, the other X chromosome will most likely carry a normal copy that acts as a backup.)

Our sense of touch and hearing are less well understood than vision, but they also depend on receptor proteins. In the case of hearing, receptor proteins in the sensory hair cells of the inner ear are stimulated by sound vibrations. This triggers a change in their shape, leading to nerve impulses being sent to the brain. Similarly, our sense of touch and muscle movement depend on receptor proteins that alter their shape in response to mechanical changes, such as stretching of the cell membrane.

We sense our environment through receptor proteins, proteins that change their shape in response to chemicals, light, sound or other stimuli. They convert information from the outside into the internal chemical language of cells, eventually leading to nerve impulses being sent to the brain.

Modifying internal patterns

In many ways, the principles by which developing organisms sense their environmental conditions are not too different from the way our own senses operate. In both cases, receptor proteins of various types play a key role in responding to external circumstances. The main difference has to do with the outcome of triggering the receptors. In the case of our senses, a nerve impulse is eventually sent to the brain, whereas for development we shall see that the outcome is often a modification of the organism's hidden colours.

We have already encountered a situation in which the triggering of receptors can lead to a change in internal patterns. Recall that in the previous chapter, we saw that cells can respond to scents from their neighbours by changing their hidden colours. The scents trigger the receptor proteins, setting off a chain of reactions in the cell which eventually lead to an alteration in the shape of a master protein, a change in hidden colour. This new colour can then be interpreted by other genes in the cell which may get switched on or off as a consequence. In a similar way, the receptor proteins triggered by the external stimuli, such as light, water or physical damage, may eventually lead to a modification of the organism's hidden colours. This change can then be interpreted by various genes, leading to a change in the proteins being produced in the cells and hence their properties. I should point out that there are still major gaps in our knowledge about how this occurs, but it is worth looking at a few examples to illustrate the likely principles involved.

How do plants respond to light? Plant cells are known to contain several different types of receptor protein that are sensitive to light. Like the receptors in the eye, these plant receptor proteins change shape when light strikes (as with the eye receptors, they have a small molecule attached to them that changes shape when it absorbs light; this influences the shape of the receptor protein). This change in shape of the receptor protein then sets off a chain of events in the cell. We still do not know the details of all these events, but one important outcome appears to be a change in the cell's hidden colour. Genes may interpret the altered colour in various ways, being expressed or not as the case may be, eventually leading to a change in cell properties.

To see how this might allow a plant to orient its growth with respect to light, look at Fig. 12.7. It shows what typically happens when a plant tip that was originally growing upwards is illuminated from only one side. The cells on the lit side grow more slowly than those on the shady side, resulting in the stem bending towards the light (in some cases, cells on the shady side may also grow more quickly). All the details of how this response occurs are not known, but let me give a simplified explanation of what may be going on. Because the plant is illuminated from only one side, the light receptor proteins on the lit side of

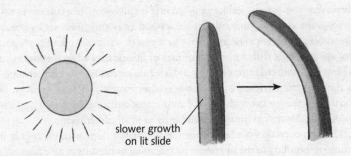

slower growth
on lit slide

Fig. 12.7 Illumination of a shoot from one side causes growth to slow down on the lit side, making the shoot bend towards the light.

the growing tip will be triggered more than those on the shady side. This sets off a chain of chemical events which may eventually result in a difference of hidden colour between the two sides, with the cells on the lit side turning from, say, cloudy-grey to sunny-yellow.* Various genes then respond or interpret the sunny-yellow colour in such a way that the proteins they produce result in growth slowing down. This means that the cells on the lit side grow more slowly, causing the shoot to bend towards the light.

I wish to emphasise that the way the system behaves depends on the particular combination of hidden colours in the responding tissue. For instance, in the previous example, the shoot started with one overall colour, cloudy-grey, which responded to the light stimulus by turning sunny-yellow. If the tissue had started with a different set of colours, as for example would be the case in a root tip, it may have responded entirely differently, say bending away from the light rather than towards it. This is because both the sensitivity (receptors) and response will depend on what the hidden colours are to begin with. This allows responses to change at different times, according to the hidden colours. A good example is in ivy-leaved toadflax, *Linaria cymbalaria*, which is to be found commonly growing on walls. The flower stems of this species initially grow towards light, carrying the flowers outwards, clear of the leaves and wall. After the flowers have been pollinated, however, the stems twist away from light, tending to bury the seeds in the cracks of the wall. The response depends on the stage of the flower stem, whether it is reacting before or after pollination. This might be explained if the act of pollination leads to a change in the hidden colours of the flower stem, so that afterwards it responds differently to light.

The notion of organisms responding to the environment through a change in hidden colours can also help us to understand what happens during regeneration. Imagine you are working on an expanding canvas, elaborating the patterns of colour in collaboration with your neighbours. Eventually you arrive at the end of the process, with the canvas fully extended and the final colours displayed. You are just feeling satisfied that it's all over, when it starts to rain. Being water soluble, your colours are washed away and you react by starting the whole painting exercise again. In a similar way, when the cells of the begonia leaf are presented with a new environmental situation, as when they are cut off a plant and plunged into soil, their hidden colours return to an earlier state. It is not that the environment actually washes away the hidden colours, but particular cells in the leaf respond to the new situation by modifying their colours so that they eventually resemble those of an earlier stage.

Limb regeneration can be explained along similar lines, with the cells in the stump responding to the operation by returning to producing an earlier set of

---

*This response is thought to be coordinated by a plant hormone, called auxin, that is produced at the growing tip—see next section.

hidden colours. In this case, however, there is a further feature that needs to be explained. Only the part of the limb that is removed tends to be regenerated. If a newt's limb is amputated near the base, almost the entire limb is replaced, but if the amputation is at the wrist, only the hand regenerates. As noted by the eighteenth-century French scientist René-Antoine Ferchault de Réaumur, who was studying the same phenomenon in crustaceans: 'Nature gives back to the animal precisely and only that which it has lost.' How is it that the dividing cells only regenerate what is missing and not more or less? The answer to this is not known, but one possibility is that it has to do with the persistence of some hidden colours in the limb. We might imagine the newt limb is divided into a series of hidden colours from base to tip during early development, and that these colours are then maintained even up to the time that the limb matures. When the limb is amputated, the cells at the tip of the stump may retain some of these hidden colours, even though others are lost or modified. This means the regenerating cells in the stump will have a different set of hidden colours when the amputation is at the wrist, compared to when it is carried out near the base. This could provide some of the information that determines how much of the limb is then regenerated.

In all these cases of regeneration, the cells respond to an injury by changing their hidden colours to resemble an earlier state; but there is also an artificial method of achieving a similar sort of result. This involves taking a nucleus from a cell in the mature body and transplanting it into an egg cell that has had its own nucleus removed; equivalent to physically transporting an artist from a mature canvas onto a fresh piece of initial canvas with the early hidden colours. Being surrounded by this new cellular environment, the nucleus might then start to interact with it in such a way that a new individual eventually develops. The new individual would carry exactly the same set of nuclear genes as the individual that donated the nucleus: it would be a clone of the original (just as a plant derived from regeneration of a begonia leaf is a clone). This is essentially the way that researchers were able to clone a sheep, called Dolly. A cell from the mammary gland of a mature sheep was fused with an unfertilised egg cell that had had its own nucleus removed. In this case, the cytoplasm from the donor cell was introduced together with the nucleus, but the effects of the donor cytoplasm were presumably swamped out by the cytoplasm in the large recipient egg cell.

Hormones

An important feature of many of these environmental responses is that they are often coordinated through long-range signalling between cells. This can be achieved through a set of internal scents, called *hormones*. Hormones are typically small molecules (small, that is, relative to most proteins), many of which are synthesised through a series of chemical reactions catalysed by proteins (enzymes). As with other types of internal scent, these hormones match the shape of particular types of receptor protein, allowing them to trigger reactions in cells which may lead to a change in hidden colour. However, unlike most of the other scents I have mentioned, they can sometimes travel relatively large distances within the organism.

Plant hormones, sometimes called growth regulators, have been shown to be involved in a range of environmental responses including stem growth, flowering time, fruit ripening, leaf shape and leaf fall. In many of these cases, their role seems to be to coordinate the responses between cells in the plant. A good example is what happens in pruning. If you remove one or more growing tips of a plant, it will stimulate the shoot-buds further down to develop and grow out. An injury caused by the environment in one part of the plant influences growth and development of structures that are very far away in terms of cell numbers. This effect is mediated by a particular growth hormone, called auxin, that is produced at the growing tips. Auxin travels from the tips down the plant, where it inhibits the growth of side buds. That is to say, the cells in the side buds respond to the auxin scent by slowing down their growth. Cutting off a tip removes a major source of auxin, liberating the side buds from its inhibitory effects, so they start to grow out (a similar sort of explanation accounts for the response of *Hydra* to surgery).

Flowering time provides another instance of long-range communication. Many plant species flower under particular day-length conditions, ensuring that they produce flowers in one season. Some species (e.g. chrysanthemums) flower when days are short, in autumn, whereas others (e.g. antirrhinums) flower when days are long, in summer. The response to day-length depends on receptor proteins in the leaves that are triggered by light. When the duration of the day reaches a certain length, it sets off a chain of reactions in the leaf cells. As a consequence, one or more long-range signals are then produced which travel up to the growing tip, triggering cells there to initiate flower development. The precise nature of these signals has yet to be determined, but some hormones have been strongly implicated.

Hormones also play a role in coordinating the way animal development responds to its environment. For example, the effect of temperature on the determination of sex in reptiles (see p. 210) is coordinated by hormones in the developing animal. Each sex produces different hormones. The cells of the animal then respond to the hormones, leading to male or female characteristics. The precise mechanism by which temperature influences sex hormones in reptiles is not known, but one model is that it changes the hidden colours of the cells in the developing gonads, which then leads to altered levels of hormone synthesis. (The sex of mammals is also coordinated by hormones, but in this case the type of hormone produced depends on the presence of a gene on the Y chromosome rather than temperature. This gene codes for a hidden colour and is expressed in the developing gonads of the male, where it leads to testis development and the production of the male sex hormone, testosterone. Females do not have a Y chromosome, so their gonads do not express this hidden colour and therefore develop into ovaries, which do not make testosterone.)

Another example of how hormones can coordinate responses involves moulting in insects. The growth of insects is largely limited by their rigid outer covering, which must be shed every so often if they are to continue growing. In the blood-sucking bug *Rhodnius*, this moult occurs after the animal has gorged itself on a large blood meal which distends the abdomen. The distension stimulates receptors in the cells of the abdomen which then send nerve impulses to the brain. This in turn leads to the insect producing a small molecule, a hormone called ecdysone, from one of its glands, which then circulates throughout the body. Cells respond to this internal scent by changing their hidden colour, switching on the genes needed for moulting. The role of the hormone is to coordinate changes in hidden colour throughout the body in response to the consumption of a large meal.

Responding in art and development

These various examples of environmental responses show that the elaboration of hidden colours, scents and sensitivities does not occur in isolation but is influenced by the general surroundings. In all of these cases, two sorts of process need to be borne in mind.

On the one hand there is the internal process of elaborating hidden colours. This involves a highly interactive series of events in which local signalling between cells, through scents and sensitivities, allows patterns of hidden colour to be built upon and refined in a coordinated way. In this case all the signals are generated by the organism itself. This is comparable to a painter interacting with a canvas, where all the colours are applied by the artist. The artist is responding to signals that he or she has put there.

On the other hand, this system of internal elaboration is continually responding to the external environment. In the case of development the mechanisms are essentially the same as those for internal patterning, the main difference being that the receptor proteins are triggered by signals from the external environment, such as light or gravity, instead of scents produced by other cells. Unlike the process of internal elaboration, the signals in this case are not generated by the organism itself but are relatively independent of it, coming from its surroundings. In a similar way, the dialogue between the artist and canvas is influenced by the general surroundings, such as the lighting conditions or landscape. The artist detects these environmental factors in the same way as the colours on the canvas, through the visual senses. The difference is that in the case of the canvas, the stimuli are a direct consequence of the artist's actions, the colours applied to the support, whereas the external environment is more independent of the artist.

For many aspects of development, particularly in animals, the environment may be held relatively constant, providing a uniform context that allows development to proceed. Heterogeneity in the surroundings plays a relatively minor role in the patterning process, as is the case with an abstract painting.

By contrast, plants are the landscape artists of the living world. Plants are continually modifying their hidden colours in response to variation in their surroundings. Like artists, they alter their patterns in response to the light and shade around them. They also establish orientations of colour in response to gravity, much as an artist might emphasise vertical or horizontal lines in a painting. Amphibious plants even modify their hidden colours according to the level of water in a pond, just as someone might depict the surface of water in a lake-side scene. The environment is written all over a plant, not as a mould over a cast, but through the particular ways that the plant responds to its circumstances. I have given some of the more obvious examples to make this point clearer but there are many other more subtle ways in which plant development is continually reacting to its surroundings. If we could read plants as well as we read pictures, perhaps we could tell as much about an external environment from a plant as from any landscape painting.

The developing brain

I have emphasised that development in plants responds more to variation in the environment than is the case in animals. Yet it seems to me that there is a very important aspect of animal development that comes much nearer to the plant style: brain development. On the face of it, it seems peculiar to make such a comparison, because brains are what we least associate with plant life. But if we look at the problem in terms of how processes are influenced by the environment, it is in brain development that responses to the surroundings start to become very significant for animals. Although we are still very far from understanding how the brain develops, it is worth pursuing this comparison further so as to bring together some of the key issues raised so far in this book.

Our brains contain many billions of nerve cells or *neurons*, which are connected with each other to form complex networks of interacting cells. The neurons work by transmitting electrical impulses along their length, which then influence other neurons that are connected to them. These connections are not direct: there is a tiny gap between each neuron and the next, called a *synapse*. When an electrical impulse from one neuron arrives at a synapse, it causes the release of a specific scent (signalling molecule), called a transmitter, which then moves across the gap, triggering a receptor protein in the neuron on the other side. This in turn sets off an electrical impulse in the responding neuron, allowing further transmission of the signal. It is the pattern and strength of all the connections between the various neurons in the brain that is believed to underlie our thought processes.

Many of the basic features of our brain's layout are established during development in the womb, through a process of internal elaboration that probably has much in common with many of the other developmental systems I have described in previous chapters. However, the patterning process does not stop there. As soon as we are born, the brain starts to get bombarded with information through our various senses. As a result, the pattern and strength of connections between nerve cells in the brain become modified in response to experiences. We normally call this learning, rather than development, because the environment plays such a large part. But we might equally say that it is simply a more plant-like form of development, in which the organism modifies its internal patterns in response to variation in its surroundings. The internal patterns here refer to the set of connections between the cells of the brain.

Let me give an example to illustrate the point. If one eye is kept covered at a critical period within the first few months of birth, so as to deprive it of any visual stimulation, the neurons emanating from that eye will fail to form connections with the main brain—so that the eye becomes almost entirely and irreversibly blind, while the other eye is still perfectly fine. In other words, without stimulation from the outside, the connections from the eye to the brain do not form. This mechanism is thought to ensure that only active or working neurons that come from the eyes are connected to the brain. By covering one eye, the neurons that come from it essentially behave as defective neurons, and so are not connected up.

This case shows, in a somewhat extreme way, how the patterns of cell connections that form in the brain may be influenced by the environmental conditions. Because the resulting change is so severe in this case, involving the loss of sight from one eye, we would probably say that the environment is affecting the development of the brain. But we might also say that the young mammal is learning about its environment, forming the appropriate connections in the brain that allow each eye to function properly. In more subtle types of response, as when the cell interactions in the brain are modified by being exposed to language or objects around us, we would call the process learning. But we could equally say that it is a form of development in which the internal patterns of cell connections are being modified through a response to the surroundings. The point is that our brain processes are not divorced from development, they are a rather special extension of it.

As with other examples of development, we can distinguish between various styles in which the brain may interact with the environment. In a uniform environment, as when we are in a quiet dark room, the brain is still very active, with one set of cell interactions leading to another; this is comparable to a developing organism elaborating its hidden colours in a relatively constant and uniform environment. We could call this pure thinking rather than learning, because of the lack of stimulating input from the outside.

When we are exposed to a stimulating environment, we talk about learning from our surroundings rather than just thinking. It is not that thinking stops but that it now also responds to numerous environmental stimuli. As with other cases of development, learning is not a matter of the environment moulding or imposing itself on the individual. Our brains are not like buckets that are passively filled up with environmental information; rather they respond in particular ways to their surroundings. Stimuli from the outside trigger receptors in our various sense organs, leading to a complex response. The response will depend on the particular history of the brain, the way it has developed and interacted with the environment on previous occasions. This is comparable to plant development, where internal patterns change in response to variation in the environment.

In a sense, both thinking and learning are creative activities: they are highly interactive processes that lead to novel ideas or levels of understanding by building on what went before. These activities are based on the history of each individual, the way his or her brain has developed and responded to previous experiences. This is true whether we are talking about a child learning how to read, or a poet making up a new verse. In all of these cases, the outcome is not anticipated in advance: the child does not know what it is like to be able to read before having learned, and the poet does not know the new lines before he or she has composed them.

The same is also true for the way that theories arise in science. As emphasised by the philosopher Karl Popper, scientific hypotheses or theories are not arrived at by following a plan or logical procedure, they arise creatively: '. . . theories are seen to be the *free* creations of our own minds, the result of an almost poetic intuition, of an attempt to understand intuitively the laws of nature.'

A scientist does not derive theories by following a formula, they are the product of a freely creative mind. By *free*, Popper does not mean to imply that the theories come from nowhere: theories depend on the scientist's past, including earlier observations and theories, going all the way back to childhood. When a scientist sitting in a bath suddenly has a new idea, it may feel as if it has come out of the blue, but it is still based on the previous history of the individual scientist. Popper's point is that a scientist does not come up with a new theory by a logical analysis of this earlier information; rather, the theory arises through a creative process in which his or her brain builds on what went before. It is only *after* the scientist has come up with a hypothesis, that logic comes into play, through testing its predictions against observations. This may lead to the rejection of the hypothesis because it is found to be logically inconsistent with the observations, or perhaps the hypothesis survives the test and is therefore corroborated. In either case, the hypothesis itself is not derived by a logical procedure but by a creative process in the brain.

Now there is a further aspect of environmental interaction that starts to become increasingly important in the case of the brain. So far I have treated the general environment as something that is relatively independent of the developing system. In most cases this is only an approximation because the individual does influence its surroundings to some extent as it develops. In the case of the brain, this can be quite significant. In addition to receiving various stimuli, it also influences the external world through the way we act. When we see an apple on a table we may respond by picking it up and eating it. Instead of the apple being stationary, we see it moving towards our mouth and start to detect its smell and taste. This modification in the environment, the movement of the apple, is a consequence of the brain's original response, the desire to pick up the apple in the first place.

If the extent of these two-way interactions between the brain and a part of its environment become particularly pronounced, then this aspect of the environment can no longer be considered as truly external or independent; it becomes integral to the interactive process. This is precisely what happens in the case of painting. The artist and the canvas are so interdependent that the canvas becomes almost an extension of the artist. Rather than the canvas being an external environment, it becomes an internal environment that is continually being modified as well as responded to. Although the canvas is physically outside the artist, it has been internalised within the creative process. It is as if the brain has established a two-way interaction to such an extent that its dynamic processes spill onto the canvas.

To my mind, human creative processes, such as painting, thinking and learning, have much in common with development, in the sense that they all depend on informed responses that continually elaborate what went before. In each case, we do not need to invoke instructions that are being followed, software as distinct from hardware, or plans that are subsequently executed. A further aspect of all these processes is that each responds in some way to a general environment. In some cases, as with pure thinking, abstract painting, and for many aspects of development, the general environment provides a relatively uniform background that allows the process to occur. In other cases, as with learning, some aspects of plant development, and landscape or portrait painting, heterogeneity in the general environment plays a more important role. Here, the creative process is continually responding to variation in its surroundings. In none of these cases does the environment mould or dictate the way things should go; rather, the creative process is modified to a greater or lesser extent in response to what is around it.

In my view, the reason for this commonality between human creativity and development is that the brain is itself a special extension of development. The interactive processes that characterise development themselves underlie the way our own brains work.

I have come full circle. In previous chapters I used comparisons with human creativity, like painting, to give an intuition of how development occurs. I am now saying that the reason this comparison works so well is that creativity is itself grounded in developmental processes: creativity is the child of development. This is not to deny the importance of the environment, because as we have seen, the environment plays a fundamental role in development. Nor do I wish to imply that the brain is the same as any other group of cells; it clearly functions in a distinctive way through the complex interactions between neurons. My point is that there is a continuity between the process of development and the functioning of the brain, which leads to them having many features in common: both are highly interactive processes that elaborate on what went before in a historically informed manner, in the context of a particular environment.

Once this is appreciated, it becomes possible to look at our own creativity in a new light. The interactive processes within the human brain continually manifest themselves to us in terms of our creative thoughts and activities. But we cannot access the processes underlying these experiences purely by introspection. We cannot unravel the mechanisms behind our thoughts simply by thinking about the problem, because each thought we have will itself be a manifestation of the underlying process. However, I have tried to show that by studying development from a scientific viewpoint, we can get a clearer view of the mechanisms lying behind a process that is comparable to human creativity. Of course, as I have just pointed out, this scientific understanding is itself dependent on our own creativity: our ability to come up with ideas and hypotheses. But these hypotheses have the merit of being testable through experiment and observation, allowing us to get a distinctive insight into the problem. That is why I believe that our scientific understanding of development can give us some useful intuitions about ourselves: we can begin to see how our own creativity might arise as an extension of an underlying developmental process, rather than as something that floats all by itself.

I shall return to the issue of how development is related to human creativity later on in the book, when dealing with the broader context of evolution. In the next few chapters, though, I want to continue with the story of development and show how the principles we have learned so far can be used to understand many of the overt features of plants and animals. In particular, I want to look at how the basic geometry and anatomy of organisms, their symmetry and form, can arise through the action of internal processes.

CHAPTER 13

Elaborating on asymmetry

We seem to live in a world dominated by symmetry. Almost everything we make is symmetrical: think of the shapes of chairs, tables, bottles, cups, windows. Symmetry also pervades the world of plants and animals: flowers, leaves, fish, butterflies. Even monsters are symmetrical; the one eye of Cyclops is carefully placed in the middle of his face. Symmetry seems to characterise the external geometry of living beings or of things made by them.

Perhaps the most famous illustration of symmetry in the living world concerns our own body: Leonardo's cartoon of a man framed within a circle and square (Fig. 13.1). In addition to the obvious mirror symmetry between the left and right halves, the picture emphasises other regularities in the human form, such as the central position of the navel or the equality between total arm span and body height. In Leonardo's own words:

> If you set your legs so far apart as to take a fourteenth part from your height, and you open and raise your arms until you touch the line of the crown of the head with your middle fingers, you must know that the centre of the circle formed by the extremities of the outstretched limbs will be the navel, and the space between the legs will form an equilateral triangle.
>
> The span of a man's outstretched arms is equal to his height.

But in spite of Leonardo's attempt to extract harmony and symmetry, the fact remains that the human body has a highly idiosyncratic shape. It is only because of our familiarity and preference for human proportions that they seem so harmonious to us. Occasionally, we get a glimpse of how peculiar we really are. I remember once waiting in an international airport for a flight home from China, when I suddenly became aware of giants with enormous noses lumbering about the lounge. I had become so attuned to the compact form of the Chinese physiognomy in the previous weeks, that Europeans and Americans suddenly seemed grotesquely out of proportion. Leonardo's canonical man is only one out of many other possible forms, as the satirist Arnold Roth emphasises in his caricature of *Leonardo at Work* (Fig. 13.2).

Our intrinsic irregularity becomes even more apparent when we think about how we develop. The shape of a fertilised human egg is pretty much spherical; yet the body that eventually develops from it is far from being the same all the

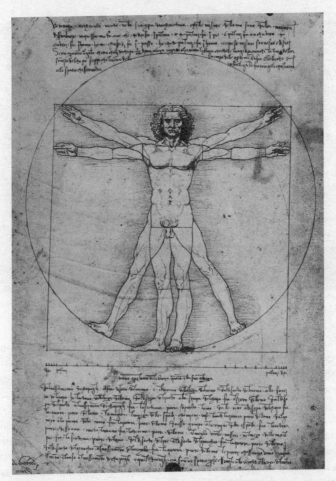

Fig. 13.1 Human proportions, from the notebooks of Leonardo da Vinci.

way round. Unlike a sphere, our bodies are very distinct from top to bottom and from front to back. It seems as if an initially perfect sphere gets pulled about and deformed in particular directions while it grows and develops into an adult.

We witness a similar sort of thing during the production of many artefacts. A glass blower might start by producing a spherical ball of glass which is then pulled or pushed in various directions to produce a more distinctive shape, such as a bottle or wineglass. Similarly, a potter working at a wheel may first form the clay into a hemispherical mound, which is then drawn out in particular directions to form a cup or vase. As with development, the point

Fig. 13.2 *Leonardo at Work* (1978), Arnold Roth.

of departure appears to be more uniform than the final outcome in these cases. In contrast to development, however, the shaping of glass or pottery depends on the guidance of external hands, rather than the action of internal processes.

Seen in this light, the problem of development can be restated in these terms: how is it that internal processes can lead to the transformation of a seemingly symmetrical fertilised egg into an organism with a much more *asymmetric* geometry? Rather than being a question of accounting for symmetry, the fundamental problem is to account for irregularity or deviations from symmetry. At first sight this may seem like a perverse way of looking at things because it is symmetry that strikes us most forcibly when we look at the living world. But we shall see that this is because of our bias or partiality towards a certain type of symmetry. Once this bias has been taken into account, we will be able to view development in terms of a progressive elaboration of asymmetry. We have already paved the way for this to some extent in previous chapters, by going through some of the basic mechanisms for elaborating patterns of hidden colour. In this and the following few chapters, I want to take the story a step further and show how these principles can be used to account for the various geometries we see around us. Before going any further, though, it is important first to establish exactly what we mean by symmetry.

Classifying symmetries

Although we might refer to many different things as being symmetrical, we have an intuitive notion that some are more symmetrical than others: a ball seems to have more symmetry than a bottle, and a bottle seems more symmetrical than a pouring jug with a handle sticking out to one side. But what precisely do we mean when we say that one thing is more or less symmetrical than another? Is there a way of defining precisely *how* symmetrical an object is?

There is a mathematical way of dealing with this problem. To illustrate how this works, we can begin with a very symmetrical object, such as a perfectly smooth sphere, and look at how many *geometrical transformations* can be applied to it which do not change its appearance in any way (Fig. 13.3). That is to say, what sort of things can we do to the sphere such that it ends up looking exactly the same as it did to begin with?

First of all, consider various types of *rotation*. We can rotate the sphere about any axis passing through its centre and still end up with an identical-looking sphere. In other words, we would not be able to tell that anything had been done to the sphere by simply looking at it after it had been turned in this way. Furthermore, for each of these axes, we may rotate the sphere by any degree we choose while still preserving its external appearance. In more formal terms we can say that the sphere has an infinite number of axes and degrees of *rotational symmetry*. Another transformation that can preserve the appearance of our sphere is *reflection*. We can reflect the sphere in any plane that passes through

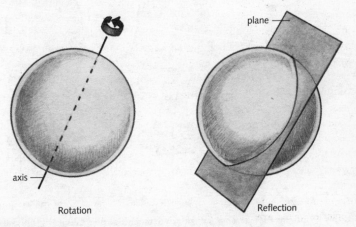

Fig. 13.3 Transformations that leave a sphere unchanged.

its centre without altering it in any detectable way. That is to say, any plane that cuts the sphere in the middle will give two equal halves that are mirror images of the other. This is true whatever angle the plane is tilted at, so long as it passes through the centre of the sphere. Putting this more formally, we can say that a sphere has an infinite number of planes of *reflection symmetry* that may pass through its centre at any angle.

The key step is that we can now use these various transformations that preserve appearances as a way of *defining* the symmetry of a certain class of objects, in this case spheres. We can say that all objects having an infinite number of axes and degrees of rotational symmetry, as well as having an infinite number of planes of reflection symmetry passing through the centre at any angle, belong to the same symmetry class, what might be called the class of *spherical symmetry*. To decide whether an object belongs to this class, you simply have to see if it satisfies the conditions of being preserved under all these transformations. In practice, it is only spheres that satisfy all these conditions.

We can now extend this approach to define other classes of symmetry. Let us take the sphere and deform it by expanding or contracting its girth to varying extents as we move along one axis, say the vertical (Fig. 13.4). We end up with

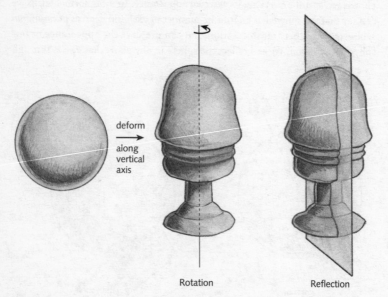

deform
→
along
vertical
axis

Rotation   Reflection

Fig. 13.4 Deforming a sphere along a single axis to produce an object displaying full radial symmetry.

a new object that is asymmetrical from top to bottom. This object now has a single vertical axis of rotational symmetry, rather than the infinite number of axes for the sphere. We can still rotate the object by any degree about the vertical axis, but this is now the only axis for which this applies; if we were to try a rotation about any other axis, the object would end up at a new angle (unless it was a trivial rotation of 360°), allowing us to tell from the end result that a transformation had been applied. Similarly, reflection symmetry is now restricted to only vertical planes rather than the infinite number of angles for the sphere. There are still an infinite number of planes but they all have to be oriented so as to include the vertical axis.

The overall result of deforming the sphere along one axis is that we have lost some, but not all, of the ways of rotating and reflecting the object into itself. We have ended up with a more restricted number of transformations that preserve appearances. We can now use this restricted set to define a new symmetry class, called *full radial symmetry*. Any object whose appearance is preserved under this same set of transformations will by definition have full radial symmetry. Many things have this type of symmetry, such as bottles, pawns in a chess game and baseball bats. All of them have a single axis of rotational symmetry about which the object can be turned by any angle, and they also have an infinite number of planes of reflection symmetry.

We can impose yet more asymmetry by taking our object with full radial symmetry and deforming it in other directions, at right angles to its main axis. For example, at each vertical level we might deform the periphery of the object, by pulling it out or pushing it in to some extent, making the outline wavy rather than circular. To keep things as regular as possible in the first instance, the same deformations could be carried out symmetrically about several minor axes around the object, say five minor axes as shown in Fig. 13.5. As with the previous object, there is a single vertical axis of rotational symmetry but instead of being able to rotate the object by any amount, there are now only five angles of rotation that bring it back into itself: turns of 1/5, 2/5, 3/5, 4/5 or 5/5. Similarly, there are now only five planes of reflection symmetry rather than the infinite number of vertical planes in the previous case. In other words, we have reduced the number of ways of transforming the object into itself even more. This sort of symmetry is sometimes called *five-fold radial symmetry* because of the five-fold nature of the rotation and reflection symmetries. Any other object whose appearance is preserved under this same set of transformations will by definition have five-fold radial symmetry. All such objects will have asymmetry along a single *major* axis in one direction (e.g. vertical in the case illustrated) as well as a repeating pattern of asymmetry along five *minor* axes coming off it at right angles (e.g. horizontally in Fig. 13.5).

We can vary the degree of radial symmetry by changing the number of minor

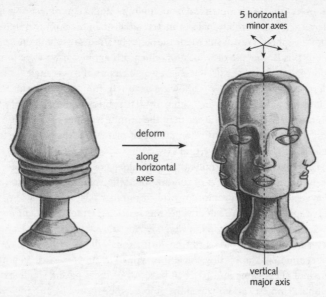

Fig. 13.5 Deformations along five minor horizontal axes produce an object with five-fold radial symmetry.

axes. If we had deformed our object at four regular intervals instead of five, giving it four minor axes, it would display a different degree of symmetry, four-fold radial symmetry. If we had deformed it along only two minor axes it would have ended up with two-fold radial symmetry. Many objects can be classified under these types of heading. If you look at chess pieces, you will see that the king commonly has two-fold radial symmetry, the rook (castle) four-fold radial symmetry, and the queen anywhere from five- to ten-fold radial symmetry, depending on the chess set. In furniture, oblong tables have two-fold radial symmetry whereas square tables or square stools have four-fold radial symmetry.

There is a particularly important symmetry class that corresponds to deforming our object along only one horizontal axis rather than several. If we carry out such deformations across the whole object, we end up with something like the head shown in Fig. 13.6, with a distinct front and back. The head has been produced by starting at one end of the object and, at each vertical level, deforming it to the same extent on each side while moving along the horizontal axis towards the other end. In the example shown, the front midline of the face has been left unchanged by the deformations: if you look at the head from the side, the frontal outline is the same as that of the fully radially symmetrical object it

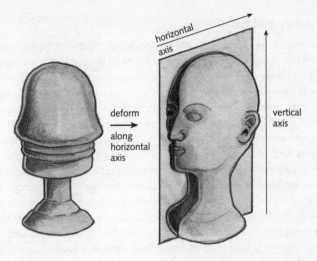

horizontal
axis

deform

along
horizontal
axis

vertical
axis

Fig. 13.6 Deformation along a second major axis generates an object with bilateral symmetry.

was derived from. Our new object now has asymmetries along two major axes: one horizontal and one vertical. In terms of geometrical transformations, there is now only one type of rotational symmetry: the trivial one involving a complete turn of 360°. There is also only one plane of reflection symmetry remaining: the one that runs vertically through the midline and divides the object in two. Although it would be reasonable to call this one-fold radial symmetry, I shall use the more common name of *bilateral symmetry* to refer to this arrangement. Bilateral symmetry, then, is characterised by a single plane of reflection symmetry, and it arises when asymmetries are elaborated along *two major axes*, in this case a vertical (top–bottom) and horizontal (front–back) axis. There are numerous objects with bilateral symmetry, including chairs, tea-cups, airplanes or knights in a chess game.

To summarise, the symmetry of an object can be defined by the number of ways it can be transformed into itself. The most symmetrical objects, such as spheres, have the highest number of these transformations. If we impose a single major axis of asymmetry on a sphere, we reduce the number of possible transformations to those of full radial symmetry, such as bottles. The extent of symmetry can be further reduced by imposing asymmetry along several minor axes at right angles to the first, giving lower degrees of radial symmetry (five-fold, four-fold, etc.). In the case of one-fold radial symmetry, more commonly termed bilateral symmetry, we end up with an object having asymmetry along two major axes.

Symmetry and function

In practice, the symmetry class of an object is often related to the way it is used. Spherical symmetry, for example, predominates in objects that are used with little or no constraint from any specific direction, as with soccer balls that roll around in any way we choose. Radial symmetry, on the other hand, is common in cases where constraints operate along one major axis. Bottles and stools, for instance, are designed to withstand the effects of gravity: bottles prevent liquids being spilt everywhere, and stools stop you from falling to the floor when you sit down. Both of these types of object have radial symmetry, with a single major axis running vertically, parallel to the gravitational force they are designed to oppose. By contrast, pouring jugs and chairs have bilateral symmetry, because they are designed to accommodate a second directional constraint in addition to gravity, the shape of a hand that comes from one side to lift them or the shape of a person who wishes his back to be supported while he is seated. One consequence of their matching the shape of the user in this way is that they can be approached from only one direction. Compare, for example, the stool of four-fold radial symmetry with the bilaterally symmetrical chair shown in Fig. 13.7. The stool can be sat upon from at least four directions whereas the chair can be comfortably occupied from only one.

Just as the symmetry of artifacts is related to their use, so the symmetry of

Fig. 13.7 Comparison of a chair with bilateral symmetry and a stool with four-fold radial symmetry.

organisms is related to the way they are adapted to their environment. Humans, for example, are to a large extent bilaterally symmetrical in terms of their external appearance. We have a single plane of reflection symmetry running down our midline, as a result of asymmetry along two major axes: the asymmetry from head to foot and front to back. The asymmetries along these axes reflect two important directional constraints on us: gravity and movement. If we lived in a free-floating world without gravity, standing would become meaningless and we would have lost much of the rationale behind asymmetry along the head–foot axis. Our second asymmetry, front–back, has more to do with movement. To generate a force that moves us in one direction, it helps to be asymmetrical along the direction of movement (i.e. the horizontal axis). Many other distinctions between front and back follow on from this. For example, the position of our eyes at the front of our head allows us to see where we are going. Most importantly, these two major aspects of our lifestyle, movement and gravity, are oriented at right angles to each other. If we normally moved parallel to gravity, only moving up or down like springs, we would have only one major direction to deal with and would no longer need to distinguish between, say, front and back. Our asymmetries are related to our lifestyle and environment: we move at right angles to the force of gravity.

For similar reasons, many other animals display bilateral symmetry. When comparing ourselves with them, however, there can be some confusion because of our upright posture. Most animals move along the direction of their head–tail axis whereas we walk at right angles to ours. The 'back' of a horse might mean its tail end (referring to the way it moves) or the part we ride on (corresponding to our back). To avoid confusion, biologists use *dorsal* to refer to our back and the upper part of most animals, and *ventral* (belly) to mean our front and the corresponding lower part of most other animals. From now on I shall refer to two major axes in animals as the *head–tail* axis and the *dorsal–ventral* axis. Humans move parallel to their dorsal–ventral axis whereas most other animals move parallel to their head–tail axis.

It is because of our familiarity with bilateral symmetry in the living world that we tend to underestimate the asymmetry that lies behind it. When you take an irregular ink blot and fold the paper in half, you end up with an ordered symmetrical pattern that looks harmonious. Yet the original irregularity of the blot has not disappeared, it has simply been balanced by a duplicate image. The single mirror plane relieves tension by giving the blot a symmetry that we are much more familiar with.

Not all animals have bilateral symmetry. A large group of soft-bodied aquatic animals, including jellyfish, sea anemones and *Hydra* (the polyp mentioned in the previous chapter), have various degrees of radial symmetry. Their lifestyle is correspondingly quite different from ours. They either drift with the water currents most of the time or stay stuck to a surface, like a rock or the sea floor. If they do move around, it is usually parallel to their major axis, as with jellyfish slowly swimming by rhythmic contractions of their bells. *Hydra* has a different solution: it can somersault its way along by bending its main axis round so that its tentacles touch the ground, and then straightening up again. It is perhaps not too surprising that these radially symmetrical animals are found in aquatic habitats, where the reduced effects of gravity are more conducive to these styles of locomotion.

Plant symmetries are also related to their lifestyle and environment. Plants explore their surroundings by growing; growth often plays a role in the lives of plants similar to movement in animals. But unlike animal movement, which is mainly horizontal, a large part of plant growth is oriented vertically, parallel to many of the environmental factors that dominate their lives, such as gravity and light. Many of their asymmetries are therefore elaborated along a single major axis. This is reflected in the typical radial symmetry of the main stems and roots. Leaves, however, usually grow out more horizontally, roughly at right angles to the direction of the light they harvest. Consequently, leaves tend to be bilaterally symmetrical: their upper surface is usually quite different from the lower and their tip is different from their base.

Flowers display a variety of symmetries, depending on how they are pollinated. Buttercups, for example, have five-fold radial symmetry, displaying five identical petals spaced out equally around the flower axis. Like stools, these flowers can be approached from several directions by their visitors. Other flowers, such as those of *Antirrhinum* or orchids, are more discriminating, accommodating selected guests more precisely through bilateral symmetry. The upper and lower petals of an *Antirrhinum* flower have quite a different shape and are united for part of their length to form a tube. The lower petals provide a sort of platform for bees to land on, prise open the flower and enter the tube where the nectar is stored (Fig. 13.8). This ensures that the flowers are only visited by certain types of insect: a bumble-bee is large enough to prise open and enter an *Antirrhinum* flower but a smaller insect would fail. The sex organs of the flower, the stamens and carpels, are carefully positioned within the tube to give and receive pollen from the bee's back. The bilateral symmetry of the flowers therefore allows their pollen to be targeted very effectively to a specific insect carrier. As with chairs, asymmetry along the second major axis of these flowers has to do with accommodating a particular type of animal.

For convenience, I shall refer to the uppermost petals of a bilaterally symmetrical flower as *dorsal*, and the lowermost petals as *ventral*, using a comparable nomenclature to that in animals. Thus the flowers of *Antirrhinum* are asymmetrical along a dorsal–ventral axis that runs from top to bottom (Fig. 13.8). Because bees enter *Antirrhinum* flowers by standing on the lower petals,

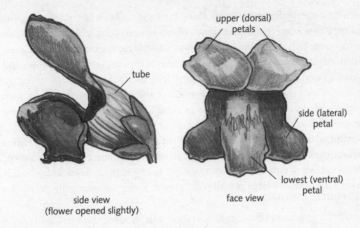

Fig. 13.8 *Antirrhinum* flower shown in side and face view.

the bee's dorsal–ventral axis runs in precisely the same direction as that of the flower (this is not true in all cases: for some plant species the animal pollinator enters the flower upside down, with its dorsal–ventral axis inverted relative to that of the flower).

Developing asymmetries

So far I have dealt with the symmetry of mature organisms in terms of geometry and the way they are adapted to their environment. I now want to look at how these manifest aspects of symmetry themselves arise; how the external geometries in the natural world arise from hidden processes.

Organisms usually display a much greater degree of symmetry early in development than later on. The embryos of many plants and animals, including humans, start off as balls of cells with a more or less spherical symmetry. It is only at later stages of development that they come to exhibit asymmetries that give them their characteristic shape and form. From a developmental perspective, then, the question is not why we are so symmetrical, but why we are not *more* symmetrical. Why are we not spherical blobs with the same symmetry as early embryos?

As we have seen, to generate something with bilateral symmetry, such as ourselves, we need to elaborate differences along two major axes. However, unlike our earlier exercise with deformations, in the case of development there can be no help from an external guiding hand: the whole process needs to be understood in terms of internal events. To some extent, we have paved the way to understanding how this might be achieved in previous chapters. We have seen how the distinctive features along the head–tail axis of fruit flies depend on the elaboration of hidden colours, eventually resulting in a set of territories from one end to the other. A comparable underlying map is also present along the head–tail axis of vertebrates. In the case of flowers, we have come across a map of concentric territories that vary from the centre outwards, along the radial axis of the flower. For each of these systems, however, we have only considered the elaboration of patterns along a single axis. I now want to bring in a second axis, the dorsal–ventral axis, to show how hidden colours can be combined to account for bilateral symmetry. As with many other developmental problems, we shall begin by turning to mutants for inspiration.

A regular monster

In 1742, a student at the Uppsala Academy in Sweden, Magnus Zioberg, was roaming over an island of the Stockholm archipelago collecting plant specimens, when he came across a strange plant he had never seen before. He dutifully pressed it and took it back to Uppsala. The specimen eventually ended up on the desk of the great taxonomist, Carolus Linnaeus, for identification. Linnaeus was the foremost expert in botany at the time, and his system of classification was destined to become the foundation of plant and animal taxonomy. In spite of Linnaeus's extensive knowledge of plants, he had never seen anything quite like this specimen before. At first he thought it was a species of common toadflax, *Linaria vulgaris*, with some strange flowers artificially stuck on to it so as to trick the specialist. On looking closer, however, Linnaeus saw that it was genuine enough. He became very excited and asked Zioberg to go back to the island and collect a living specimen so that he could study it in more detail. Linnaeus was so enthralled by the plant that he wrote a small dissertation on it in 1744.

It became clear to Linnaeus that this plant was identical to common *Linaria* in every way except for the symmetry of the flowers. Like its close relative *Antirrhinum*, the flowers of *Linaria* have five petals which are joined together for part of their length to form a tube (Fig. 13.9, left). *Linaria* flowers normally have bilateral symmetry, with distinctions between petals along the dorsal–ventral axis. This is particularly striking in *Linaria* because the ventral (lowest) petal has a distinctive outgrowth at its base, called a spur, where the nectar collects. By contrast, the plant that so interested Linnaeus had flowers with five-fold radial symmetry: all five petals were identical (Fig. 13.9, right). Instead of varying along the dorsal–ventral axis, each of the five petals closely resembled the ventral petal of the normal form, giving a flower with five symmetrically arranged spurs. It was as if the ventral petal had become repeated all the way round. We might compare this to deriving a stool from a chair by repeating

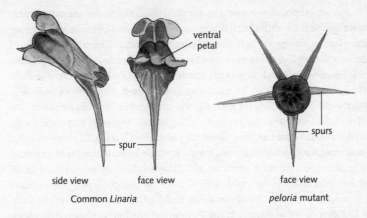

ventral
petal

spur

spurs

side view   face view

Common *Linaria*

face view

peloria mutant

Fig. 13.9 Common form of *Linaria vulgaris*, with bilateral symmetry, compared to peloria mutant with five-fold radial symmetry.

only the front of a chair all the way round, as shown in Fig. 13.7. The stool can be thought of as a chair that has lost its front–back asymmetry.

The reason that Linnaeus found this plant so interesting was that he had based his whole system of plant classification on flower structure. Accordingly, the unusual plant with a fundamentally altered flower symmetry should have belonged to a totally new species, yet the overall appearance of the plant was obviously that of common toadflax. Linnaeus had to conclude that this peculiar plant had arisen by some sort of transformation of common toadflax. This was well before Darwin's time, and the idea that one species might be transformed in some way to resemble another was very radical. Species were thought to be timeless acts of creation, forever occupying a fixed position in nature. Monstrosities of various sorts were well known but they were usually defective in some way that prevented their being taken seriously as new or modified species. What startled Linnaeus was that this plant with symmetrical flowers seemed to be a perfectly acceptable species: the flowers were beautifully formed and regular, and they could produce seed. It overturned the dogma that species could not be transformed or tampered with. Linnaeus wrote:

> Nothing can, however, be more fantastic than that which has occurred, namely that a malformed offspring of a plant which has previously always produced irregular flowers now has produced regular ones. As a result of this, it does not only deviate from its mother genus but also completely from the entire class and *thus an example of something that is unparalleled in botany* so that owing to the difference in the flowers no one can recognise the plant any more . . . This is certainly no less remarkable than if a cow were to give birth to a calf with a wolf's head.

Linnaeus eventually named the plant *Peloria*, Greek for monster. Paradoxically, the flowers themselves were quite attractive; it was their disturbing biological implications that made them monstrous. The phenomenon of peloria was noted in many other species following Linnaeus's report, but little progress in understanding was made until 1868, when Charles Darwin published some intriguing observations on *Antirrhinum*. The observations were perhaps more remarkable for what Darwin missed than for what he saw.

Darwin's theory of evolution by natural selection depended on parents being able to transmit some of their characters to their offspring. It only worked if characters giving an advantage during the struggle for existence could be passed on to later generations. Otherwise, selection would be ephemeral and could not contribute to a lasting and gradual transformation of species over many generations. For years Darwin amassed an enormous amount of information on heredity to try and support his theory of evolution. Amongst his experiments, he described some crosses between two varieties of *Antirrhinum*. As with *Linaria*, there are some varieties of *Antirrhinum* that have radially symmetrical flowers, instead of the normal bilaterally symmetrical flowers. These mutant forms are called *peloric* (Fig. 13.10).

Darwin noted that when he crossed a peloric variety with the common form he obtained hybrids with normal flowers; the peloric trait had disappeared. However, when the seed from these hybrids was sown, the peloric form reappeared in 37 out of the 127 progeny. That is, although the more symmetrical form seemed to vanish in the first generation of the cross, it resurfaced in about one-quarter of the progeny in the following generation. Anyone familiar with Gregor Mendel's work, published a few years earlier in 1865, would have concluded from Darwin's results that the peloric trait was probably determined by a single hereditary factor, what we now call a gene. To see why, recall that

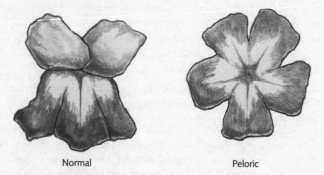

Normal   Peloric

Fig. 13.10 Normal and peloric flowers of *Antirrhinum* in face view.

every individual carries two copies of each gene, one coming from each parent. Suppose that Darwin's parental plant with peloric flowers carried a mutation in both copies of a particular gene, whereas this gene was unaffected in the normal-flowered plant. When the parents were first crossed, the hybrid would have inherited a normal copy of the gene from one parent and the mutated copy from the other. Having one normal copy of a gene is usually enough to ensure proper development, so the hybrid looked normal. But in the next generation, when the hybrid itself reproduced, these two versions of the gene would be shuffled. Each individual would then have a one-half chance of inheriting the mutated copy from its father and a one-half chance of getting it from its mother. The chance of *both* copies being mutated would therefore be one-quarter this generation, the same as the chance of getting two tails when a pair of coins are tossed. Mendel's theory of inheritance was based on precisely this sort of result, observing about one-quarter mutant progeny in the second generation of a cross. The observations made by Darwin fit this (within the bounds of statistics), implying that the peloric trait depends on a single gene.*

Darwin, though, was not familiar with Mendel's theory and was thinking along different lines. Ironically, in the same book in which he described his *Antirrhinum* crosses, Darwin went on to propose his own theory of inheritance, which turned out to be quite incorrect (we shall return to this theory in a later chapter). He based his theory on an enormous body of data, of which his results with *Antirrhinum* formed only a minor part. He therefore missed the vital clue they offered. Had Darwin placed more emphasis on his experiments with *Antirrhinum* and followed them up with further breeding studies, perhaps he would have elucidated the principles of heredity as well as those of evolution.

A celestial colour

We now know that the peloric forms of *Antirrhinum* carry a mutation in a gene called *cycloidea* (from the Greek *cyclo-*, meaning circle, referring to the more circular outline of the peloric flower), or *cyc* for short. It was most probably this gene that Darwin had unwittingly revealed in his crosses. How does the *cyc* gene influence the symmetry of a flower?

It will help first to schematise the petals of an *Antirrhinum* flower. The five petals in a normal flower can be classified into three types, each with a distinctive size and shape: two dorsal petals, two lateral petals and one ventral petal (Fig. 13.11, left). The flower has a single plane of reflection symmetry running

* The story is more complicated than this because out of his 127 progeny, Darwin also obtained 2 with a condition intermediate between peloric and normal. We now know that additional genes can influence the extent of peloria, although it is difficult to know precisely what was going on in Darwin's case given his limited amount of breeding data.

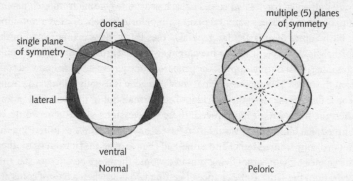

Fig. 13.11 Schematic diagram of petals of normal and peloric flowers, shaded to indicate the different petal identities. In the peloric mutant, all petals resemble the ventral type of normal flowers.

down the middle (see dotted line in left diagram of Fig. 13.11). By contrast, in peloric flowers, all the petals look similar to each other, so the flower has five-fold radial symmetry rather than bilateral symmetry (Fig. 13.11, right). The petals of the peloric flowers are not of a completely new type; they closely resemble the *ventral* petal of a normal flower. That is, instead of having petals with three different identities, dorsal, lateral and ventral, the peloric flowers exhibit a single ventral type all the way round. It is as if whatever normally establishes dorsal and lateral identities has been lost, and ventral identity is reiterated by default.

To see how this might be explained, I need to describe where the *cyc* gene is normally expressed. The *cyc* gene is first switched on at a very early stage of flower development, at a time when the bud is just a tiny bulge of cells with no obvious dorsal–ventral asymmetry. Most importantly, *cyc* is only switched on in a very specific part of the flower-bud: the dorsal (upper) region. This is shown shaded in Fig. 13.12, where the floral buds are shown in section as bumps on the periphery of the growing tip. So even though the flower-bud itself looks symmetrical from top to bottom at this stage, a hidden asymmetry is already there in terms of *cyc* expression. You can think of *cyc* as providing a distinctive hidden colour to the dorsal region of the bud. I shall name this hidden colour *celestial-blue*, reflecting its location in the higher regions of the bud. It is the production of celestial-blue in the early flower-bud that leads to the manifest asymmetry of the flower later on along the dorsal–ventral axis. In peloric flowers, this colour is missing because of a mutation in *cyc*, and the flower develops with radial symmetry.

Many of the details of how this celestial-blue colour can give rise to the

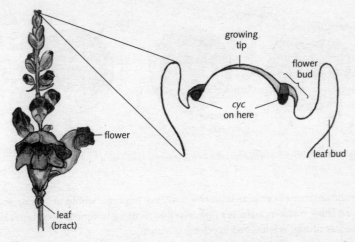

Fig. 13.12 Pattern of *cyc* expression in very young *Antirrhinum* floral buds on the periphery of the main growing tip. Each flower grows in the angle of a small leaf (called a bract). Note that *cyc* is only switched on in the upper (dorsal) region of the flower-bud, nearer to the growing tip.

different petal types in the flower are still not known, but I will try to sketch a reasonable scenario. To do this, I will need to introduce some more hidden colours. First I want a colour that we associate with the lower depths, say *sulphur-yellow*. We will start off with the flower-bud, schematised as a disc in face view, having a uniform distribution of sulphur-yellow (Fig. 13.13, left). When the *cyc* gene gets switched on, celestial-blue will be produced towards the top of the bud, so we end up with two colours: celestial-blue in the dorsal region and sulphur-yellow in the remaining part (Fig. 13.13, middle; for simplicity, I am assuming that the blue colour outcompetes or predominates over yellow). Now recall that a hidden colour can lead to a specific scent (signalling molecule) being produced, and that cells may respond to this scent by changing their hidden colour (Chapter 11). Suppose that celestial-blue cells start to produce a scent, a heavenly perfume, to which neighbouring cells respond by turning from sulphur-yellow to a more earthy colour, like *terra-cotta*. Cells further away, in the lower parts of the bud, might be too far away to detect the heavenly scent and therefore remain sulphur-yellow (Fig. 13.13 right). In this way, we have elaborated three regions along the dorsal–ventral axis: celestial-blue, terra-cotta and sulphur-yellow. The identity of the petals will depend on the hidden colour of the region they come from: the petals from the celestial-blue region will develop with dorsal identity, those from the terra-cotta region will have lateral identity, and the petal from the sulphur-yellow region will have ventral identity.

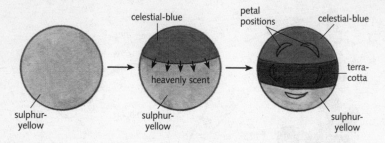

Fig. 13.13 Elaboration of hidden colours in the flower (*face view*).

In the absence of celestial-blue, this sequence of events would not occur and everything would remain as sulphur-yellow, leading to ventral petal types all the way round, as witnessed in peloric mutants.

Asymmetry along a major axis in the mature *Antirrhinum* flower—the distinction from top to bottom—therefore arises through the elaboration of hidden colours. But you might well ask how the initial asymmetry in colour is set up. Why is it that the gene for celestial-blue, *cyc*, is expressed only in the dorsal region of the early flower-bud?

Now although the detailed answers to this question are not yet known, we must not fall into the trap of believing that they will take us to some absolute starting point that anticipates everything else. For example, if you look again at the pattern of *cyc* expression shown in Fig. 13.12, you will see that the dorsal region of the flower-bud lies nearer to the main growing tip than the rest of the flower-bud. Perhaps *cyc* is switched on in this region because it is exposed to a particular scent being given off by the cells at the growing tip. This in turn would mean that the growing tip has a distinctive hidden colour, leading to the production of the scent. Thus, the asymmetric distribution of celestial-blue in the flower would itself depend on an earlier distinction between the growing tip and its surrounding buds. This distinction in turn could be traced back to when the growing tip was first being formed, in the embryo within the mother plant. We might continue in this way, trying to trace things further and further back, but we should not imagine that we will find an initial plan for flower asymmetry. The asymmetries in the early embryo do not anticipate what the flower will eventually look like, they simply provide a prelude that is interpreted and built upon. Eventually, as the plant grows and floral buds arise from the main tip, the *cyc* gene is able to add its own interpretation to these patterns. This in turn acts as a prelude to further elaborations. As we have seen before, development is a process of continual interpretation rather than one of anticipation.

Playing with combinations

We are now in a position to get an overview of how the bilateral symmetry of a flower is established. Recall that to produce bilateral symmetry, we need asymmetry along two major axes. In the case of *Antirrhinum* flowers, asymmetry along one of these axes, the *dorsal–ventral axis*, reflects a pattern of hidden colours from top to bottom, denoted by celestial-blue, terra-cotta and sulphur-yellow. Asymmetry along the other major axis corresponds to a map of concentric territories of hidden colour described previously (Chapters 4 and 6) that vary from the centre outwards, along the *radial axis* of the flower. We symbolised these as various types of overlapping red colours, with claret-red at the centre, burgundy-red further out and amarone-red outermost. Each whorl of floral organs had a different combination of reds, corresponding to its distinctive sepal, petal, stamen or carpel identity.

The key to putting these asymmetries together is that hidden colours can act in *combination*. This is because when master proteins (hidden colours) bind to a gene, it is their combined effect that determines whether it is switched on or off (Chapter 5). To see how two patterns of hidden colour might work together, we can therefore simply imagine combining or superimposing them, as shown schematically in Fig. 13.14. We now end up with an overlapping patchwork of 10 different colour combinations, with a plane of mirror symmetry running down the middle. The ability of hidden colours to act in combination has allowed us to multiply the number of territories.

The type of organ that develops from each territory will reflect its combination of hidden colours. To make this easier to see, I have outlined where the petals and stamens will form in Fig. 13.14. Dorsal petals, for example, come from a region with celestial-blue, amarone-red and burgundy-red. This combination is distinguishable from that of ventral petals (which have sulphur-yellow instead of celestial-blue) or dorsal stamens (which have claret-red instead of amarone-red). The hidden patchwork can therefore be interpreted to give dorsal petals their distinctive characteristics.

I have emphasised dorsal–ventral distinctions between petals, but the same logic can be applied to other whorls of organs. For example, the dorsal stamen of a normal *Antirrhinum* flower is much smaller than the other stamens: it is arrested early on in development and remains vestigial. The adaptive significance of this may be that this stamen is in an awkward position for transferring pollen to a visiting insect because it lies just behind the carpels, which obstruct easy access. The arrested growth may therefore prevent the development of a relatively useless stamen. Now a distinctive fate for the dorsal stamen can easily be accommodated by our patchwork because the region it comes from has a unique colour combination: celestial-blue, burgundy-red and claret-red. This

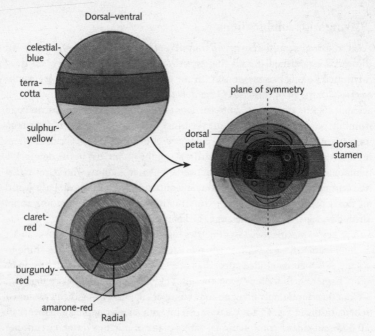

Fig. 13.14 Combination of hidden colours along the radial and dorsal–ventral axes of an *Antirrhinum* flower.

could then be interpreted in such a way as to lead to an arrest of growth (the other stamens would not be arrested because they lack celestial-blue, and the dorsal petals would not be arrested because they have amarone-red instead of claret-red). This makes a very simple prediction: if we could remove celestial-blue from the flower, the dorsal stamen would no longer be distinguishable from the other stamens and would therefore develop like them. This is precisely what is seen: the stamen in the dorsal position does indeed develop fully in *cyc* mutants. That is, peloric flowers are not only radially symmetrical with respect to petals, they are also radially symmetrical with respect to stamens.

The external geometry and symmetry of a flower therefore reflects early patterns of hidden colour, established along different axes of the floral bud. In combination, these colours form a hidden patchwork that is then interpreted and eventually becomes manifest in the mature flower.

Symmetrical flies

I have given an example of how bilateral symmetry can be produced in the case of plants, but do the same principles hold for animals? To look at this problem

I want to turn to the fruit fly, *Drosophila*, and see how its dorsal–ventral axis of asymmetry is established.

In previous chapters I described how differences along the head–tail axis of flies are elaborated during development. We saw that mutations affecting the early steps in this process caused lethality; the embryos did not make it to adulthood. Now the very same principles apply to the second major axis of asymmetry. For a fly to move, feed or do anything useful, it needs distinctions from back to belly. Mutations that eliminate differences along the dorsal–ventral axis of an embryo will therefore lead to its premature demise. For this reason, these sorts of mutation were missed for a long time, as scientists traditionally looked only at adult flies. It was only in the 1970s, when Christiane Nüsslein-Volhard started looking at defective embryos, that radially symmetrical mutants started coming to light. While working in Freiburg, she came across a female that seemed to produce symmetrical embryos:

> I had this first female and she produced eggs and I really went back every hour, every hour, to see what they were going to do. And then one made these folds, and the next ones also made these folds, and *they were symmetrical all round*.

Normally, the dorsal and ventral regions of fruit fly embryos have very distinctive features. The belly, for example, has a characteristic set of thorn-like outgrowths (denticles). In this mutant, however, the ventral features were missing. Instead, the embryos only had dorsal characteristics all the way round. Because of this, Nüsslein-Volhard named the mutant, and the gene it affected, *dorsal*.

It turns out that the *dorsal* gene produces a master protein, conferring a hidden colour that I shall call *devil-red*. As you may guess from the name I have given, the devil-red master protein acts mainly in the lower or ventral regions of the embryo. Surprisingly, though, if you look at very early stages of development, this master protein is not restricted ventrally, but is present *throughout* the embryo. That is, the embryo starts off with a uniform distribution of the master protein and it only becomes asymmetric later on. I now want to explain how this happens.

Recall that following fertilisation, the nucleus in a fruit fly egg starts to go through several rounds of division, giving rise to many nuclei immersed in a common cytoplasm (Chapter 9). At this stage the master protein made by the *dorsal* gene is uniformly distributed. But there is a key additional feature: the master protein is only found in the cytoplasm, not in the nuclei. This is because it is bound or tethered to another protein that prevents it from entering the nuclei and keeps it in the surrounding cell fluid. Now for a master protein to do anything, to switch genes on or off, it has to be able to enter the nucleus, where the DNA is located. So although the *dorsal* master protein is found

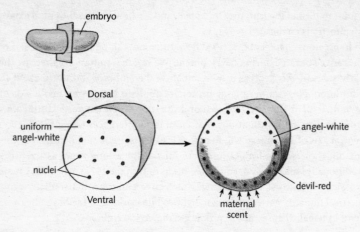

Fig. 13.15 Development of dorsal–ventral asymmetry in fruit fly embryos. Initially, the dorsal protein is inactive and the embryo is uniform angel-white. Later, the protein is released in response to a maternal scent, such that it now confers devil-red on ventral regions.

throughout the embryo, it is unable to confer a hidden colour. By being tethered to another protein in the cytoplasm it is effectively rendered colourless. The hidden colour of the embryo at this stage is therefore not affected by *dorsal* at all, but is set by other master proteins. These collectively give the embryo a uniform background colour that I shall refer to as *angel-white* (Fig. 13.15, left).

Shortly after this, something remarkable happens: the *dorsal* master protein enters the nuclei specifically in ventral regions (this happens at about the stage of development when the nuclei have started to migrate to the periphery of the embryo). That is, the protein is released from its tether in the cytoplasm and allowed to confer its devil-red colour, but only in the region that will eventually form the belly of the fly (Fig. 13.15, right). The reason is that this region is triggered by a scent given off by the surrounding cells of the mother. A group of maternal cells, located along one side of the embryo, are involved in making a particular scent (signalling protein) that stimulates the release of *dorsal* master protein from the cytoplasm and allows it to enter the nucleus. Regions further away from the scent will not be triggered and therefore do not acquire devil-red.

The embryo now has two hidden colours, devil-red in ventral regions and angel-white in dorsal regions. These colours, though, are not the end of the story; they are a prelude to a series of steps that further refine the dorsal–ventral pattern. For example, a particular scent is produced from the angel-white region, an angelic perfume, in response to which new colours might be interposed, such as *earth-brown*. The process may continue in this way, building

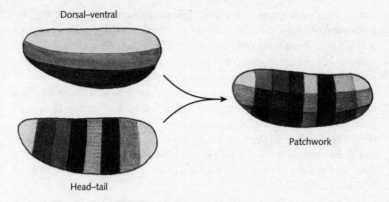

Fig. 13.16 Patchwork of hidden colours in the fruit fly embryo derived by combining patterns along the dorsal–ventral and head–tail axes.

upon and refining distinctions as the embryo develops, eventually giving rise to parts with distinctive identities along the dorsal–ventral axis. In a mutant that lacks devil-red (i.e. *dorsal* mutant), the embryo remains uniformly angel-white and therefore develops with dorsal features all the way round.

We can now begin to see how the overall bilateral symmetry of the fly can arise. There are two overlapping patterns of hidden colour. On the one hand, there is the pattern elaborated along the dorsal–ventral axis, which appears as a series of horizontal stripes when the embryo is viewed from the side, as shown in the top left of Fig. 13.16. On the other hand, there is the pattern elaborated along the head–tail axis, such as the various greens that provide distinctions between segments, seen as vertical stripes in the bottom left of Fig. 13.16. These two patterns act in *combination* to give a patchwork of distinctive territories. This means, for example, that not only can each segment be distinguished, but so can the dorsal and ventral regions within each segment. Now because the patchwork is derived from a combination of patterns along two major axes, the territories on the left and right halves of the embryo will be mirror images of each other. The final symmetry of the fly reflects the bilateral symmetry of the patchwork.

Some early connections

So far, I have considered the patterns elaborated along the two axes of the fly as arising independently of each other. Recall that the head–tail axis can be traced back to an early asymmetry in a black hidden colour, derived from RNA produced by the mother at only one end of the egg (Chapter 9). The dorsal–ventral axis stems from a quite different asymmetry in the mother, the production of a scent along one side of the embryo that triggers the appearance of devil-red. If you trace these even further back, though, you do find some connections between them. The details are rather complicated and need not concern us here, but I will try to give the general gist of the story.

To build a snowman, you first pile two or three large snowballs on top of each other, defining the head–tail axis. You might then stick a carrot in the middle of the head for a nose, defining which part is ventral. As long as the carrot is somewhere in the middle of the head it doesn't really matter where you put it round the circumference. But once you have put it in, the dorsal–ventral axis is fixed, and you put in the remaining decorations, such as the eyes, mouth and buttons, accordingly.

In a similar way, the head–tail axis of a fly is established first, by transporting the RNA needed for black (*bicoid*) to the head end. Once this head–tail axis has been initiated, further interactions eventually lead to the mother cells along one side of the egg producing the scent that triggers devil-red (*dorsal*). As in the snowman, the dorsal–ventral axis is only produced after the head–tail axis has first been put in place, allowing the two axes to be oriented properly with respect to each other. But the decision as to which particular region around the egg's circumference makes the maternal scent seems to be a matter of chance (it actually depends on where the nucleus happens to be in the egg cell). So just as with the placing of the carrot around the snowman, it does not matter which region becomes ventral, so long as once that has been established, all subsequent elaborations are based upon it in a consistent manner. And as we have seen, this follows quite naturally from the way that hidden colours build on each other.

The flower and the fly

The symmetry of flowers and flies is based on remarkably similar principles. To see this more clearly, I want to show them from a common viewpoint, by pressing a fly flat out such that its segments form concentric circles, like the whorls of a flower (Fig. 13.17). Seen this way, the two major axes of the fruit fly *Drosophila* are directly comparable with those of the *Antirrhinum* flower. The radial axis of the flower corresponds to the head–tail axis of the fly, and the dorsal–ventral axes line up with each other (whether you think of an axis as radial or longitudinal just depends on how much it is compressed or stretched out).

In both cases, the axes reflect a set of hidden colours that are gradually elaborated during development. In the case of flies, elaborations along the head–tail axis lead to a series of green territories that provide distinctions

Fig. 13. 17 Schematic flower and fly to show relationship between major axes of asymmetry.

between segments. In flowers, elaborations along the radial axis lead to over-lapping red colours that give each whorl of organs a distinctive identity. In both cases, these colours act in combination with those elaborated along the dorsal–ventral axis to establish a patchwork of territories. These are further interpreted and built upon, eventually leading to the manifest geometry of the mature flower or fly.

Going belly-up

Where do vertebrates, such as ourselves, fit into this scheme? We have seen how vertebrates have a map of green territories along their head–tail axis, similar to that found in flies. But what about the other major axis? Are there also common elements underlying the dorsal–ventral asymmetry in these different types of animal?

Some intriguing observations relevant to this question started to emerge in the 1990s, from studies on the frog *Xenopus*. As with flies, the dorsal–ventral axis of frog embryos is established at a very early stage of development. In this case, the axis is set at the time of fertilisation, according to where the sperm happens to enter the egg: sperm entry provides the equivalent of placing a carrot on a snowman. This is because sperm entry triggers a chain of events from which the dorsal–ventral axis is elaborated. During the early 1990s, a key signalling protein (scent) involved in this process of elaboration was identified. This protein was found to be expressed only in ventral regions of the frog embryo, where it acted to promote ventral identity. The protein turned out to be very similar in sequence to a protein involved in dorsal–ventral asymmetry of fruit flies. A common scent was involved in the dorsal–ventral axes of flies and frogs. But there was a curious twist in the story.

Recall that during the elaboration of dorsal–ventral asymmetry in flies, the angel-white region produces a particular signalling protein, an angelic perfume. Well, it is this signalling protein that turned out to be similar to that isolated from frogs. The twist is that in frogs this scent is expressed in *ventral* regions, whereas in flies it is in *dorsal* regions: frogs seem to be upside down relative to flies!

Although this may sound rather peculiar, a similar notion had already been put forward one hundred and seventy years earlier, by the French anatomist Geoffroy Saint-Hilaire. As we saw in Chapter 7, Geoffroy was trying to establish a common plan for vertebrates and insects. In addition to comparisons from head to tail, he also tried to draw correspondences along the dorsal–ventral axis. But in trying to do this, he found that he needed to *invert* insects relative to vertebrates. For example, the nerve cord runs along the dorsal side of vertebrates, whereas it runs along the ventral side of insects. It was as if one of them was the other way up. As with his other ideas, the proposal that insects are upside down relative to vertebrates met with disbelief from most quarters.

It now looks as though Geoffroy may have not been too far from the truth. As the mechanisms of development have started to be unravelled, common elements have been found that promote dorsal identity in insects but ventral identity in vertebrates. From an evolutionary point of view, this is perhaps not too difficult to explain. For a primitive aquatic ancestor, the difference between swimming on its back or on its front might not have been so great. The eventual orientation of the dorsal–ventral axis relative to the ground could have been largely a matter of chance. By contrast, the head–tail axis appears to have been more conserved. It seems that once the head–tail axis was established in our ancestors, they always kept moving in the same direction, but whether their belly faced down or up was more variable. So in Kafka's story of a man waking up to find himself transformed into an insect lying on its back, the man had probably fallen asleep with his head on the pillow but had been lying on his belly.

To summarise, development can be viewed as a process whereby asymmetries are elaborated as a fertilised egg divides and grows into a mature organism. This depends on hidden colours, scents and sensitivities continually building upon each other. In cases of bilateral symmetry, this process occurs along two different axes to give a combined patchwork of hidden colours. This hidden patchwork is further interpreted and elaborated upon, eventually becoming manifest in the visible structure of the organism. These processes allow organisms from humans to daisies to develop with their own idiosyncratic asymmetries.

Throughout this chapter, we have been concerned with the external geometry of organisms: their surface appearance. But organisms are not just empty shells;

they contain complex internal arrangements of tissue. By concentrating on the outer surface, I have given a rather two-dimensional picture. In the next chapter, I want to go deeper and explore how asymmetries of internal anatomy are also elaborated, to give a more complete three-dimensional view of how organisms develop.

CHAPTER 14

Beneath the surface

Humans are superficial creatures. We tend to judge things by their outward appearance rather than by what is inside them. The outer surface of an object can be seen or touched but its inner qualities are largely inaccessible to us. Even if we break it open, we can still only examine an object by feeling or looking at its inner surfaces rather than directly sensing what is within. There is one notable exception to this rule: our own bodies. We can, for example, sense internal pain or muscular tension by getting direct information from our insides. But even here, our senses are very poorly developed. We are entirely oblivious to most things that go on inside our bodies. We do not know the state of our internal blood vessels, or what our kidneys are up to, or where our thoughts come from. This is because most of the senses of which we are conscious are oriented outwards: they point to the outside world rather than to our insides.

To get a detailed idea of our internal structure, we therefore need to expose it to view. Leonardo, for instance, dissected more than ten human bodies so as to get an appreciation of human anatomy. The majority of us, though, would not relish the idea of contemplating our insides. Even the most attractive man or woman would seem abhorrent if we were to look at the tissue just beneath the skin. Notwithstanding this natural aversion, some artists have explored the aesthetics of raw flesh and bone. Rembrandt, for example, chose a slaughtered ox as a subject for a portrait (Fig. 14.1), and similar subjects have inspired other artists (e.g. Chaim Soutine, Francis Bacon). Even in these cases, though, the depiction of flesh can seem disturbing. This is not because there is something intrinsically ugly about an organism's insides; we find the grain and texture of wood positively attractive, even though this comes from the inside of trees. Rather, it seems to me that the explanation has more to do with our bias towards the surface of living animals, including ourselves. Our insides are only exposed to us on pain or death, and it is these unpleasant associations that are most probably the source of our discomfort. If we were naturally endowed with an ability to see into animals, with something like X-ray vision, skeletons and tissues might lose their morbid associations and become part of the beauty of life.

In the previous chapter, we dealt with the outward appearance of organisms. We saw how external geometry, such as bilateral symmetry, can arise through

Fig. 14.1 *Slaughtered Ox* (1655), Rembrandt. Louvre, Paris.

the elaboration of asymmetries along two major axes. But a third asymmetry was implicit in this analysis: an *inside–outside axis*. Even a perfect sphere has a distinct inside and outside; it would look and feel very different if you were inside rather than outside a sphere. But because of our bias towards external appearance we tend either to overlook this asymmetry or take it for granted. As far as organisms are concerned, though, the inside–outside axis is the most fundamental distinction of all. You can imagine a spherical organism, but you can't conceive of an organism without distinctions between inside and outside. Such an organism would be indistinguishable from its environment, so it would be impossible to think or talk about it in any meaningful way. In this chapter I want to explore this issue further, by considering the process of development from a more three-dimensional perspective.

Moving in

One of the primary distinctions along the inside–outside axis of animals is the subdivision of the early embryo into different *germ layers*. These are concentric

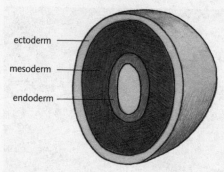

ectoderm

mesoderm

endoderm

Fig. 14.2 The three germ layers of an embryo.

cell layers, each of which gives rise to particular tissues in the body. Vertebrate embryos, for example, comprise three germ layers (Fig. 14.2). The outer layer of cells, or *ectoderm*, gives rise to the external covering of the animal, the skin (it also gives rise to the nervous system). Cells in the middle layer, or *mesoderm*, give rise to most of the underlying flesh and bone (e.g. muscle, blood, heart, kidneys, skeleton). Finally, cells in the innermost layer, or *endoderm*, produce the inner linings of the body, such as the gut and lungs. These three layers therefore provide one of the earliest distinctions between the bodily tissues from inside to outside. Understanding how they are first established will provide a good entry point for exploring the origins of internal asymmetry.

The different germ layers of the animal embryo arise through a series of cell movements, known as *gastrulation*. In most animals, the early divisions of the fertilised egg give rise to a ball of cells enclosing an internal space. The process of gastrulation occurs when some of the cells on the outer surface of the ball start to move internally, ending up inside the ball. It is the inward movement of these cells that leads to the distinctive layers of cells from outside to inside. Without gastrulation, our internal body tissues would not form and we would end up as empty shells. The biologist Lewis Wolpert once emphasised this by saying that the most important event in your life is not birth, marriage or death, but gastrulation!

To see how gastrulation works, I want to return to the fruit fly, *Drosophila*. Look at the left part of Fig. 14.3, which shows a cross-section through a *Drosophila* embryo when it has a single outer layer of cells. Gastrulation begins when a group of cells running along the ventral region of the embryo (shown shaded in Fig. 14.3) start to move inwards, creating a fold or furrow. As this inward movement continues, these cells gradually become internalised within the embryo, forming an internal layer, the mesoderm (Fig. 14.3, right).

Why is it that only the cells in the ventral region move in to form the

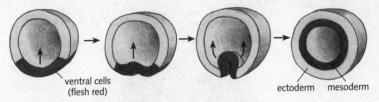

ventral cells
(flesh red)

ectoderm   mesoderm

Fig. 14.3 Formation of mesoderm during gastrulation in fruit fly.

mesoderm? What distinguishes them from the other cells around the embryo? In the previous chapter, we saw that the ventral region of the early embryo has a distinctive hidden colour, devil-red (produced by the *dorsal* gene). This colour provides an early frame of reference that can be interpreted by other genes, leading to them being expressed in particular regions. Two of these genes (called *twist* and *snail*) are switched on specifically in a ventral group of cells. It is precisely this pair of genes that distinguishes the cells that will move in to form the mesoderm. These genes produce master proteins, giving the shaded cells in Fig. 14.3 a distinctive hidden colour. Because these cells eventually give rise to most of the internal flesh of the animal, I shall name this colour *flesh-red* (actually the flesh of insects is not red but this does not matter for our purposes). The presence of flesh-red in the ventral cells leads to particular genes being switched on or off, which in turn results in the cells moving inside the embryo. In mutants that lack flesh-red, the ventral cells do not move in, but remain on the outer surface.

The overall result is that the embryo ends up with an internal layer of cells with a distinctive hidden colour, flesh-red (Fig. 14.3 right). This can then be interpreted by other genes, eventually leading to the development of characteristic mesoderm tissues, like muscles or heart. Without flesh-red, embryos will not form these structures.

Similar principles apply to the formation of the innermost germ layer, the endoderm. Again, cells with a distinctive combination of hidden colours move in from the outer surface of the ball, eventually forming a third layer within the embryo (in this case, the cells that move in start off at either end of the embryo).

The overall result of gastrulation is that the embryo ends up with a pattern of hidden colours that varies along the inside–outside axis, from one germ layer to the next. It is this pattern of colours that provides the basis of inside–outside asymmetry at this stage. This can then interact with the patterns along the other two axes of the embryo, head–tail and dorsal–ventral, to give a three-dimensional overlapping patchwork of hidden colours, shown schematically in Fig. 14.4. You will notice that this patchwork still has bilateral symmetry— the left and right halves are mirror images of each other—because the

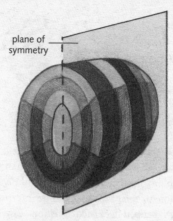

plane of symmetry

Fig. 14.4 Cross-section of schematic patchwork composed of overlapping patterns of hidden colour along three axes.

inside–outside axis acts in a similar manner on each side of the embryo. In other words, it is symmetrical about the mirror plane set by the other axes. This early patchwork provides an in-depth frame of reference that is further interpreted and elaborated, eventually leading to the complex internal and external anatomy of the animal.

Telling in from out

In describing gastrulation, I skipped over a vital step. Although the dorsal–ventral pattern can distinguish those cells (flesh-red) that will form mesoderm, it does not account for why they move *in* rather than in any other direction. Why, for example, don't the flesh-red cells move outwards? There has to be some asymmetry that leads the cells to move specifically in an inward direction.

We can address this problem by looking in more detail at how the early cell movements occur. One of the first signs of inward movement is that the flesh-red cells become wedge-shaped, by a contraction of their outer surface, as shown in Fig. 14.5. By changing shape in this way, the cells naturally form an inward arch, starting the process of gastrulation. Now the direction of this movement clearly depends on which surface of the cells contracts: if the outer surface contracts, as illustrated in the figure, the cells arch inwards; but if the inner surface were to contract, the opposite would happen, and the cells would arch outwards. Why, then, is it only the outer surface that normally contracts? It turns out that the contraction is caused by a change in the shape of a special group of protein fibres, called contractile proteins. These proteins come to lie

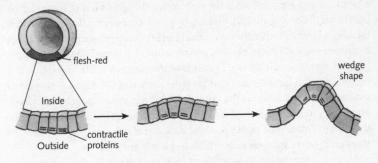

Fig. 14.5 Inward movement of ventral cells caused by localised contraction.

near to the outer surface of the cells, so it is this region that preferentially contracts (Fig. 14.5). We have therefore traced the direction of movement to an asymmetry in the distribution of contractile proteins.

The next logical step would be to ask how the contractile proteins come to be mainly in the outer regions of these cells. That is, how come the surface cells are themselves polarised along the inside–outside axis? The detailed answer to this is not known, but it is thought to trace back to earlier events that occur while the embryo is still a single cell. Recall that following fertilisation, the nucleus of a fruit fly egg undergoes several rounds of division to give many nuclei immersed in a common cytoplasm (Chapter 9). At this stage there is only one membrane around the cytoplasm so the embryo comprises a single cell. The nuclei then migrate to the outer regions of the cell, where they eventually become enclosed by membrane to form separate cells. This means that even before these individual cells are formed, the embryo has a defined structure from outside to inside: a membrane on the outside, cytoplasm containing nuclei further in, and then a central core. It is likely that some aspect of this earlier asymmetry between inner and outer regions of the embryo, when it is still a single cell, eventually leads to the preferential location of contractile proteins near the outer surface of the individual cells that form later on.

I have gone through this chain of events in some detail to show how the inside–outside axis can be traced back to properties of the egg cell. Basic differences in the egg, from inside to outside, lead to distinctions between inner and outer regions of the early embryo. These can then be further elaborated, in combination with the other asymmetries, such as the dorsal–ventral pattern of flesh-red, leading to the inside–outside pattern of hidden colours that distinguishes the germ layers.

I have concentrated on fly development, but many of the same considerations apply to other animals. In vertebrates, for example, an inside–outside axis is established very early on, when the embryo is a clump of a few cells (unlike the fly, the fertilised egg divides directly to give a clump of cells, without first forming a large cell with multiple nuclei inside). At this stage, particular types of protein molecules may become preferentially distributed towards the inner or outer regions of the clump. Exactly how this happens is not known but it is believed to stem from the fact that the outer surface of the clump is exposed to the surroundings, whereas the inner areas are not. Some aspect of this basic asymmetry leads to an altered distribution of molecules from inside to outside the clump. This is then built upon and elaborated, although precisely how the three germ layers are eventually established is still not clear.

There is a general lesson to be learnt from these examples. Because the early embryo is a coherent structure, a single cell or a clump of cells, there is an automatic distinction between inside and outside, between the embryo and its surroundings. This basic asymmetry can lead to particular molecules being distributed unevenly from inside to outside, eventually leading to the more elaborate inside–outside axis of the adult. Unlike the other axes of asymmetry I have mentioned, the origin of the inside–outside axis is so basic that it can almost seem trivial. We can imagine an embryo starting off without head–tail or dorsal–ventral distinctions but if it is at all coherent, it already has to have differences from inside to outside. These may be as simple as the difference between a membrane on the outside versus the cytoplasm and nucleus within, or an outer surface of a clump of cells as distinct from its interior. The particular aspect of asymmetry that is elaborated may vary from case to case, but the potential almost has to be there to begin with. In this sense, the inside–outside axis is the most fundamental of them all.

The same point can be vividly illustrated with plants. Unlike the cells of animals, plant cells are surrounded by rigid cell walls that cannot move easily relative to each other. When a fertilised plant egg cell divides, a new cell wall is laid down within it, partitioning the cell in two (Fig. 14.6). Each of these cells then divides, leading to the introduction of further partitioning walls. With each further division, new internal partitions continue to be added in this way. But throughout this process, the original wall of the egg cell continues to

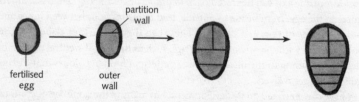

Fig. 14.6 Early development of a plant embryo. The outer cell wall is shown thicker for clarity.

surround the embryo. Although new internal walls are added, the outer foundation remains. Of course, the outer wall has to grow and expand to accommodate the enlarging embryo, but it maintains its original external location. This means that the outer surface of a mature plant traces back to the wall of the egg cell it came from (except for losses like the shedding of bark from trees and the loss of surface tissue from roots). So in this case, a basic asymmetry between outside and inside for the egg—the cell wall as distinct from its contents—becomes embodied in a corresponding asymmetry for the mature organism.

Of course, mature plants are not just a bag of nondescript cells surrounded by an outer wall. Like animals, they have a complex pattern of internal tissues. Although many of the details of how this pattern develops are not yet known, it most likely reflects the elaboration of hidden colours along the inside–outside axis that traces back to a very early stage, when the embryo comprises one or a few cells. Unlike in animals, however, the elaboration process does not involve cell movement, because plant cells are held in relatively fixed positions by their cell walls. Rather, it relies on cells signalling and responding to each other.

Handedness

So far I have described how patterns can be elaborated along three different axes to give a complex patchwork that retains a plane of mirror symmetry. However, there are some aspects of plant and animal development that display a further degree of asymmetry, leaving no mirror plane at all. I now want to look at how this might come about.

Figure 14.7 shows a self-portrait by Vincent van Gogh with a bandaged ear. It was painted within a short time of his having had a mental breakdown, during which he cut off part of his ear. According to *The Oxford Companion to Art*, the picture shows that van Gogh must have mutilated his *right* ear, because it is clearly this one that is bandaged in the picture. A moment's reflection, though, tells you that it actually must have been his *left* ear that was cut, because the picture is a self-portrait, carried out with him looking in a mirror and therefore inverting left and right. (You can also see this from the fact that the buttonhole is on the right of his jacket in the painting—the reverse of the conventional arrangement in which the buttonhole for male garments is on the left.)

Van Gogh's self-portrait makes the point that anything with asymmetry from left to right is distinct from its mirror image. But it turns out that for us to be able to distinguish an object's left and right, it also has to have asymmetries along two other axes. In the case of van Gogh's picture, for example, these are the asymmetries from top to bottom and front to back. We have no problem

Fig. 14.7 *Self-portrait with Bandaged Ear* (1889), Vincent van Gogh. Cortauld Institute, London.

knowing which way up to hang the picture because of the many obvious differences from top to bottom; similarly, there is no problem in distinguishing the back of the canvas from the front. If we were to eliminate differences along either of these axes, we would no longer be able to tell which ear he had cut off. Suppose, for instance, we got rid of differences from front to back, by taking a photographic slide of the portrait that could be looked at either way. Looking at the slide from one side we would conclude that he cut his left ear, but looking at it from the other side we would come to the opposite conclusion, that it was his right ear. Without knowing which way round we ought to look, we simply cannot tell. I have used the bandaged ear to make this point, but the same could be said of all the features in the painting that vary from left to right, such as the position of the window frame or the picture shown hanging on the wall behind.

We can summarise by saying that for an object to be distinguishable from its mirror image, it needs three axes of asymmetry, labelled as top–bottom, front–back and left–right in Fig. 14.8. Such an object can come in two possible forms that are mirror images of each other. For convenience, I shall refer to one of these as the left-form or *L-form*, and the other as the right-form or *R-form*. In each case it will be a matter of convention as to which way round these labels

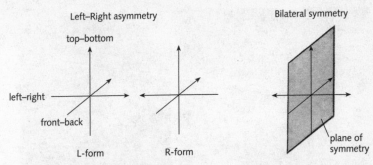

Fig. 14.8 Objects with a left–right axis can come in two possible forms and have no internal plane of reflection symmetry; whereas objects with bilateral symmetry have a single plane of reflection symmetry and come in only one form.

are applied. If we decide that the original van Gogh portrait should be called the R-form, then its mirror image would be the L-form. Such objects have no internal planes of reflection symmetry: it is not possible to find a plane running through the object that divides it into two halves that are mirror images of each other. Compare this with the situation of bilateral symmetry, shown on the right of Fig. 14.8. Here there are also three axes, but one of them, the horizontal axis in this case, acts symmetrically with respect to the other two. The object could therefore be superimposed on its mirror image, and it retains a single plane of reflection symmetry. This is the situation we encountered earlier on in this chapter with the patchwork of hidden colours, because the inside–outside axis acted symmetrically with respect to the other two axes (dorsal–ventral and head–tail).

For objects with left–right asymmetry we can ask whether the two possible forms, L and R, have the same chance of occurring, or whether there is a bias that favours one over the other. Take the case of van Gogh cutting his ear. Was is just a matter of chance that he chose to cut his left rather than his right ear? Probably not. Based on a portrait of van Gogh at work by his friend Paul Gaugin, we can be fairly confident that van Gogh was right-handed. Now it seems to me that if a right-handed person wants to cut a bit of his ear off, he would most naturally choose his left ear: this ear is much easier for him to cut with the right hand while holding it with the left hand. There would therefore have been a bias that favoured one form over the other because of the handedness of van Gogh.

Such a bias arising from the handedness of artists can be witnessed in other ways. Compare the two sketches shown in Fig. 14.9, by van Gogh and Leonardo. Van Gogh's main lines of shading run diagonally from the top right of the picture to the bottom left, the way a right-handed person would naturally tend

Fig. 14.9 Sketches by Leonardo (left) and van Gogh (right) showing opposite directions of shading.

to shade things. By contrast, Leonardo, who was left-handed, naturally shaded in the opposite direction, from top left to bottom right. The direction of shading is not random but is biased by the preferred hand of the artist. We might call van Gogh's diagonals representative of the R-form of shading whereas Leonardo's would be typical of the L-form.

For any individual artist there will therefore be a bias towards producing certain forms over others because of his or her hand preference. If you were to consider the works of many artists together, this bias might to some extent be averaged out because some are right-handed and others are left-handed. But the key point about handedness is that there is not an equal proportion of each type in human populations: there are more right-handed than left-handed people (I shall come back to why later). This means that there will be an overall bias towards certain forms, such as the direction of shading, over others. Our own handedness therefore provides a very good example of something that creates a bias towards one mirror image over the other. From now on I shall use *handedness* as a general term for any system in which there is a consistent bias towards an L- or an R-form.

Left–right asymmetry in plants and animals

There are many examples of left–right asymmetry in the living world. In some cases, the L- and the R-forms occur in equal numbers, and it is largely a matter

Fig. 14.10 Male fiddler crab with enlarged left claw.

of chance whether any individual will develop as one form or the other (i.e. there is no overall handedness). Male fiddler crabs, for example, often have an enlarged claw on one side (Fig. 14.10). Whether this is on the left (L-form) or right (R-form) depends on the previous chance experiences of the animal. Young males start off with two large claws, but if one of these is lost through injury, a small claw will regenerate in its place, resulting in a crab with a large claw on only one side. Unlike for van Gogh, in crabs there seems to be no bias as to which side gets injured, so the left and right forms typically occur in equal numbers. (If both claws are lost, the crab ends up with two small claws; whereas if neither is lost, the adult ends up with a large claw on each side.)

Another example of chance asymmetry is the arrangement of leaves in some plants. For example, if you trace the consecutive leaves of a tobacco plant up the stem, they form a spiral or helix. The direction of the spiral can be most easily seen by looking down on the growing tip, from where the leaves further down appear progressively larger (Fig. 14.11). In some individual plants, the spiral runs clockwise from top to bottom (R-form), whereas in others it runs anticlockwise (L-form). These two forms are mirror images of each other. In a survey of over twenty thousand individual tobacco plants, the two forms were found to occur in approximately equal numbers. Furthermore, if either form was self-pollinated, it gave progeny which again were a 50:50 mixture of each type. It seems, then, to be largely a matter of chance whether an individual develops as one form or the other. Perhaps the direction that the plant will take is very finely balanced early on in development and a minor random perturbation is enough to flip it one way or the other.

In these examples, the distinction between left and right forms is a matter of chance. But there are other cases in which there is a consistent bias towards one

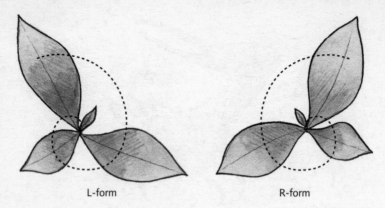

L-form   R-form

Fig. 14.11 Two tobacco plants viewed from the top showing spiral arrangement of leaves.

form over the other. In flowers of greater periwinkle (*Vinca major*), for example, the petals are arranged as fan blades that all seem to turn in the same way (Fig. 14.12, left). If you start from the centre of the flower and trace the edge of a petal outwards, you will be turning anticlockwise (the flower has five-fold rotational symmetry but no internal planes of reflection symmetry). This is a case of handedness because all individuals of this species have flowers arranged in the same manner (the L-form). By contrast, plants of oleander (*Nerium oleander*) have flowers that consistently exhibit the mirror image arrangement or R-form (Fig. 14.12, right).

Vertebrates typically exhibit bilateral symmetry on the outside, but several aspects of their internal anatomy show handedness. Our heart and stomach are usually on the left and our liver is on the right. This difference between internal and external asymmetry probably has to do with the way we move. Walking from A to B in a straight line is faster than moving in a left-handed curve, as would tend to happen if your left leg was shorter than your right. The only time we move in consistent curves is in the womb, where the foetus usually turns to its right, giving the vessels of the umbilical cord a corresponding twist (Fig. 14.13). Our external bodies are symmetrical for the same sort of reason that cars are outwardly symmetrical. Internally, though, we can put the steering wheel on the left or right of a car without influencing the direction it moves in. Similarly, we could argue that in order to pack our insides effectively, some organs need to go to one side or other. It is important that this is done in a consistent way. It is no good putting the steering wheel on the right of a car if the accelerator pedal is on the left. For similar reasons, individuals in which one or two organs are inverted from left to right while the other organs remain normal may have medical complications. About one in ten thousand people

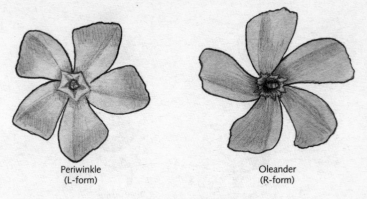

Periwinkle
(L-form)

Oleander
(R-form)

Fig. 14.12 Flowers of greater periwinkle (*Vinca major*) and oleander (*Nerium oleander*) showing handedness.

Fig. 14.13 Human umbilical cord showing twisted arrangement of blood vessels.

have *all* of their internal organs inverted, a mirror image of the normal arrangement, a condition termed *situs inversus*. Nevertheless, because the organs can still function properly with respect to each other, these individuals can live quite happily.

Although moving in left or right curves is not very common in animals, it is precisely the way that twining plants grow. When a twining plant climbs up a pole, the growing tip tends to curve consistently either to the right or left, resulting in a spiral or helix. I should briefly mention how the direction of a helix is defined. Helices are conventionally classified as either right-handed (R-form) or left-handed (L-form), according to whether or not they match the appearance of a right-handed corkscrew (Fig. 14.14). An easy way to recognise

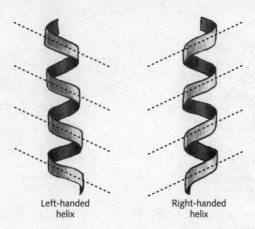

Left-handed helix   Right-handed helix

Fig. 14.14 Two types of helix: left-handed and right-handed. Note how the diagonals of the helices slope in the same way as the shading of a person with the corresponding handedness.

these forms is to look at the diagonal pattern made by the edges nearest to you. If they slope the same way as the shading of a right-handed person, the helix is right-handed; if they match the shading of a left-hander, the helix is left-handed (if you apply this to the umbilical cord shown in Fig. 14.13 you should easily see that it forms a left-handed helix). Note that whether the helix is right or left-handed is intrinsic to a helix and doesn't depend on which way up it is. You can see this by turning Fig. 14.14 upside down and observing that the direction of the diagonals remains unchanged (helices have rotational symmetry even though they have no internal plane of reflection symmetry).

Most species of twining plant show handedness, either consistently forming right- or left-handed helices when they climb. Runner beans (*Phaseolus*) and bindweeds (*Convolvulus*), for example, make right-handed helices, whereas hops (*Humulus*) and honeysuckles (*Lonicera*) produce left-handed helices (Fig. 14.15). The handedness depends on the behaviour of the growing tip. The tip of a twining plant tends to rotate once every few hours in a consistent direction as it grows, either clockwise or anticlockwise. It is the direction of this rotation that determines whether the plant will form a right- or left-handed helix as it climbs (Darwin made a fascinating study of this type of rotation, called circumnutation, by closely observing the tip of a potted plant in his room while he was confined there by illness).

Another common example of handedness involving spiral growth is the direction of coiling in shells. Most snail or sea shells have right-handed spirals. If you look into such a shell, while holding it with the tip of the spiral pointing away from you, the opening is on the right (Fig. 14.16, right). The direction of

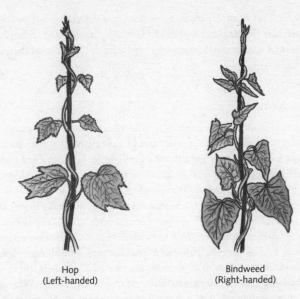

Hop
(Left-handed)

Bindweed
(Right-handed)

Fig. 14.15 Helices made by hop (left-handed) and bindweed (right-handed).

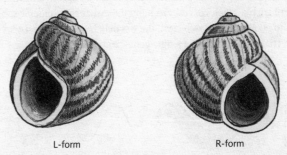

L-form

R-form

Fig. 14.16 Left-handed and right-handed snail shells.

the spiral depends on how the animal grows from a very early stage. In some right-handed species, rare mutant individuals can be found that grow in the opposite way and give left-handed shells.

In all these cases of handedness, it is not at all clear, from the viewpoint of adaptation, why one mirror image should be favoured over the other. There is no obvious disadvantage in having all our internal organs arranged in the precise mirror image of the common form. Similarly, why should a twining plant or shell consistently spiral one way rather than the other? The answer is probably not that one mirror image works better than the other but that the basic molecular mechanisms that orient left–right asymmetries often happen to do so in a handed or biased manner. I now want to look at what these mechanisms might be.

The worm's turn

Handedness is challenging to explain because there has to be some way to orient the left–right axis relative to the other axes in a consistent manner. As children it takes us a long time to distinguish left from right, so how can mere molecules or cells achieve this? We still do not know the definitive answer to this problem, but we can get a clue by looking at some mutant worms.

The tiny nematode worm (*Caenorhabditis elegans*) normally moves along a solid surface by lying on its left or right side and undulating its body by a series of muscular contractions, leaving a wavy track (Fig. 14.17). However, mutant worms have been found that, instead of snaking smoothly along, roll around and move in circles. These *roller* mutants are twisted along their length so that their dorsal and ventral sides are helically arranged (Fig. 14.17). When such a mutant attempts to move, the head and body rotate like a screw, and the animal tends to travel in a circular path. Of most interest for our present purposes, *roller* mutants tend to circle around in a constant direction, showing handedness. There are two classes of *roller* mutant: those that circle clockwise and those that go anticlockwise. For each class, the worm behaves in a consistent manner. How is it that a mutation in a single gene can make a worm always turn one way?

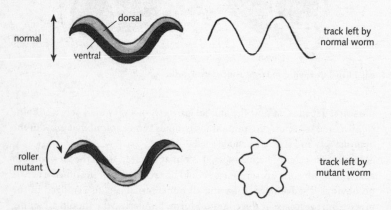

Fig. 14.17 Movement and structure of a normal worm compared to a *roller* mutant.

Fig. 14.18 Structure of collagen.

It turns out that the gene affected in the *roller* mutants codes for a protein called *collagen*. Collagen is a long stiff protein that forms a major part of the hardened cuticle or outer covering that protects the worm (collagen is also found in humans, where it contributes the major fibrous element of skin and bone). A key feature of collagen proteins is that they are formed of three long twisted fibres that are always arranged as *right-handed helices* (Fig. 14.18). Now imagine that many of these collagen proteins were wrapped around the worm in parallel with each other. Because they all twist in the same way, their twists might well be compounded to give the covering of the worm an overall twist as well. And because the collagen proteins are handed, the overall twist would be biased predictably in one way or the other. In normal worms, this potential twisting may be averted by ensuring that the various collagen molecules (and other handed fibres) are oriented in particular ways with respect to each other. In other words, things are normally carefully arranged in order to *avoid* an overall twist. This means that a mutation that alters one type of collagen protein could disturb the balance, giving an overall right- or left-handed twist to the worm. The worm would then move in endless circles, clockwise or anticlockwise depending on the hand of the twist. This appears to be the sort of thing that happens in *roller* mutants. The consistent circular movement of mutant worms in one direction or another seems to derive from the handedness of its component collagen proteins. Why then are collagen proteins always arranged as right-handed helices? How can a protein distinguish left from right?

Recall that every protein is made up of a chain of subunits, called amino acids. Now each amino acid in turn consists of a carbon atom with four different molecules or atoms attached to it,* forming a tetrahedron (Fig. 14.19). But notice that this arrangement has no planes of reflection symmetry: the mirror image of an amino acid cannot be superimposed on the original whichever way we try and turn it. There are therefore two potential forms for each amino acid, an L-form and a R-form (Fig. 14.19). If these were present in a 50:50 mix, there would be no overall handedness to amino acids. However, at an early stage of evolution, only one form of amino acid, the L-form, became predominantly used by organisms and essentially all amino acids involved in protein synthesis ended up being of this type. Exactly how this came about is

---

*One amino acid, glycine, is an exception to this rule because two of the atoms attached to the carbon are the same (hydrogen).

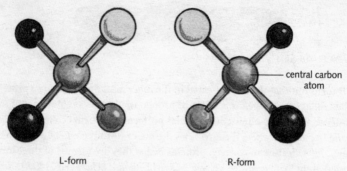

L-form   R-form

Fig. 14.19 Structure of two forms of amino acid.

still a matter of debate, but one important consequence has been that proteins can be made in a consistent shape because the same form of amino acid is always used. In principle, the R-form could have been used just as successfully as the L-form, and it may have been largely a matter of chance which form survived, just as whether we drive on the left or right side of the road is a matter of historical accident. It doesn't matter which side we drive on so long as everyone else does the same. For some countries (e.g. England) the convention is to always drive on the left whereas in others (e.g. America) it is the right, but a mixed strategy of sometimes driving on the left or right would be disastrous. In a similar way, interactions between molecules during evolution may have ensured that only one type of amino acid eventually came to predominate in the living kingdom (according to one view, this choice may itself have been biased by a handedness in physics, involving the properties of subatomic particles).

When L-amino acids are strung together to make proteins, they often form helices. And because the amino acids are handed, so are the helices that they form. In the case of collagen, three strings of amino acids wind around each other to form a right-handed spiral: each of the individual fibres of collagen shown in Fig. 14.18 is a string of L-amino acids. The handedness of collagen therefore reflects the handedness of its component amino acids (DNA is a right-handed double helix for a similar reason: its subunits are handed molecules).

By stringing together small handed molecules, the L-amino acids, we end up with a handed helical protein of collagen. The collection of handed collagen proteins can in turn give a handed twist to the cuticle, leading to a worm wandering around in circles in a consistent direction. The final handedness we see in the mutant worm can be traced to a basic molecular handedness, which in turn may have been set by chance events early on in the evolution of life.

Handed painting

The case of the turning worm shows how a handed twist of the body might follow from molecular handedness, but what about more complex arrangements, such as the left–right asymmetry in our internal organs? In 1995, Michael Levin, Cliff Tabin and colleagues, working at the Harvard Medical School in Boston, made a remarkable observation. They were studying some genes for hidden colours (master proteins), scents (signalling proteins) and sensitivities (receptor proteins) in the early chick embryo, when they noticed that some of these genes were expressed mainly on the left side of the embryo, whereas others were on the right. In other words, there was a left–right asymmetry in the distribution of these gene activities at an early stage of development. Similar types of observation were then made for other vertebrate embryos, such as those of mice and frogs. This means that there are left–right distinctions in the early map of hidden colours and their associated scents and sensitivities.

This asymmetry is related to the left–right differences of internal organs that develop later on. For example, there is a mutant type of mouse in which the organs are always in a mirror image of the normal arrangement (*situs inversus*). The early embryos of such mice also tend to have the left–right distinctions in their hidden map reversed, so that genes expressed normally on the left are now found to be expressed on the right. These sorts of observation have led to the following model.

In addition to distinctions along the three axes previously mentioned (head–tail, dorsal–ventral, inside–outside), an asymmetric pattern of hidden colours is also established very early on from left to right. This results in distinctive colour combinations in the left and right halves of the embryo's patchwork that can be further interpreted and elaborated, eventually leading to left–right distinctions in internal anatomy. Our internal left–right asymmetry traces back to an asymmetry in the patchwork of hidden colours.

This begs the question of what the left–right differences in the patchwork themselves depend upon. The answer is not yet known but, as we have seen with the twisting worm, whatever they trace back to must itself be handed. Like the worm, the most obvious candidates are the handed building blocks of life, such as the L-amino acids. But instead of these molecules contributing to the handed movement of the whole animal, we need to imagine that they lead to differences in hidden colour between the left and right halves within the embryo. For example, handed proteins (made of L-amino acids) might lead to a particular scent being preferentially transported from the left to the right side of the embryo. Once left–right differences are established for one type of scent molecule they can then be responded to and further elaborated. (Essentially this sort of model was proposed by Nigel Brown and Lewis Wolpert in 1990, and it has recently received some experimental support from the finding that one of the genes affecting handedness in mice codes for a protein, dynein, involved in molecular transport.) The details are still far from clear but the key point is that the handedness of the patchwork most probably reflects the handedness of basic molecules, such as L-amino acids. Perhaps all forms of biological handedness, from twining plants and spiral shells to the arrangement of our visceral organs, trace back, in one way or another, to the handedness of molecular building blocks like amino acids.

Might the same principles also underlie the bias towards right-handedness in humans? An early theory of human handedness was that it reflected the internal arrangement of organs. We now know that this is not the case because the vast majority of people, including the ten per cent or so of left-handers, have the same internal anatomy, with the heart on the left and the liver on the right. Hand preference is therefore not correlated with visceral anatomy. Instead, the bias has more to do with the way our brain works. We tend to think of our brain as a single structure, but it is actually divided into left and right halves or hemispheres, which can function differently. For example, in most people, speech is localised to the left hemisphere whereas visuo-spatial abilities are concentrated in the right hemisphere. The hemispheres also control our basic body movements in a cross-wise fashion: the left hemisphere controls the right side of the body (e.g. right arm, right leg) whereas the right hemisphere controls the left side. Because of this, predominance of right-handedness is thought to be due to a dominating effect of the *left* hemisphere on manual activities.

Our hand preference may therefore reflect an asymmetry in the way our brain develops: there is a bias that leads to manual dominance more often developing in the left hemisphere than the right. Nobody knows what the origin of this bias is, but it seems to me that one possibility is that it stems from molecular handedness. Just as handed molecules might bias the development of certain organs towards one side of our body, so they might also bias the way our brain develops. We would expect at least some of the molecules involved to be different for brain and visceral left–right asymmetry because these can be inverted independently of each other. Furthermore, the bias in the case of the brain may be weaker because cross-cultural studies indicate that about one in ten people are left-handed*, whereas only about one in ten thousand people have fully inverted visceral organs. Nevertheless, the ability to distinguish right from left in both cases might trace back to the same sort of molecular handedness. If so, it would mean that the handedness of internal painting, the elaboration of

*The bias is probably greater than this implies because some individuals may be genetically predisposed to be left-handed.

hidden colours, and the handedness of human painting would have the same basis.

To summarise, the most fundamental type of asymmetry is along the inside–outside axis. All organisms express this in some form or other. In multicellular organisms, asymmetry along this axis can be elaborated during development, to give a pattern of distinctive hidden colours from inside to outside. This may interact with patterns along two further axes to give an overall patchwork with bilateral symmetry. In some cases, there is an additional left–right pattern of hidden colours that can be superimposed to give the patchwork an overall handedness. As with other cases of biological handedness, this most probably traces back to the handed nature of molecular building blocks.

We have seen how the overall geometry and symmetry of multicellular organisms depends on the elaboration of hidden colours along various axes. However, there is another fundamental aspect to the geometry of many organisms: internal repetition. A striking feature of many plants and animals is that they contain many repetitive elements or themes, such as the numerous leaf-like organs of a plant, the segments of an insect, or the vertebrae of a mammal. In the next chapter, I want to explore the nature of this internal repetition and show how it is related to the other patterns and asymmetries we have encountered.

CHAPTER 15

Themes and variations

While I was at school in Liverpool, I was sometimes given 'lines' to write out as a form of punishment. Following some misdemeanour, I would be told to write a sentence like 'I must not use my ruler as a catapult' one hundred times. I would begin by writing the phrase at the top of a page and then write it again and again on the lines beneath, continuing down until the page was covered. In performing this task, a strange sort of satisfaction would start to emerge. There was something gratifying about seeing the same series of pen strokes running regimentally down the page, forming a regular pattern. In fact it was often quicker to form this pattern early on, by writing each word in turn as a vertical series. Written out in this way, the sentences lost their meaning and the exercise became one of making an almost abstract pattern of regular marks.

It seems that simply repeating something many times can lead to a certain type of symmetry. The same principle underlies the appeal of many friezes and wallpaper patterns. Even a basic design can be rendered quite attractive by the simple act of repetition (Fig. 15.1).

Fig. 15.1 Simple repetitive design.

As we shall see, repetition is also a fundamental feature of the living world. In this chapter I want to look at how repetition can arise during the development of organisms and how it relates to the types of symmetry covered in previous chapters.

Translational symmetry

The type of symmetry displayed when a sentence or design is repeated is known as *translational symmetry*. Remember that the symmetry of an object can be defined by the number of different transformations that can be applied to it that leave it unchanged (Chapter 13). We have already come across rotational symmetries, in which an object is left unchanged when it is turned; and reflection symmetries, in which appearances are preserved following reflection in a plane. An object with translational symmetry doesn't change when you move or shift it along in a straight line by a certain amount. Imagine, for example, that the design in Fig. 15.1 carried on for an infinite distance to the left and right. If you were to shift the pattern along by one length of the repeating unit, you would end up with exactly the same pattern again. In other words, you would not be able to tell that such a transformation, called a translation, had been applied by simply looking at the outcome. Similarly, an infinitely long page of written 'lines' would not be altered by shifting it up or down by one line. Strictly speaking, only infinite patterns show this sort of symmetry. If the repeating pattern is of a limited length you could tell that it had been shifted along by observing the movement of its ends. Nevertheless, for our purposes it will be useful to apply translational symmetry in a looser sense to include finite as well as infinite repeating patterns. After all, you can always *imagine* converting a finite pattern into an infinite one by extending it indefinitely.

With this notion of translational symmetry in mind, we can distinguish between two basic types of repeating pattern. Look at the simple linear design at the top of Fig. 15.2. As well as showing translational symmetry along the horizontal axis, the pattern has a vertical plane of reflection symmetry running down the middle. That is to say, the left and right halves of the design are mirror images of each other. If the pattern was repeated indefinitely, there would be a similar plane at regular intervals all the way along (the interval between the planes of symmetry would be half the length of the repeating unit). Such patterns can be said to show *no polarity* because they look the same either way round. A band of two alternating colours also has this sort of symmetry, with the mirror planes running down the middle of each colour block (Fig. 15.2, bottom). Another good example is the Greek palmette motif shown in Fig. 15.3 (taken from Hermann Weyl's book on symmetry).

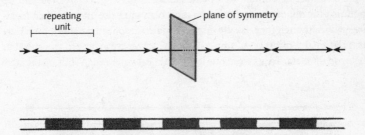

Fig. 15.2 Translational symmetry with no polarity.

Fig. 15.3 Palmette design showing translational symmetry with no polarity.

The second type of repeating pattern has no such planes of symmetry. For example, if the pattern shown at the top of Fig. 15.4 is reflected in any vertical plane, the direction of the arrows will be reversed, allowing you to tell that the transformation has been applied. It therefore has a lower degree of symmetry than the previous examples because there are fewer ways of transforming it into itself. Such patterns can be said to exhibit *polarity*. A banding pattern of three different colours, shown at the bottom of Fig. 15.4, has this sort of symmetry because any reflection at right angles would invert the order of the colours. Other examples are the design in Fig. 15.1 and written 'lines' on a page.

All cases of translational symmetry fall into one or other category, either having polarity or not. A stack of similar coins oriented in the same way, say heads up and tails down, would have translational symmetry with polarity; each coin would be related to the next by a vertical shift of one coin's thickness, and the direction each coin was facing would be inverted by reflection in any plane at right angles to this (a horizontal plane in this case). By contrast, a stack of double-headed coins (coins with heads on both sides), would form a pattern showing translational symmetry without polarity.

A good example of translational symmetry in the living world is provided by the segments of centipedes and millipedes (Fig. 15.5). You can see from the orientation of the legs that these repeating patterns also have polarity: if you were to reflect the animal in a plane across it, the segments would end up pointing the opposite way. Leaves on most plants are also arranged in a repetitive way, although they usually display a more complicated pattern than pure translational symmetry. In some plants the leaves are arranged as a spiral or helix up the stem. To get from one leaf to the next you need to shift up the stem

Fig. 15.4 Translational symmetry with polarity.

Fig. 15.5 Centipede (Scolopendrid) showing translational symmetry of segments.

(translation) and move part way round its circumference (rotation). That is to say, the pattern remains unchanged following a combined translation and rotation. In other cases, as with the begonia shown in Fig. 15.6, subsequent leaves are related by a translation up the stem followed by reflection in a longitudinal plane. Most types of leaf arrangement involve a combination of translation, rotation and/or reflection symmetry (they also show polarity because the upper surface of a leaf is usually quite different from the lower surface).

All these examples involve repetition of the same or very similar units. But there is also a broader category of patterns in which repeating units *vary* to some extent from one to another. Suppose, for example, that instead of identical coins we stack a mixed set of coins of different denominations. The coins can still be stacked with the same polarity (heads facing up and tails down) but there is now no translational symmetry because each coin is distinct from its neighbour. Furthermore, each coin may have a slightly different thickness so that there is no standard length for shifting them up or down. Nevertheless, all the units in the stack share some features because they are all easily recognisable as coins. One way of describing the situation would be to say that there is a basic *theme* that is repeated up the stack—the general notion of a coin—but there are also important *variations* on this—the particular denominations.

The same point can be illustrated with a queue of people. If all the people in a queue are identical and stand at perfectly repeated intervals, the queue would show translational symmetry. But in real queues, each person is different from the next, so there is no strict symmetry of this type. Nevertheless, we can still recognise a repetitive structure in the queue, which might be expressed by

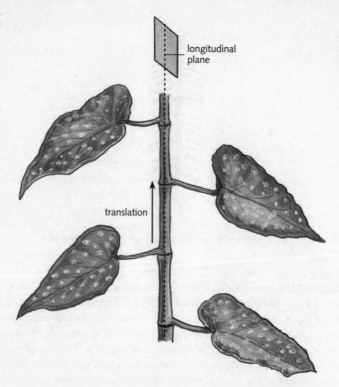

longitudinal
plane

translation

Fig. 15.6 Arrangement of leaves on a begonia plant, in which subsequent leaves are related by a combined translation along the stem followed by a reflection in a longitudinal plane.

saying there is a basic theme—the general notion of a person—that is repeated along the queue, together with numerous variations upon it—the particular characteristics of each individual in the queue. In all these cases a basic type of unit, such as a coin or person, is repeated in a linear sequence but the repeating units are not identical.

A fruit fly can also be thought of as exhibiting variations on a theme. We talk about a fruit fly as being made up of repeating segments even though they are clearly not all the same: some have legs, some also have wings, yet others have no appendages. It is as if the fly displays a repeated theme, the segment, combined with variations upon it. The same might be said about the different whorls of organs in a flower, or the fore and hind limbs of a vertebrate: they are all examples of variations on a theme. In previous chapters we have looked at how such variations might arise. We saw, for example, that the variations between the segments of a fly reflect a series of hidden colours (various types of green) that vary from head to tail. But I said nothing about what the segments have in common, the nature of the underlying *theme*. I now want to look at this problem, but before doing so it will help to distinguish between two types of theme.

Explicit and implicit themes

In 1957, Pablo Picasso produced a series of paintings inspired by one of the masterpieces of art, *Las Meninas* (The Maids of Honour), painted about three hundred years earlier by Diego Velázquez. Velázquez's painting shows a girl of the Spanish royal family, attended by her maids of honour (Fig. 15.7, top). The girl has come to watch the Court painter, Velázquez himself, at work in his studio. Picasso was fascinated by the picture and produced a series of paintings based upon it. Some of Picasso's paintings are based on the whole of Velázquez's picture (Fig. 15.7, bottom); others concentrate on a detail such as the girl alone (Fig. 15.8). Once you know the history of the paintings, you can almost see the original Velázquez within them, even though on the surface they look so different.

This series of paintings by Picasso might be said to represent variations on a theme. The theme is the original painting by Velázquez whereas the variations are the different versions of it that Picasso came up with. Although the theme and its variations are all paintings, there is no problem in distinguishing between them. The theme clearly preceded the variations and was their source of inspiration rather than the other way round.

But there is also another theme that unites Picasso's paintings: they have a common style. They are all examples of Picasso's art and vision. In a sense this is a more important theme because anyone can look at a Velázquez and produce a series of paintings, but only Picasso could have produced the works that we see in his *Las Meninas* series. Yet defining this theme, Picasso's style, is not so easy. You might look at many examples of paintings by Picasso and recognise some unity, but it would be difficult to say precisely what they all have in common, what makes them all identifiable as works of Picasso.

Picasso's *Las Meninas* series illustrates two basic types of theme. On the one hand, there is the theme provided by the Velázquez. We can point to Velázquez's original painting as distinct from the variations based upon it. We might say that in this case the theme is quite *explicit* and precedes the variations. By contrast, the other type of theme, the stylistic unity of Picasso, is *implicit*. It seems to be locked within each of the variations rather than being a separate painting we can point to. Unlike an explicit theme, there is no implied direction here: any two Picasso paintings are *equivalent* expressions of Picasso's style, rather than one being derived from the other.

Fig. 15.7 *Las Meninas* by Velázquez, above (1656, Prado, Madrid) compared to version by Picasso, below (1957, Museo Picasso, Barcelona).

Fig. 15.8 Detail from *Las Meninas* by Velázquez, left (1656, Prado, Madrid) compared to picture by Picasso, right (1957, Museo Picasso, Barcelona).

The leaf

Much confusion about the nature of repetition in the living world has come from confounding implicit with explicit themes. A good illustration of this concerns Goethe's idea that all the appendages of a plant are equivalent to leaves. Recall that he proposed that all the organs of a plant were variations on a theme: sepals, petals, stamens, carpels and foliage leaves may look different from each other but they share an underlying similarity. Goethe wanted a term to describe the theme that unified the various organs of a plant and he chose to use the word *leaf*:

> It goes without saying that we must have a general term to indicate this variously metamorphosed organ, and to use in comparing the manifestations of its form; we have hence adopted the word *leaf*. But when we use this term, it must be with the reservation that we accustom ourselves to relate the phenomena to one another *in both directions*. For we can just as well say that a stamen is a contracted petal, as we can say of a petal that it is a stamen in a state of expansion. And we can just as well say that a sepal is a contracted stem-leaf . . . as that a stem-leaf is a sepal.

Goethe was using the word *leaf* in an unusual way, to refer to the *implicit theme* behind all organs. There was no implied direction: a comparison between any pair of organs, like a stamen and a petal, or a sepal and a foliage leaf, could always be made both ways. This equivalence could be summarised by saying that they are variations on the same theme, all different types of *leaf*. Unfortunately, in choosing the word *leaf* to denote the underlying theme, he caused major confusion because we also use leaf to indicate one of the variations, namely foliage leaves. By saying that all organs were essentially leaves, he did not mean to imply that leaves, in the narrow sense of green foliage leaves, were an explicit theme upon which everything was based. He was using *leaf* in a more implicit sense, as a theme common to foliage leaves and floral organs. His notion of *leaf* was more like the stylistic unity of Picasso's *Las Meninas* paintings than the distinct Velázquez that preceded and inspired them.

Even though he was very careful to point this out, it did not stop people misinterpreting his theory. Many botanists believed Goethe to be saying that the different parts of a flower were derived from leaves in the common sense of foliage (this mistake is still sometimes made even today). The green leaf of a flowering plant was taken to be the template that everything else was based upon. After Goethe died, this same misinterpretation surfaced again but in a slightly different way, under the guise of evolution. As people became convinced of Darwin's theory of evolution in the late nineteenth century, they started to consider what the ancestors of flowering plants might have looked like. If the floral organs were indeed modified foliage leaves, as they thought Goethe was saying, it would seem that leaves came first and flower parts evolved from them secondarily. The implication was that ancestral plants must have had only foliage, and flowers then evolved through the modification of this ancient greenery.

The trouble is that if these ancestors had only had foliage leaves, they would have been completely sterile. A rose leaf cannot produce sperm or egg cells, no matter how long you wait. Nobody was around to witness the ancestors of flowering plants but it is extremely unlikely that they were sexually sterile. Sex is one of the oldest and most ubiquitous inventions of evolution and greatly predates the arrival of flowering plants. It is therefore highly improbable that the early ancestors of flowering plants had only sterile organs. They are more likely to have been similar to modern fern plants. Ferns do not have flowers but they do reproduce sexually. They achieve this through the production of sexual spores from their green fronds or 'leaves'. These spores will eventually give rise to sperm and egg cells which can fuse to produce the next generation of fern plants. In other words, the fronds on a fern resemble both foliage leaves, being green and spread out to collect the energy from the sun; and flowers, being the route to sexual reproduction. In flowering plants there is a separation between organs involved in nourishment (foliage) and sex (flowers). It seems likely that this is a later specialisation, and that both functions were carried out by the single type of organ in their early ancestors.

Some botanists therefore concluded that Goethe must have been wrong when he said that the parts of the flower were modified or metamorphosed leaves. It seemed to imply that ancestral plants were just sterile foliage, which was obviously incorrect. As the botanist F. O. Bower said in 1911: 'There is no need now for any theory of metamorphosis to explain the origin of the Flower. As first stated by Goethe it was a theory of mysticism, which was doomed to dissolution.' But this is because Goethe's notion of *leaf* was taken to be an explicit theme from which other organs were derived. Goethe was actually using *leaf* to denote an implicit theme that was simply manifested in different ways in the various parts of the plant. According to this viewpoint, the leaves of ancestral plants would have represented yet other variations on the same theme.

Even when taken on his own terms, though, Goethe's view seems to lead to an impasse. What exactly is a *leaf*, this common theme that is continually being depicted by the plant? There seems to be no meaningful way to pursue the problem other than by saying it is that which is common to all plant organs. This just seems to be evading the issue. It was for such reasons that Goethe's ideas were often considered to be mystical idealism rather than serious science.

Is there any way of getting beyond this impasse? Although the stylistic unity of Picasso's paintings is difficult to describe, we know where its origin lies: with Picasso himself. The implicit theme of his paintings reflects the way Picasso's brain worked and interacted with the canvas as each painting was produced. This means that one way of getting a deeper insight into the implicit style of his work might be to try and understand more about how Picasso's paintings arose, what led Picasso to apply the paint in one way rather than another. In a similar way, to define the implicit theme that is common to petals, stamens and foliage leaves, it is no good looking at more and more examples of fully developed plant appendages. We need to look deeper, for the common internal processes that lie behind the development of these organs. To illustrate how this can work, I want to turn to a very well studied example of this problem in the insect world: the nature of segments in fruit flies.

Repeating colours

If you look at the embryo of a fruit fly at a relatively late stage of its development, certain features can be seen to be repeated along the length of the body. Most obvious is the pattern of thorny outgrowths or denticles. As shown in Fig. 15.9, the denticles are most abundant towards the head end of each segment. The segments display a consistent repeating pattern, with a thorny head region and a smooth tail region, as you travel along the length of the body, like a stack of coins oriented in the same way. Ignoring the variation between segments for the time being, the pattern shows translational symmetry because you can shift

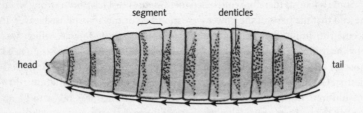

Fig. 15.9 Fruit fly embryo at late stage of development, viewed from the belly, showing repeating pattern of denticles on the segments. Only 11 of the 14 segments can be seen clearly because the 3 head segments are tucked inside the mouth at this stage.

it into itself by moving it along by one segment. The pattern also has polarity because of the consistent orientation of the head–tail differences of each segment: if the embryo looked at itself in the mirror, it would see another embryo looking in the opposite direction with the polarity of each segment flipped around to point the other way.

As we have seen in previous chapters, a good way of trying to understand biological patterns is to study mutants in which they are altered. In this case, we would be looking for mutants with alterations in some aspect of their translational symmetry. In the late 1970s, Eric Wieschaus and Christiane Nüsslein-Volhard were looking through thousands of fruit fly embryos, trying to find exceptional cases that deviated from the norm in interesting ways (Chapter 9). As they went through these screens, they noticed a striking class of mutants. These had the normal number of segments but instead of having a head–tail polarity, each of the segments was symmetrical about its middle (the segments were also shorter). For instance, the thorny outgrowths, typical of the head region of each segment, might be repeated as a mirror image, replacing the tail region, so that the embryo became equivalent to a stack of double-headed coins (Fig. 15.10). It was as if the polarity of the pattern had been lost; Nüsslein-Volhard and Wieschaus therefore named these *segment polarity mutants*.

It is important to bear in mind that these mutants do not affect the variation between the segments: segment 4 can still be distinguished from segment 5 even though each segment is now made up of two symmetrical halves. In a similar way, you could imagine replacing a stack of coins with different denominations by a stack of double-headed coins, while leaving the denominations unchanged. Segment polarity mutants affect the theme but not the variations. I first want to deal with how the repeated theme arises, and then I shall return to the question of how it is related to the variations.

Many of the segment polarity genes are expressed (switched on to make RNA

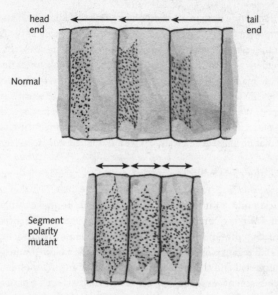

Fig. 15.10 Three segments from a normal fruit fly embryo at a late stage compared to a segment polarity mutant showing double-headed segments.

copies) in a repeating pattern in the early embryo, corresponding to the pattern of segments that will form later on. Figure 15.11 shows a typical example, with a segment polarity gene being expressed in stripes at regular intervals along the embryo. The basic repeating pattern is illustrated below the embryo. The region depicted by each of these stripes will eventually form part of a segment, say part of its tail end. Another segment polarity gene might have an expression pattern that is shifted along slightly, such that each stripe corresponds to a different region of each segment, say part of its head end. In other words, before the segments become visibly manifest, the embryo already contains a series of gene

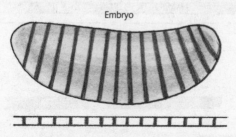

Fig. 15.11 Expression pattern of a segment polarity gene in an early fruit fly embryo, with the basic repeating pattern shown below.

activities that divide it up in a corresponding repetitive pattern. This pattern is normally hidden from view but can be revealed by looking at the expression pattern of the segment polarity genes.

As with many of the other genes we have come across that affect fundamental symmetries, many of the segment polarity genes code for hidden colours (master proteins), associated scents (signalling proteins) or sensitivities (receptor proteins). We can think of the normal role of these genes as establishing a repeating pattern in the early embryo. To see how this might be related to the segments that form later on, I want to give a simplified story using three hidden colours.

Imagine the early embryo contains repeating stripes of three different coloured regions (Fig. 15.12, top). One set of three colours makes a repeating unit, corresponding to the length of one segment. To make things easier to remember, I will use colours with the same initial letter as the corresponding region of each repeating unit: *hazel* (head region), *mauve* (middle region), and *tangerine* (tail region). These hidden colours provide a basic frame of reference that is interpreted and elaborated, eventually becoming manifest in the anatomy of each segment. As we saw earlier in this chapter, such a pattern of three colours has translational symmetry with polarity: it can be moved into itself by shifting it along by a repeating unit, but has no planes of symmetry at right angles to this.

Now suppose we have a mutant in which one of these hidden colours, say tangerine, is lost (Fig. 15.12, bottom). We end up with a repeating pattern of only two colours, hazel and mauve. As we saw earlier, a pattern of only two alternating colours still has translational symmetry but lacks polarity: a plane of reflection symmetry passes through the middle of each region of colour. By removing one hidden colour, tangerine, we have increased the degree of

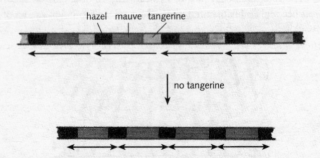

Fig. 15.12 Basic repeating pattern of three hidden colours in a normal embryo (*top*) compared to a mutant in which one colour (tangerine) has been lost (*bottom*).

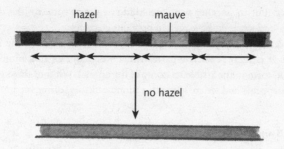

Fig. 15.13 Mutant in which hazel is lost as well as tangerine to give a pattern of uniform mauve.

symmetry because there are now more ways of transforming the pattern into itself. The same symmetry will also become evident in the anatomy of the segments that develop by interpretation of this pattern. This is what happens in the mutants with double-headed segments: they lack a key element in the basic pattern of hidden colours and so develop with the corresponding symmetry.

What would happen if a second hidden colour was lost? This is illustrated in Fig. 15.13, where hazel has been lost as well as tangerine, so now there is only mauve all the way along. You might expect such a mutant, with a single colour, to give rise to an animal with uniform segments. However, the change is even more fundamental than this. This is because the anatomy of the animal, including the production of visibly distinct segments, depends on the interpretation and elaboration of hidden colours. By eliminating the repetitive pattern of hidden colours we would also get rid of any anatomical repetitions, such as the development of a visible segment boundary at regular intervals. In other words, there would no longer be any segments to speak of because the boundaries between segments are themselves a later manifestation of the repetitive pattern of hidden colours. This is essentially what happens in mutants in which all of the segment polarity genes are inactive: the embryos lack clear repetitive structure and remain unsegmented.

In mutants with only one hidden colour all the way along, it may look as if there is no repetition left. Strictly speaking, though, we have ended up with a *higher* degree of translational symmetry. A uniform line can be transformed into itself by shifting it along by any amount (assuming it is infinitely long). It is also unchanged by a reflection in any plane at right angles to it. The loss of hidden colours from our original pattern therefore leads to an increase in symmetry: loss of one colour leads to more reflection symmetry (lack of polarity), loss of two colours to an increase in both translational and reflection symmetry. Put in another way, the hidden colours increase the degree of *asymmetry*, much as we have seen in previous chapters.

The segment polarity genes therefore provide the basic theme behind a segment. It is their repetitive pattern that we recognise in a modified and elaborated form in the visible anatomy of the animal. Without these genes, the theme disappears and we are left with a monotonic organism.

Combining themes with variations

So far I have been concerned with the repeating theme of hidden colours that are common to segments (hazel-mauve-tangerine). But we have seen in previous chapters that variation between segments also depends on a set of hidden colours: the various types of green that distinguish segments from the head to the tail of the embryo (Chapter 6). To see how these two aspects, themes and variations, interact with each other, we need to combine their respective colours. I have illustrated this for three segments in Fig. 15.14, where the two patterns of hidden colour are simply superimposed to give an overall patchwork. By combining the colours in this way, each segment gets a distinctive overall type of green, while at the same time there is a common pattern to all segments set by the repeating sequence of hazel-mauve-tangerine. Mutants lacking the green colours have segments, but all of them are identical (i.e. they have a theme without variation). Conversely, mutant embryos that lack the repeating colours do not have segments, but still show variation from head to tail (i.e. they have variation without a theme).

The notion of themes and variations seems much more straightforward when

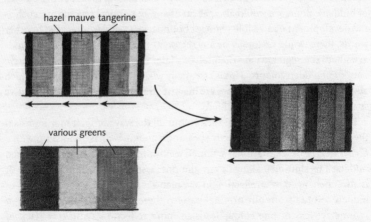

Fig. 15.14 Combining theme and variations for three segments.

understood in terms of overlapping hidden colours than when we look at an adult fly. This is because the adult is a much later manifestation of this underlying pattern. The patchwork provides a frame of reference that is interpreted and elaborated, eventually becoming manifest in the anatomical features of the fly. However, during this process, the simplicity of the underlying pattern can be extensively transformed so that we perceive it much more dimly and indirectly in the final anatomy. In some cases the transformations can be so extensive that the segments become hardly recognisable at all: the head of a fly seems very different from the rest of the animal, yet much of it is derived from three of these repeating units. The themes and variations we see in the adult are a highly distorted view of a much earlier simplicity.

From rainbows to zebras

For this system to work, a theme and its variations need to be tightly coordinated. The two overlapping sets of hidden colour have to be interwoven in a precise way so that the variations and theme can be correctly superimposed. How is this precise coordination achieved?

I have already described how the various types of green are established (Chapter 9). Recall that the green colours themselves depend on an earlier pattern of rainbow colours that divide the embryo from head to tail into broad regions of red, orange, yellow, green, blue, indigo and violet. The rainbow gives a progressive change from head to tail, and this information is interpreted to give various types of green from one end to the other (apple-green, bottle-green, cyprus-green, etc.). This seems relatively straightforward because we are using one pattern that progresses from head to tail, the rainbow, to generate another pattern that progresses in a similar way, the various greens.

What about the repeating pattern of hazel-mauve-tangerine; where does that come from? This also depends on interpreting the rainbow pattern of hidden colours, but in a rather different way. To illustrate how a progressive rainbow can be used to give a repeating pattern, I will give a picture that is rather simplified but that nevertheless explains many of the basic principles involved.

Like a true rainbow, the hidden colours of the early embryo are not separated by sharp boundaries, but gradually merge from one to the other. The hidden colours overlap with each other, so that where yellow overlaps with green, for example, you get a combined yellow-green colour, because both the yellow and green master proteins are there together. This means that in addition to seven regions of what might be called pure colour, there are also six regions of mixed colour, where the colours from adjacent regions overlap. To see how this fuzzy rainbow of colours can be interpreted to give a repeating pattern, I first need to remind you briefly about how genes interpret hidden colours.

Recall that each gene is divided into two parts: a regulatory region containing sites where master proteins can bind, and a coding region carrying the information used to make a protein (Chapter 5). The way a gene interprets a pattern of hidden colours depends on the combination of binding sites in its regulatory region. This will determine where the gene is switched on or off.

To make a repeating pattern, we will need an interpreting gene with binding sites for the rainbow colours. Recall that a binding site is named according to the hidden colour that recognises it: red master protein binds to an R-site, orange to an O-site, yellow to a Y-site, etc. I want to consider an interpreting gene with *all* of these sites strung together in the regulatory region: an R-site, O-site, Y-site, G-site, B-site, I-site and V-site (Fig. 15.15). We also need a rule for how the master proteins affect the activity of the gene. In this case, the rule will be that the gene only gets switched on *if two master proteins are bound at the same time* (recall that such rules depend on how these master proteins interact with each other and with other proteins bound to the regulatory region). This means that the gene will only be active in regions where the colours overlap with each other: the gene would not come on, for example, in the pure yellow or pure green regions because they only contain one master protein, but it would come on in the yellow-green region because it would have

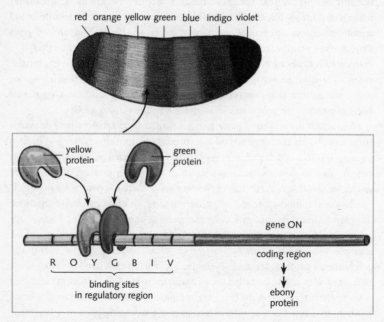

Fig. 15.15 Interpretation of a rainbow by a gene for ebony.

both the yellow and green proteins bound to it. The overall result would be that this gene would come on at every one of the six regions where the colours overlap. (A key piece of experimental evidence in support of this type of model is that if many of the binding sites are deleted, to give a regulatory region with a more limited number of sites—say only a Y-site and a G-site—the interpreting gene only gets switched on in one stripe, where yellow and green overlap.)

Now suppose that this interpreting gene itself produces a hidden colour, say *ebony*, from its coding region. This means that ebony will be produced in all the regions where the rainbow colours overlap: we will end up with a repeating pattern of ebony stripes at regular intervals along the embryo (Fig. 15.16). The progressive rainbow pattern has been interpreted to give a repetitive pattern of hidden colour. In a similar way, we may introduce a second gene that codes for a different hidden colour, say *ivory*. Like the gene for ebony, it also has binding sites in its regulatory region for all the rainbow proteins. But in this case, the rule is different: the gene for ivory only gets switched on when *only one master protein is bound*. If more that one protein is bound, the gene for ivory gets switched off rather than on. This gene will therefore only be switched on in the seven regions of pure colour, and will be off in the regions of mixed colour. It will paint a pattern of ivory stripes alternating with ebony stripes. The rainbow has been interpreted to give a zebra.

The genes for ebony and ivory correspond to another group of genes that

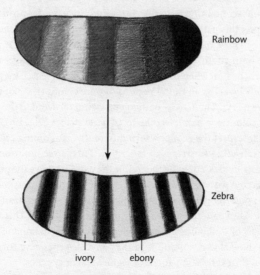

Fig. 15.16 Interpretation of rainbow to give a zebra pattern of alternating stripes in the embryo.

affect early embryo development in fruit flies, *pair-rule* genes (so called because they affect alternate segments along the embryo). The discovery of these genes was a major breakthrough because they provided a way to link the progressive pattern of the rainbow with a repeating pattern. When these genes were isolated, the zebra pattern of alternating stripes could be revealed by staining the embryo (there are actually seven stripes of both ebony and ivory rather than the six stripes of ebony in my simplified model). I vividly remember the first time I witnessed these alternating stripes. I was attending a symposium in 1983 and towards the end of one session, the chairman announced that there were some exciting new results. An unscheduled speaker, Ernst Hafen (working in Walter Gehring's lab in Basel, Switzerland), jumped to the stage and showed a slide of an elliptical embryo with 7 beautiful zebra-like stripes of gene activity running down it, glowing in the dark. There was an audible gasp in the audience. For the first time, we could see that what seemed like a featureless embryo, showing no physical signs of repetition, was actually subdivided into regular repeating regions of gene activity.

It is relatively straightforward (in principle) to get from the alternating pattern of ebony-ivory in the embryo to the stripes of hazel-mauve-tangerine. The main difference between these patterns is that there are twice as many hazel-mauve-tangerine repeats (14) as individual ebony or ivory stripes (7). Getting from one pattern to the other depends on further interpretations and elaborations. Imagine that the ebony and ivory regions overlap to give a series of grey regions, containing both the ebony and ivory master proteins. A gene might interpret this pattern such that it was only switched on in the regions of pure ebony or ivory, but not in the grey regions in between (that is, the gene would have both ebony and ivory binding sites in its regulatory region, but would be switched on when only one of these was occupied). This gene would therefore only be active in the 14 regions of pure colour (i.e. 7 of pure ebony and 7 of pure ivory). The activity of this gene may in turn lead to the production of a hidden colour, say hazel, leading to 14 hazel stripes (Fig. 15.17). Another gene might interpret the zebra pattern differently, say coming on only in the grey regions, leading to stripes of tangerine. Further elaborations, through scents and sensitivities, could further refine the pattern to give the precise hazel-mauve-tangerine pattern. This is essentially the way the expression pattern of the segment polarity genes is established, through interpretation and elaboration of the zebra pattern established by the pair-rule genes. (This was verified experimentally by Philip Ingham and colleagues working in London in the mid-1980s, who showed that mutations in the pair-rule genes gave corresponding alterations in the expression pattern of segment polarity genes.)

To summarise, two different patterns can be traced back to the rainbow. On the one hand, interpretation of the rainbow via the zebra pattern leads to the

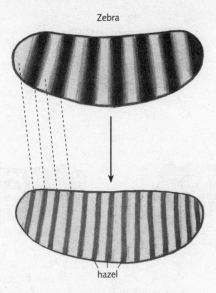

Fig. 15.17 Interpretation of zebra to give a pattern of 14 hazel stripes.

repeating pattern of hazel-mauve-tangerine stripes (Fig. 15.18, right). On the other, the rainbow provides the frame of reference for the different types of green, providing variation between segments (Fig. 15.18, left). That is to say, the rainbow pattern is interpreted along two routes. One gives a repeating pattern of hazel-mauve-tangerine stripes, the common theme that is depicted in every segment. The other route gives the progressive change in segment identity from the head end to the tail, providing a distinct green colour for each segment.* Although the two routes are shown as separate, they actually occur concurrently in the embryo as it develops, with some cross-talk between them that ensures theme and variation are kept in register with each other. The key point is that because both the variation and theme are an elaboration of a common fundamental pattern, the rainbow, their superimposition is automatically coordinated.

Implicit themes and language

We are now in a position to look again at implicit biological themes. By examining an adult fly, we get an overall impression of a repeating structure or

*You may notice that the different greens cannot account for all the variation between segments, because there are fewer types of green than segments. Some of the variation between segments depends on other factors, such as the intensity as well as the type of green.

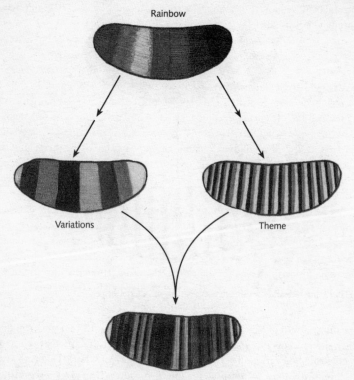

Fig. 15.18 Common source of theme and variations.

theme, the segment, together with many variations upon it. Yet it is difficult to define the precise nature of this theme independently of the variations, because any individual segment will always be a combination of the two. If we try to define the theme explicitly, as a visible segment, we would have to say what sort of segment it was, but in doing so we would end up by describing only one particular variation, not the general theme.

There is no way out of this impasse so long as we try to define the theme purely in terms of the appearance of the organism. But as we have seen, if we look at how the segments arise during development we can define the theme in a different way: as a repeating pattern of hidden colours that provides an early frame of reference for all segments, irrespective of their variations. This theme provides a common pattern that is then interpreted and elaborated in combination with other colours, eventually becoming manifest in the anatomy of the organism. We can only see the more deeply seated theme by looking through a special window that reveals the activity of particular genes and their products. If we look at the adult, this theme can only be discerned implicitly, through its elaborate manifestation in the visible organism.

Similar considerations apply to other repeated themes, such as the different limbs of a vertebrate or the various organs of a plant. For example, related patterns of gene activity have been found in the developing wings and legs of birds. Similarly, researchers have started to find some common hidden colours in various developing plant organs, such as foliage leaves and floral organs.

To my mind, this brings us much nearer to understanding what lies behind Goethe's notion of a *leaf*. By comparing different plant organs, Goethe discerned that each was a different manifestation of an underlying theme. But the nature of this theme was inaccessible to him. It was implicit rather than something that could be pointed to. As a result, his notion sounded mystical and obscure.

But we can now see that this follows automatically from the nature of development. The visible structures on a plant are based on the interpretation and elaboration of hidden patterns; patterns that can only be revealed by studying the activity of genes. In some cases, as with many of the foliage leaves on a plant, the same hidden patterns can be repeated and interpreted with little or no variation, so we end up with a series of very similar-looking organs. But in other cases, as with different types of foliage and floral organs, variation is superimposed on the repeating patterns so that they become manifest in different ways. This variation is itself another pattern of hidden colours, such as the various types of red that distinguish the whorls of flower organs. In these cases, repetitive themes may only be dimly perceived by looking at their visible manifestation in plant organs. That is to say, the theme is not a plant organ of any description; it is something that can only be clearly conceived of at a different level, in terms of processes that are normally hidden from view.

It seems to me that once this is appreciated, we can look at other types of implicit theme from a new perspective. Take the meaning of words, for example. What do we mean by a word like 'chair'? You might say that 'chair' means an object with a seat, back and legs. But this would simply raise the question of what we mean by 'seat', 'back' and 'legs'. Instead of one word, we now have to define the meaning of three. Each of these would in turn have to be described in more words, so we end up with an infinite regress. One way out of this problem might be to say that we know the meaning of 'chair' not through a verbal definition but by forming a picture of it in our mind. If you were asked what a chair is, you quickly imagine what one looks like. But the problem here is that when you try to picture something in your mind, such as a chair, you always see one particular example, a certain type of chair. Yet the notion of a chair applies to all chairs, not just one, so how do we know the meaning of the word 'chair' in general as opposed to individual cases? How can we imagine the shape that is common to all chairs, rather than one particular instance? The problem is the linguistic equivalent of Goethe's difficulty with defining the general notion of *leaf*, as the philosopher Ludwig Wittgenstein alluded to in his book *Philosophical Investigations*:

> So if I am shewn various different leaves and told 'This is called a "leaf"', I get an idea of the shape of a leaf, a picture of it in my mind.—But what does the picture of a leaf look like when it does not shew us any particular shape, but 'what is common to all shapes of leaf'?

In other words, as soon as we try and define the meaning of the word 'leaf' more precisely, by conjuring up a mental picture of one, what we see in our mind will be one particular leaf, not the general meaning. You might picture a simple broad flat green apple leaf, or a leaf of grass, or a pine needle, but these are just examples, variations, not the general notion itself. Wittgenstein was using the example of a leaf to make a general point about language: whenever we try and conjure up a picture to represent a word that we use, the picture will always be only one example rather than the general concept. Wittgenstein could have chosen any common object, such as a chair, to make the same point, but he probably chose a leaf because he saw the relationship between this problem and Goethe's ideas on plants.

Although Wittgenstein was concerned with the nature of the word 'leaf' in the *mental* rather than biological sense, we can get a general insight into the problem he raised based on what we have learned in this chapter. We have seen that in the case of plants, the common unity of plant organs is not a visible leaf at all, but something that can only be described in other terms, such as an underlying pattern of hidden colours. By analogy, we might expect that common mental themes, our general notion of what 'leaf' or 'chair' means, could not be pictured in our mind's eye, but would lie hidden from us. In this case, the theme might reflect a common pattern of brain activity, rather than hidden colours. For example, every time we think of or use the term 'chair', a common set of brain connections could be triggered, irrespective of the individual type of chair involved. This common pattern would itself have been established and elaborated upon during previous responses of our brain to the various situations in which the word 'chair' was used, during which we eventually learned the meaning of the word. The important point is that we would be oblivious to this underlying pattern and would only be conscious of it indirectly, through its manifestation in particular cases, such as when we try to imagine what a chair looks like. The particular type of chair a person imagines may vary according to his or her state of mind, but all such chairs may nevertheless share some underlying patterns of activity in the person's brain. So although we have an implicit notion of a common meaning of the word 'chair', we cannot access the underlying unity directly by introspection. As soon as we try to analyse it, by, say, visualising something, we are stuck with a particular case rather than the underlying theme—much as when we look at a mature plant organ we see one particular manifestation rather than the common theme. It seems to me, then, that by getting a clear understanding of how repeated themes are elaborated during biological development, we can get a useful intuition into how analogous processes might operate in other cases, including the workings of our mind.

We have seen how both the idiosyncrasies and repetitions evident in the anatomy of organisms depend on hidden processes. Development involves the continual interpretation and elaboration of hidden colours, scents and sensitivities, giving an internal patchwork that underlies the final geometry we see. If one or more of these hidden components is lost through mutation, there is a corresponding change in the geometry of the organism that develops. But there is still the issue of how this process of internal painting eventually becomes manifest in the visible structure of each plant or animal. How does a pattern of regional distinctions actually influence the shape and form of the organism as it grows? This is the problem I wish to explore in the next chapter.

CHAPTER 16

Shifting forms

A striking feature of paintings of the Byzantine period is their curious sense of proportion. Look, for example, at the Madonna and Child of Fig. 16.1. The head of a child is normally bigger in relation to its body than that of an adult; yet in the painting the child is shown with adult proportions, making its head seem unnaturally small. The reason for the difference in proportions between child and adult has to do with the way humans grow. A child's head grows much more slowly that the rest of its body, so that by the time adulthood is reached, relative head size is greatly diminished. The Byzantine artist seems to have taken little account of this, so that the child looks strangely mature for its age.

Not only does relative head size change during growth, but so do facial proportions. Eyes, for example, hardly enlarge at all as a child grows, so they become progressively smaller in relation to the rest of the face. This change in relative size is emphasised by Magritte's painting, *The Spirit of Geometry* (Fig. 16.2) in which he swaps around the head of a baby with that of its mother, keeping overall head size to scale. The head on the mother now seems to have a disconcertingly large pair of eyes, highlighting the distinctive facial proportions of a baby.

These differences in proportion between baby and adult illustrate how the regions of an organism grow to different extents as it develops. Were it not for this, the adult would end up looking just like an enlarged version of the embryo, or going even further back, like an inflated fertilised egg. If we compare the growth of a developing organism to an expanding three-dimensional canvas, we therefore need to imagine that some regions of the canvas grow more quickly than others. It is a dynamic canvas of continually changing size and proportion. But I have also been saying throughout this book that the shape of an organism depends on the interpretation of hidden colours on the canvas. There seems to be a contradiction here. On the one hand, I am saying that an organism's shape reflects the growth of the canvas; on the other, that it depends on the hidden colours on the canvas. How can both of these views be correct?

This brings us to an important distinction between human and internal painting. For an artist, the canvas provides an independent support to which colours are applied. However, for internal painting, the canvas is not an independent support but something that continually changes and interacts with the hidden colours. We shall see in this chapter how this interaction provides the

Fig. 16.1 *Enthroned Madonna and Child* (detail). Probably painted in Constantinople about AD 1200. National Gallery of Art, Washington, DC.

fundamental link between hidden colours and the manifest anatomy of multi-cellular organisms. I should say in advance that this is perhaps the hardest aspect of development to think about and one that is still poorly understood. This chapter will therefore tend to be more speculative than the others. Nevertheless, we shall be able to look at some of the key issues involved.

Deformations

In order to understand how an organism grows, we need to be able to relate the successive shapes it goes through as it develops. One of the best ways of

Fig. 16.2 *The Spirit of Geometry* (*c.*1936), René Magritte. Tate Gallery, London.

conveying the relationship between two shapes is to imagine ways of *deforming* one of them, by say stretching or compressing it in various directions, so that it comes to resemble the other. To give a very simple example, a square may be transformed into a rectangle by stretching it out in one direction. The relationship between these two shapes, square and rectangle, can therefore be summarised by this basic deformation, a linear stretch.

An early illustration of how this approach can be applied to more complex shapes, of the sort encountered in biology, is found in Albrecht Dürer's *Four Books of Human Proportion*, published in 1528. In this work, Dürer describes how the human body should be depicted. He first gives the detailed proportions of what may be considered to be a standard human being. Having established this canonical form, he goes on to give a general method for producing deviations:

> Herein will I show how the proportions given above may be altered and changed according to each man's will, the parts being lengthened or shortened so that the figure can no longer be recognised and becometh quite different from its original form.

Fig. 16.3 Standard head (*left*) compared to three different deformations (after Dürer).

The key to his method was to alter human proportions by stretching or compressing particular regions. Dürer illustrated his method by showing how a normal head inscribed in a grid could be deformed in various ways by changing its relative proportions, as if the picture were on a piece of deformable rubber (Fig. 16.3). By following his approach, he believed artists might be able to explore a whole new repertoire of facial types and expressions, ranging from the beautiful to the ugly or monstrous. We can recognise a less formal version of this method in many caricatures in which certain facial features, such as a nose or mouth, seem to have been greatly stretched out or magnified.

The use of deformations for relating shapes was applied more generally in the early twentieth century by the biologist D'Arcy Thompson, in his monumental book *On Growth and Form*. Look at the top left of Fig. 16.4, which shows a species of fish inscribed within a rectangular grid. By deforming the grid in different ways, Thompson could obtain shapes that came pretty close to other fish species that existed in nature, such as those shown in the rest of the figure. The importance of this method was its simplicity and economy. Without the grid, all the fish species in Fig. 16.4 would look very different on several counts, such as the shape of the different body parts and the various types of fin; but with the benefit of a deformable grid, all of these aspects are seen to flow from much simpler overall deformations. The method allows seemingly unrelated and complex differences between two species to be brought under the umbrella of a single geometrical operation. Thompson applied the same method to illustrate the relationships between many other forms, such as the shapes of different crab shells, or the skulls of related crocodiles. In each case, he inscribed one shape on a regular grid and obtained related forms by simply deforming the grid, bringing apparently complex differences under a unified scheme.

Thompson realised that his method worked best when comparing closely related forms, where the correspondences between different animals could be easily identified. He therefore applied it to cases where it was possible to make easy and meaningful comparisons, like closely related fish species, or the skulls of various crocodiles. The method could not be used in a convincing way to compare more remote shapes, like those of fishes with birds, where there were too many differences to be accounted for by a simple deformation. He pointed

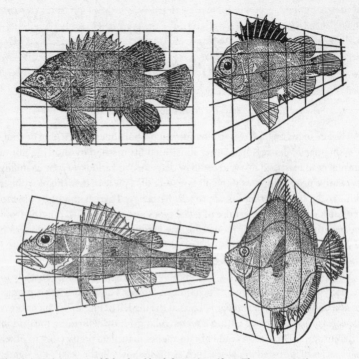

Fig. 16.4 Various types of fish related by deformations (from Thompson 1942).

out, though, that even with these more distantly related forms, a certain type of correspondence could still be achieved by an extension of his method:

> In these transformations of ours every point may change its place, every line its curvature, every area its magnitude; but on the other hand every point and every line continues to exist, and keeps its relative order and position throughout all distortions and transformations. A series of points, *a, b, c,* along a certain line persist as corresponding points *a', b', c',* however the line connecting them may lengthen or bend; and as with points, so with lines, and so also with areas. Ear, eye and nostril, and all the other great landmarks of cranial anatomy, not only continue to exist but retain their relative order and position throughout all our transformations.

He was saying that an essential feature of his method of deformations was that the relative order of parts remained the same. A simple stretch or compression does not change the order of landmarks on a skull, even though it may alter the distances that separate them. So when you come to compare more distantly related forms, you may still be able to discern the constant relations of their landmarks, even though you cannot draw a simple diagram to relate them. There is no simple way of deforming the skull of a mammal into that of a fish, bird or frog, but there is still something that links them: the relative order or arrangement of their landmarks, like ears, eyes and nose. We have come across this idea before: it is a restatement of Geoffroy's Principle of Connections (Chapter 7). Geoffroy's principle stated that the relative order of bones—the way they were connected to each other—was preserved in different skeletons, as in a human arm, horse's leg or bird's wing. This is precisely what Thompson was saying, except that he was using the more general term of *landmarks* rather than bones.

Now although Thompson's approach provides a very useful and immediate way of appreciating relationships, it has an important limitation. Because he is comparing different *adult* forms, the deformations he shows are indirect. The various fish species in Fig. 16.4 were not actually derived by physically stretching or compressing one form into the other. They are different adults that developed independently. As the biologist J. H. Woodger pointed out:

> For no one supposes that an adult crocodile's skull has ever been transformed into another adult crocodile's skull, or that an adult fish of a certain shape has ever been transformed into another adult fish of a different shape, even though the two shapes can be shown to be related in a particular way with the help of some mathematical transformation.

To get around this problem it is no good simply comparing adults, you need to look at how each adult itself develops. Fish have different shapes because of a difference in the way that each fish species grows from a fertilised egg. Several scientists in the 1930s and 40s who wished to apply Thompson's method concluded that it might be more useful for comparing different developmental *stages* of the same species rather than adults from different species. Look at Fig. 16.5, for example, which shows various stages of human growth inscribed on a simple grid, taken from a paper by Peter Medawar at that time (1945). By adjusting each stage to the same height, the diagram shows how the upper part of the human body, including the head, grows much more slowly in relation to the lower parts. It is a more formal way of describing the changes in proportion we encountered at the beginning of this chapter. Medawar is using the method of deformations to relate various developmental stages within one species, humans in this case, rather than to compare adults from different species. This means that the deformations he shows are those that actually occur as a human grows up. As far as Thompson's method was concerned, Medawar wrote: 'There can be no doubt . . . that its true field of application lies in development, and not in evolution; in the process of transforming, and not in the *fait accompli*.'

Even when applied to development in this way, however, the method of

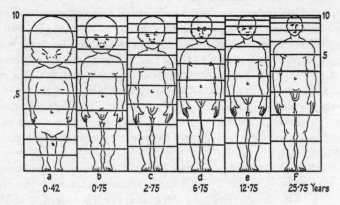

Fig. 16.5 Different stages of human development, adjusted to the same size (from Medawar 1945).

deformations still has a fundamental limitation. It is relatively easy to follow the growth from foetus to adult, because the essential landmarks, such as ears, eyes and nose, are already present at the foetal stage. All we have to do is measure the relative distance between these pre-existing features as the baby grows. But what about earlier stages, before the landmarks have appeared? Without ears, eyes or nose, it is obviously not possible to make any measurements on their relative positions: comparisons become meaningless. You cannot extend the human measurements all the way back to the fertilised egg because the landmarks are simply not there. Much of development is about the origin of landmarks, the production of greater complexity as the organism grows. Thompson's method assumes a constant level of complexity and therefore ignores a fundamental aspect of development: the increase in geometrical complexity. As Medawar himself concluded: 'the most important problem of all is so far wholly out of reach.'

Growth and hidden colours

It seems to me that many of the problems encountered with the method of deformations stem from the fact that it was applied to visible features rather than to underlying processes; it gives a limited description of the outcome of development, without addressing the mechanisms behind it. It is instructive to look at how D'Arcy Thompson tried to address this issue. He sought analogies between development and the way humans modify or transform shapes, such as a glass-blower gradually working a piece of glass into a particular form. Another example he gave was the way humans could influence biological forms, such as the shape of a gourd (a fruit of the same family as melons and pumpkins):

> The little round gourd grows naturally, by its symmetrical forces of expansive growth, into a big, round, or somewhat oval pumpkin or melon. But the Moorish husbandman ties a rag round its middle, and the same forces of growth, unaltered save for the presence of this trammel, now expand the globular structure into two superposed and connected globes. And again, by varying the position of the encircling band, or by applying several such ligatures instead of one, a great variety of artificial forms of 'gourd' may be, and actually are, produced. It is clear, I think, that we may account for many ordinary biological processes of development or transformation of form by the existence of trammels or lines of constraint, which limit and determine the action of the expansive forces of growth that would otherwise be uniform and symmetrical.

By this example, Thompson wanted to show how a simple physical restraint, like tying a rag round the middle of a growing fruit, can result in a transformation in shape: from a sphere to a dumb-bell. The constriction imposes an asymmetry on the growing system which results in a more elaborate form. Now the key asymmetry here is imposed from *outside* the organism, by a Moorish husbandman who ties the rag and therefore imposes a restriction on the gourd. It seems to me that Thompson chose an external asymmetry because during his time there was no clear understanding of how internal asymmetries could arise or be elaborated. There was no picture of hidden colours, scents or sensitivities that might have helped him to link observable deformations in geometry with internal processes. I now want to try and show how such a link might be made.

We shall begin with a hypothetical layer of cells with a very simple patchwork of hidden colours (Fig. 16.6). Each of the hidden colours will be interpreted by genes, leading to particular genes being switched on or off in the various

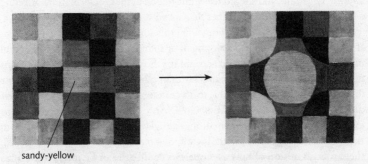

sandy-yellow

Fig. 16.6 Deformation of a patchwork by interpretation of hidden colours.

regions. I want to concentrate on one particular region in the centre of the patchwork, with a hidden colour that I shall call *sandy-yellow*. Now suppose that the sandy-yellow colour specifically leads to the activation of genes that promote cell growth and division. That is, the proteins produced by these genes modify the chemical reactions in a cell in such a way that it grows and proliferates more quickly. This would mean that the sandy-yellow patch would start to expand more than the surroundings. To accommodate the expanding island of sandy-yellow, either some of the surrounding regions would have to be compressed or the island might bulge out of the plane. Either way, the patchwork would become deformed, as shown on the right of Fig. 16.6.

This type of deformation is vividly illustrated by M. C. Escher's lithograph, *Balcony* (Fig. 16.7, top). Below is shown Escher's original sketch of the scene before he applied the deformation. The central balcony, the fifth one from the bottom, seems tiny in comparison with the deformed version shown above (the final lithograph is a mirror image of the original sketch because of the inversion caused by lithography). The overall visual effect of the deformation is that the centre of the picture seems to bulge out towards you.

The changes in shape of a sheet of cells show how, in principle, a deformation can result from the interpretation of hidden colours. In the example I gave, the deformation only affected the internal regions of the sheet, leaving its perimeter unchanged. But you could equally well imagine a hidden colour influencing the shape of the outline as well. For instance, if one of the squares at the edge of the sheet had a hidden colour leading to rapid growth, this region would grow more than its neighbours, tending to displace one edge outwards, deforming the outline. The key point is that the sheet of cells, the canvas, both supports and is modified by the colours. It would be as if every time an artist put a colour on a canvas, the newly coloured region might start to grow or shrink relative to the rest of the picture. The final shape of the painting could end up being rather complicated, simply because of the hidden colours that are applied. Of course, you should really imagine all of these deformations affecting a three-dimensional canvas rather than a two-dimensional sheet.

It is important to be clear about what this sort of explanation has achieved. If you were to observe the development of an organism from a purely external point of view, you might try to account for the various deformations in terms of some regions growing more quickly than others. But you would be at a loss to explain why this was so; unable to account for what makes one region behave differently from another. That is why D'Arcy Thompson was driven to use analogies with artificial deformations caused by imposing asymmetries from the outside, like a glass-blower at work, or a man tying a rag around a gourd: there was no notion of how asymmetries could arise or be interpreted from within. But we have seen in previous chapters how distinctions between regions

Fig. 16.7 *Balcony* (1945), M. C. Escher, compared to preliminary sketch (*below*).

are generated during development through the elaboration of hidden colours. These colours provide an internal frame of reference that underlies the geometry we see: if one or more of these colours is lost through mutation, the external appearance of the organism undergoes a corresponding change. Furthermore, we have seen how this hidden patchwork can be interpreted by genes through their regulatory regions, allowing each gene to be expressed in a specific regional pattern. Once we have this internal view of the organism in mind, it is no longer difficult to see why one region might grow differently from another: it reflects the way the hidden patchwork is interpreted by genes that influence growth.

As well as accounting for differential growth, the patchwork of hidden colours also provides a set of landmarks: each region of colour can be thought of as a hidden landmark, equivalent to an anatomical feature used by D'Arcy Thompson. The patterns of growth may displace and deform the boundaries between these hidden landmarks, but their relative order remains the same, just as with anatomical landmarks. Unlike anatomical landmarks, however, which can only be traced back to the time when they first become visible, we have seen how the patchwork of hidden landmarks is anchored in earlier events, through the process of internal painting. If we simply observe the development of a fertilised egg from an external point of view, anatomical landmarks such as ears, eyes or nose seem to appear from out of the blue, representing a mysterious increase in complexity. But this is not the case when we look at development from within: the internal landmarks of hidden colour are firmly grounded on earlier processes of interpretation and elaboration. Indeed, the anatomical landmarks are themselves a later manifestation of these internal processes: the emergence of ears, eyes or nose in particular positions depends on the interpretation of the patchwork of hidden colours. This means that the relatively constant pattern of anatomical connections noted by D'Arcy Thompson and Geoffroy Saint-Hilaire is itself a reflection of the preserved order in the underlying pattern of hidden colours.

Nevertheless, in spite of this considerable advance in our understanding of development, there is still an important gap in our knowledge. Although many genes affecting the growth and division of cells have been identified, it is not clear precisely how they respond to or interpret patterns of hidden colour, scents and sensitivities, so as to change the size and shape of particular regions in the developing organism. I believe that one reason for our present ignorance may be that these genes act in a quantitative way, slightly increasing or decreasing the rate of growth here or there; making them more difficult to pin down than the genes that establish the patchwork. One of the major challenges facing biologists is to try and understand precisely how genes influence growth in response to the developing patchwork of hidden colours, scents and sensitivities.

Direction of growth

In my hypothetical example, I assumed that each region of the cell layer grew in a symmetrical fashion, but it is also possible for growth to be oriented in particular directions. A good illustration of this comes from a study on the growth of gourds carried out in the 1930s by Edmund Sinnott at Columbia University. Gourds come in many different shapes and sizes. A commonly grown type, known as the bottle gourd, has a broad base and narrow waist, resembling the shape of a bottle. Sinnott was comparing two different races of bottle gourd: a *miniature* form that was only about 10 cm long at maturity, and a *giant* form that was more than twice this length (Fig. 16.8, top). Although both races have an overall bottle shape, the giant form is wider in proportion to its height than the miniature form. This is shown more clearly in the bottom part of Fig. 16.8, where the gourds have been adjusted to the same height.

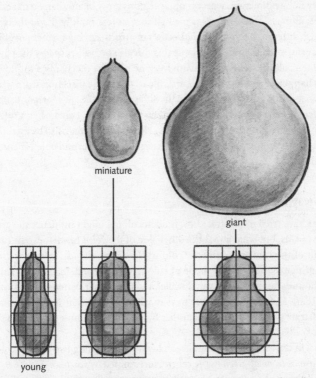

Fig. 16.8 Top shows a comparison of miniature and giant gourds. Bottom shows how their shapes compare to a gourd at an earlier stage of development (adjusted for size).

To see why there was such a difference in relative width between these races of bottle gourd, Sinnott looked at the way each shape developed, making measurements at various stages of growth. He found that young gourds were very narrow in proportion to their height, but gradually made up for this with time because they grew faster in width than length. I have illustrated this in the bottom part of Fig. 16.8 by inscribing a young gourd in a grid (left) and comparing it to the mature forms, adjusting for overall size. The gourd gradually gets fatter with time due to its growing more quickly in width than height. This means that the bigger the gourd, the greater its relative width will be. The reason the miniature is relatively narrower than the giant form is simply that it reaches maturity at a smaller size; were it to continue growing it would eventually end up having exactly the same shape as the giant form.

This example shows how growth can be oriented preferentially in certain directions. In this case, the whole gourd was growing faster in width than height; but you could also imagine some regions of an organism growing preferentially in one direction, whereas other regions grow in another. So in addition to certain parts of the canvas growing more or less rapidly than others, they may also differ in their preferred direction of growth, giving further possibilities for transforming shape. Presumably this can also be influenced by the patterns of hidden colour, scents and sensitivity. One possible mechanism for how this could happen might be that cells are able to detect the direction from which a scent is coming, by responding to slight differences in the concentration of scent at different ends of a cell. This might then allow cells to orient their growth in relation to directions of internal scents. However, it remains to be established whether this and/or other mechanisms are actually responsible for directed growth.

Two-way interactions

So far I have tried to illustrate how hidden colours might influence the growth of the canvas, but what about the other side of the coin: to what extent can the growth of the canvas modify the colours? We have already seen how growth may influence the size and shape of coloured regions, deforming or shifting their boundaries in various ways, while preserving their pattern of connections (Fig. 16.6). But there are further ways in which growth may influence matters. To illustrate this, I want to give another hypothetical example involving a single layer of cells.

We will start with a very simple patchwork, comprising two hidden colours: an island of *tropical-green* in the centre, surrounded by *sea-blue* (Fig. 16.9, left). We will assume that the sea-blue cells are giving off a particular scent, the smell of the sea air. Although all the cells of the island are initially tropical-green, they

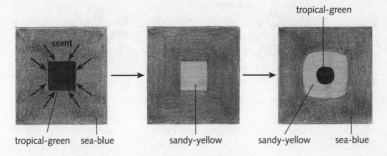

Fig. 16.9 Deformation of a patchwork leading to the production of an additional hidden colour.

eventually turn *sandy-yellow* in response to the sea scent (Fig. 16.9, middle). The scent penetrates all the way into the island, so all of its cells change colour in this way. Now once the island turns sandy-yellow, it starts to grow more quickly (genes promoting growth are switched on by sandy-yellow), deforming the patchwork just as we saw before. But in doing this, the island may expand so much that its most central region is no longer exposed to much sea scent. That is, the island has become so big that the scent no longer penetrates all the way to its interior. Being sheltered from the scent, the central region of the island might therefore revert to its original colour, tropical-green, as shown on the right of Fig. 16.9. We get a final pattern of three hidden colours: sea-blue, sandy-yellow and tropical-green.

Now the key point is that the pattern of colours we end up with depends on the growth of the canvas: if the island of sandy-yellow cells had not grown, there would have been no internal region far enough away from the sea to produce tropical-green. There is a two-way interaction between the canvas and the colours: the colours modify the growth of the canvas, while the growth of the canvas can in turn influence which colours appear.

The examples I have given are mostly hypothetical because we still know very little about the detailed interactions between hidden colours and the rates or direction of growth. I have given a few possibilities but they are by no means the only ones that could be envisaged. Nevertheless, they serve to illustrate how hidden colours and canvas are not independent but can continually build upon and respond to each other during development.

Cell death

So far I have emphasised the role of growth, but there is another way that the shape of the canvas can change: through parts of it withering away and dying.

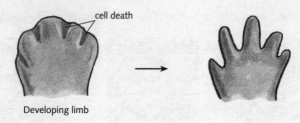

cell death

Developing limb

Fig. 16.10 Cell death leading to the separation of digits.

A good example is the way digits form in most mammals, birds and reptiles. In the embryos of these animals, the digits are initially joined together and only become separate later on, through the death of cells that lie between them (Fig. 16.10). These cells do not die because they have been damaged by an external agent, but through a controlled form of cellular suicide. In duck embryos, these deaths are less extensive so the adults develop with webbed feet.

A similar type of process is responsible for the holes in leaves of *Monstera deliciosa* (Swiss cheese plant, Fig. 16.11). The young leaf buds of this plant comprise a continuous expanse of tissue but, later on, cells in particular regions start to die, leading to the holes in the mature leaf.

Cell death can also result in the loss of entire structures from an individual. The most striking example is the fall of leaves in autumn. Before leaves are shed from a deciduous tree, the cells in the leaves undergo senescence, during which many of their valuable contents are recovered by the plant. Only after this has happened are the remains discarded by leaf fall. Similarly, the tail on a tadpole is eventually lost through the death of its cells, although in this case the cell

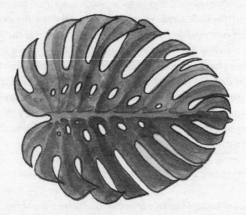

Fig. 16.11 Leaf of *Monstera deliciosa* showing holes caused by local cell deaths.

corpses are resorbed by the animal rather than falling off it. The technical term for these various forms of controlled cell death is *apoptosis*, a Greek word originally used to describe the 'dropping off' of petals from a flower or 'falling off' of leaves from a tree.

Major advances in understanding cell death have come from studies on the small worm, *Caenorhabditis elegans*. The main body of this animal is made of about 1000 cells. John Sulston and colleagues, working in Cambridge in the 1970s, were able to define precisely the pattern of cell divisions and deaths that occur as this worm develops from a fertilised egg. They showed that of the 1090 cells formed during the development of an individual worm, 131 always die. These cell deaths are not random but involve exactly the same cells in every individual.

A clue as to what was going on came from mutant worms in which this pattern of cell deaths was altered. During the 1980s, Robert Horvitz and colleagues in Cambridge, Massachusetts were able to identify and analyse several mutant worms in which cell deaths did not occur: most of the 131 cells that normally die now survived instead. These mutants lack particular types of protein that are normally involved in the act of cellular suicide. In normal worms, these proteins can set off a chain of events in a cell that lead to its self-destruction. However, in the mutants without these proteins, cells are unable to kill themselves, so the 131 that would normally commit suicide escape their grave fate. Similar proteins have now been shown to be involved in cell deaths in other animals, including vertebrates. Most cells in an animal's body appear to contain these cell death proteins but they are not in the right form to start the process of self-destruction. Only when the cell death proteins assume an active form do they become lethal. It is as if every cell carries a gun for shooting itself, but the trigger is only pulled in certain cases.

What controls the trigger? Why is it that the cell death proteins are activated in some cells but not others during development? The detailed answers are not known, but hidden colours, scents and sensitivities have been shown to be involved in several cases. To illustrate how this might work, let us return to our sheet of cells comprising an island of tropical-green surrounded by sea-blue. As before, the sea-blue cells give off a scent to which the tropical-green cells are sensitive. But in this case, the chain of reactions set off by the scent leads to a change in form of the cell death proteins (Fig. 16.12). As a result, these proteins become active and all of the tropical-green cells die, creating an island of dead cells in the centre. The cell corpses could either fall away from the sheet or be absorbed by the surrounding sea-blue cells. Either way, the sheet would now have a hole in it. Thus, the process of internal painting can influence not only the way the canvas grows but also which bits of it survive or die, allowing further types of modification to the shape of the organism.

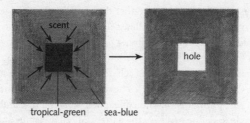

Fig. 16.12 Localised cell death triggered through a combination of hidden colour, scent and sensitivity.

Cell movement

In addition to growth and death, there is another way that the canvas can be modified that is particularly important for animal development: movement. Unlike the cells of plants, which are held in relatively fixed positions by their walls, those of animals are able to undergo extensive movements and migrations within the developing body. We have already seen how the elaboration of the inside–outside axis of animals involves the inward movement of cells (Chapter 14). There are also many other instances where cell movements play a key role. The cells of the developing nervous system, for example, undergo extensive migrations. Rita Levi-Montalcini, a pioneer in the study of nerve cells, recalls her impressions during the 1940s of their remarkable behaviour:

> At the beginning of spring, from the window of my small room in our cottage, I contemplated the ducklings following their mother in single file . . . In specific areas of the embryonic nervous system, cells in the first stages of differentiation detach themselves from cellular clusters of cephalic nuclei and move singularly, one after the other like the little ducklings, toward distant locations along rigidly programmed routes . . . In other sectors of the developing nervous system, thousands of cells move about like colonies of migrating birds or insects—like the Biblical locusts that I was to see many years later in Ecuador.

What leads particular cells to start migrating and to move along defined paths? We still do not know many of the details, but it has been shown that hidden colours, scents and sensitivities play an important role. To illustrate the likely principles involved, I want to start with a hypothetical embryo with an internal landscape of various hidden colours. We shall concentrate on a group of cells in one region of the embryo with a particular colour, say *racing-green*. I want to show how these racing-green cells might be able to migrate along a particular route to another region of the embryo.

The first step is to get the racing-green cells to become mobile. This can be achieved if we assume that racing-green activates genes that encourage a cell to start moving. These genes might, for example, produce proteins that change their shape in certain regions of a cell, causing local contractions or extensions that make it move along. The racing-green cells therefore start wandering about.

The next problem is getting the racing-green cells to migrate along a defined route to a new location. The trick is to exploit the pattern of scents in the embryo. Because there are many hidden-colours in the embryo, there will also be a corresponding pattern of scents: an orange region produces the scent of oranges, a lemon region the scent of lemons, etc. In other words, the embryo will be decorated with an internal landscape of diverse scents, each being emitted from a particular region. The racing-green cells will be able to respond to some of these scents through its receptors. Now suppose that the racing-green cells respond to a nearby scent, say orange, in such a way that their movement becomes oriented towards the source of the scent. In other words, instead of wandering about aimlessly, the racing-green cells would start to modify their pattern of contractions and extensions so as to move nearer to where the orange scent was coming from. All the racing-green cells would then head off in the direction of the orange region. Having arrived there, they might start to pick up another scent, say lemon, that was previously too far away to have been detected, so they now veer off towards the lemon region. In this way they might move through a sequence of several different regions. The racing-green cells are navigating the embryo using their sense of smell, blindly finding their way by following the landscape of scents. They take a predictable route because of their specific responses to the sequence of scents they encounter. Eventually they may arrive at their final destination: a region where the high concentration of a particular scent leads them to give up their migratory behaviour and settle down.

I have given this example to illustrate two key points about cell movement. First, the pattern of cell movement depends on the landscape of hidden colours and their associated scents and sensitivities. In other words, movement and rearrangement of parts of the canvas depend on the colours applied to it. Secondly, the process of cell movement in turn influences the pattern of hidden colours. This is because the act of migration can lead to the introduction of a colour, racing-green in this case, at a new location. The racing-green cells have acquired new neighbours with different colours from those that originally surrounded them, changing the overall pattern on the canvas. Unlike differential growth, cell movement can influence the order or connections between the colours, providing further ways of elaborating internal patterns.

The story so far

In previous chapters we saw how a patchwork of hidden colours can be gradually elaborated through a highly interactive process involving continual interpretation and refinement. This developing patchwork provides a frame of reference that underlies the geometry of the organism, its complex and idiosyncratic asymmetries that distinguish it from the fertilised egg that it came from. Some of the colours are so basic that mutants without them develop with fundamentally altered symmetries, lacking distinctions along major axes. However, the mechanism by which these hidden patterns eventually become manifest in the final geometry and anatomy of the organism was left open. In this chapter, I have tried to bridge this gap by showing how interpretation of the hidden patchwork may lead to changes in shape: regions of the canvas growing in particular ways, dying off or moving about from one region to another. In doing this, we have come across another level of interaction: all these changes in the canvas can themselves influence the pattern of hidden colours. There is therefore a two-way relationship between hidden colours and the canvas. It is as a result of these interactions that geometry can become gradually more elaborate as an organism develops.

We have arrived at an overall picture of how development occurs. But this is itself part of a broader canvas: evolution. The development of multicellular organisms from fertilised eggs is a process that has evolved over many millions of years. To appreciate the nature of development and its fundamental relationship to other types of making, we will need to place it in a broader evolutionary context, as we shall see in the remaining chapters.

The story of colour

The art of genes has a long history. Hidden colours, scents and sensitivities are not recent inventions; they have an ancient origin, tracing back to the time when all life-forms were single cells. In this chapter, I want to describe the history of internal painting, how colouring has changed during evolution from relatively simple beginnings to the production of complex patterns in the plants and animals of today.

Before embarking on this history, I need to be clear about what I mean by an evolutionary progression. We use the word *progress* in many different ways, leading to many misunderstandings. To avoid this problem, I want to clarify at the outset the sense in which evolution can or cannot be said to involve progress.

Progression and compromise

One of the earliest ways of looking at the living world, going back to Aristotle, was by arranging organisms along a scale or ladder. Plants were at the bottom of the ladder, as they only exhibited the most basic requirements for life: nutrition, growth and reproduction. Next came animals, which had the additional qualities of being able to move in accordance with their feelings or sensitivity. Above the animals came man, with his further faculties of thought and reason (in some versions of the ladder, the rungs continued beyond man, reaching up to the angels and finally God). It is a ladder of progress, with every organism earning its place according to how many qualities it has. Organisms towards the bottom are considered to be somehow defective when compared to those above them because they lack one or more qualities: plants lack movement and feelings compared to the animals above them, and animals in turn lack the rational thought of the superior humans. The ladder still permeates our everyday language. A person who has lost all power of movement and thought is referred to as a vegetable or cabbage. Obviously this is not because of a physical resemblance to vegetation but because such people retain only the basic qualities of life, bringing them down to the same level as plants at the bottom of the ladder. In a similar vein we use *veg out* to refer to human inactivity. Students of behaviour have extended this usage to fruit flies: mutant flies that are defective in learning have been given names like *radish* and *rutabaga*.

The theory of evolution by natural selection has made the scale of nature dispensable. If you were to have a criterion of success from an evolutionary point of view, it would not be your ability to move or how brainy you are, but how well you reproduce. An organism that leaves no offspring or relatives is a failure from an evolutionary standpoint, no matter how sensitive or intelligent it might be. Rather than classifying living organisms along a scale of qualities, the most meaningful evolutionary arrangement depends on common ancestry or descent. Instead of a ladder, we have to imagine an extensive family tree, with living species occupying the tips of branches. A species at one tip need be no better or worse than a species at another tip: they all occupy equivalent stations on the tree. Nevertheless, there is another sense in which we might want to search for progress in evolution, not so much in the comparison of living forms at the tree tips but in their relationship to more primitive ancestors that lie deeper down in the earlier branches. Perhaps evolution involves progress in time, with organisms becoming better as you travel up the family tree, from the trunk of ancient organisms, through the branches and twigs, eventually arriving at the tips of today.

According to the theory of natural selection, we might expect organisms to become better and better adapted to their environment during evolution. This is because natural selection will favour heritable variations that increase the chances of leaving offspring. Over many generations of selection, organisms should become better able to survive and reproduce, more adapted to their environment. The features of a lion that allow it to hunt and catch its prey so efficiently have arisen from countless generations of selection in the lion's ancestry, gradually improving its adaptive features. It is tempting to conclude that natural selection will therefore lead to an ever increasing number of adaptive qualities, taking organisms on a road towards perfection. This would be misleading for several reasons.

The first reason could be called the problem of the 'moving goal posts': the environment of organisms is not fixed but continually changes during evolution. When we say that a lion is well adapted, we are assuming a particular physical and biological environment. If you were to place a lion in a different environment, such as that of one billion years ago, it would not fare very well. This is not just for climatic reasons, but because all the organisms around it would be microscopic single cells, totally inappropriate prey for lions to hunt and eat. With respect to this environment, the modern lion is much less well adapted than its ancestors of one billion years ago. In saying that a lion is well adapted, we are taking particular climatic and biological conditions for granted, including the other organisms around it such as the grazing herbivores. The important point is that the environment is bound to change during evolution as organisms change, so it is simply not possible to measure adaptation against a fixed standard. The environmental goal posts are always on the move so there can be no such thing as a perfectly adapted state towards which organisms are evolving.

We can illustrate the same problem with progress in art. The history of western painting has sometimes been portrayed as a progression from primitive symbolism through to realism and perspective which in turn led to the discovery of atmospheric colour and Impressionism. It seems like a steady march of progress, in which paintings get better and better. One problem with this view is that not only do paintings change with history, but so do the ways we look at them. When the pictures of Impressionists were first exhibited in the nineteenth century they were met with hostility and incomprehension from the audience. Indeed the term *Impressionism* was coined by an art critic as a derogatory label, implying that the paintings were mere impressions rather than profound works of art. As one critic wrote after seeing an exhibition in 1876:

> An exhibition has just opened at Durand-Ruel which allegedly contains paintings. I enter and my horrified eyes behold something terrible. Five or six lunatics, among them a woman, have joined together and exhibited their works. I have seen people rock with laughter in front of these pictures, but my heart bled when I saw them. These would-be artists call themselves revolutionaries, 'Impressionists'. They take a piece of canvas, colour and brush, daub a few patches of colour on them at random, and sign the whole thing with their name.

Of course, few people would now think of Impressionism in this way; we have become so used to Impressionist pictures, to the point of being bombarded with them on postcards and chocolate boxes, that they seem very conventional, and it is difficult to see what all the fuss was about. The goal posts have moved because our visual environment, the paintings and images to which we have been exposed during our lives, is quite different now from what it would have been one hundred years ago. This means that the visual environment and aesthetic standards are continually changing as a result of the paintings themselves. As paintings change, so do our ways of looking at them, so the notion of progress towards perfection in art becomes meaningless because there can be no fixed reference point from which to judge.

Apart from the problem of moving goal posts, there is another reason for perfection never being attainable in biology. Why are there several species of large carnivores in Africa, such as lions, cheetahs and leopards? Why doesn't the best one win and eliminate the others? The simplest answer is that they are specialised in different ways. For instance, although the cheetah might run fastest, the lion is more powerful and the leopard can climb trees, so each

Fig. 17.1 *Portrait of B. van Orley* (1521), Albrecht Dürer. Dresden.

carnivore occupies a slightly different niche in the environment. But why doesn't natural selection favour a carnivore that combines the best features of a cheetah, lion and leopard to give a super-cat that would out-compete everything else? In a similar way, why doesn't a super-herbivore evolve that combines the best in zebras, antelopes and gazelles? Taken to its extreme, we might end up with a single super-organism that dominates the whole world. There is a rather obvious fallacy in this line of reasoning: if you took what might seem like the best adaptive qualities from a cheetah, lion and leopard you would not end up with a super-cat at all, you would get a hotchpotch that would be slower than a cheetah, less powerful than a lion and not as good at climbing trees as a leopard—a jack of all trades but master of none. Rather than out-competing the other species, our generalised cat would be beaten on all fronts. There is a limit to adding up adaptive qualities because being good at doing one thing often means being not so good at doing another. The elegant streamlined

Fig. 17.2 *Portrait of a Man*, Frans Hals. Leningrad.

construction of a cheetah that enables it to run so fast is incompatible with the power and strength of a lion. To put it another way, in a population of cheetahs, individuals with a heavier build, giving them more power and strength, would not be favoured by selection because they would not be as fast as the other cheetahs and not as powerful as lions; they would fall between two stools and be less able to survive and reproduce. In the context of a cheetah, what may seem at first sight to be an adaptive quality, increased power and strength, is in fact non-adaptive. Adapting in one way means sacrificing other adaptive options. There can be no perfect plant or animal because perfection in one direction carries built-in imperfections in other directions.

There is a parallel to this in art. Paintings come in a variety of styles. Heinrich Wölfflin, for example, distinguishes between the *linear* style, exemplified by Dürer (Fig. 17.1), in which outlines are very definite so that you can almost feel exactly where each object starts and finishes; and the *painterly* style, which is much more diffuse and less tangible, with outlines merging imperceptibly into the background as shown in a painting by Frans Hals (Fig. 17.2). Each style has its own atmosphere and aesthetic merits: we can appreciate and enjoy both the definite quality of the linear and the more ambiguous features of the painterly pictures. Even so, there is an incompatibility between these styles: an artist who paints in a linear style does so at the expense of painterly diffuseness. Similarly, you cannot work in a painterly fashion without sacrificing hard linear qualities. If you were to try to combine the two styles, rather than adding together the strength of linear and painterly, you would end up with some sort of compromise between the two. There is no such thing as a perfect style that sums up the linear and painterly because progress in one direction is at the expense of the other. We encounter a similar incompatibility in many other elements of style, such as depth as opposed to flatness, brightness as opposed to subdued colour, realism versus abstractness. You cannot combine the best of everything in one painting style any more than you can in one plant or animal. The history of art is no more a march towards perfection than is the evolution of organisms.

Perhaps if organisms cannot be perfected by simply adding together different adaptive qualities, perfection lies in being extremely specialised. But this also carries its price. The more specialised an organism, the narrower the range of biological and physical environments it can exploit. It is possible to be over-specialised, so finely attuned to a particular way of life that you are vulnerable to competitors that can survive and reproduce in a broader range of conditions. Adaptation therefore involves a compromise between specialisation and generalisation, between having too narrow a lifestyle on the one hand and being too much of a non-specialist on the other. Adaptation is about compromise rather than perfection.

I have taken pains to describe the case against perfection, so as to be clear about a more limited sense in which I think evolution can be said to involve a progression. Although combining different adaptive qualities need not always lead to an improved ability to reproduce in a particular environment, in some cases it clearly can. Were it not for this, we could not account for the complex adaptive features of organisms that have evolved over many generations of selection. The human eye, to give a popular example, is a compound structure, made up from a whole series of adaptations that have been built on each other. If every new adaptation undid the benefits of previous adaptations, if adaptations were always incompatible with each other, there could be no effective evolutionary change. Evolution is progressive in the sense that it is based on what went before: each adaptation assumes and is imposed upon a whole series of previous adaptations. In the long term, this does not lead to perfection for the reasons I have already mentioned, but it may lead to novel compromises, organisms that have new or more effective strategies for surviving and reproducing in a particular environment.

Now because every new possibility is tied up with what went before, the whole process is enormously constrained by history. The organisms we see today can only be understood in the light of their past. The same could be said about paintings. Impressionism or Cubism are only comprehensible in relation to their antecedent styles. Leonardo could not have produced a Cubist version of the *Mona Lisa* in the early sixteenth century because Cubism only makes sense in the light of a historical progression. Artistic styles do not appear in a vacuum: each style is informed by its predecessors, incorporating the lessons of the past, whether it is a reaction against them or an extension of them. It is in this sense of historical progression that I now want to trace some of the elements that have allowed complex patterns of development to emerge during evolution.

Unicellular chameleons

For most of this book, I have been describing how hidden colours, scents and sensitivities are used in multicellular organisms as a way of elaborating internal patterns during their development from a single cell, the fertilised egg. The rationale behind the various colours was to establish regional differences between cells, distinguishing cells in one area of an individual from those in another. Yet, as I shall describe, it is very likely that colours, scents and sensitivities were present in the unicellular ancestors of plants and animals. Now, on the face of it, there appears to be little point in a single-celled organism having a system of colours. If the *raison d'être* of hidden colours is to produce differences between cells, what possible significance could they have had for an ancestral organism that comprised only one cell? They would have nothing to differentiate between. It would be like having a box of paints to decorate a canvas that could only fit one colour on it.

Although we cannot answer this question by going back billions of years and looking at the behaviour of our unicellular ancestors, we can do the next best thing and look at the properties of unicellular organisms alive today. To be sure, they will not be exactly the same as the ancestral forms because unicellular organisms have also changed during evolution. Evolution has not stood still for unicellular life any more than for multicellular life. Nevertheless, the unicellular organisms of today can tell us about the significance of colours, scents and sensitivities for a single-celled lifestyle.

Fortunately we do know a great deal about some unicellular species because their small size and ability to reproduce rapidly make them ideal subjects for genetic studies. For instance, much of our basic understanding of how genes work has come from studies on a bacterium, *Escherichia coli*, which normally lives in the human gut. It is very straightforward to breed large numbers of *E. coli* in the laboratory, where it can reproduce about once every twenty minutes. Now not only are hidden colours to be found in *E. coli*, it was through studying *E. coli* that they were discovered in the first place.

In the 1950s, François Jacob and Jacques Monod were working in the Pasteur Institute, Paris, on the problem of how *E. coli* bacteria manage to utilise different types of sugar. Like ourselves, bacteria use sugars as a source of energy and chemical components. *E. coli* is quite versatile and can metabolise a whole range of different sugar molecules, such as glucose, galactose and lactose. This is because it can produce a range of different proteins, enzymes, that can catalyse the breakdown of each type of sugar molecule. There is one enzyme for breaking down glucose, another for galactose, yet another for lactose. The key problem that attracted Jacob and Monod was that *E. coli* does not produce all of these enzymes all of the time; it is able to *regulate* which enzymes it produces according to what is available in the environment. If, for example, there is no lactose in the environment, *E. coli* does not bother to produce the lactose-digesting enzyme. But if you put the bacteria into a situation with lots of lactose, the relevant enzyme needed to digest lactose is now produced. This makes a great deal of adaptive sense: it avoids a needless waste of valuable energy on making an enzyme when there is nothing for it to work on. But this raises a problem: how do the bacteria sense whether or not lactose is around, and regulate themselves accordingly? How do they detect and act upon what is going on in their environment?

A good way to try and answer this question is to study mutants that are unable to regulate themselves: mutants that cannot respond to their environment and effectively ignore what is going on around them. By working out what is wrong with such deregulated mutants, you may be able to understand how regulation normally occurs. Using this approach, Jacob and Monod showed that in order for *E. coli* to regulate its response to lactose, it needed a particular gene, called *repressor*. Mutants without this gene simply ignored whether there was any lactose around and went ahead to make the lactose-digesting enzyme anyway. The repressor gene was therefore essential for *E. coli* to regulate itself.

How does the repressor gene work, and affect whether or not bacteria make an enzyme? The story is slightly convoluted but the final message is simple. We can think of the repressor gene as making a hidden colour, say *red*. Like the other hidden colours I have mentioned, red corresponds to a master protein— a protein that can directly bind to genes and switch them on or off. In this case, the gene it binds to is the one responsible for producing the lactose-digesting enzyme. When the red protein is bound to the regulatory region of this gene, it stops it from being active, and no enzyme is made (Fig. 17.3, left). The red

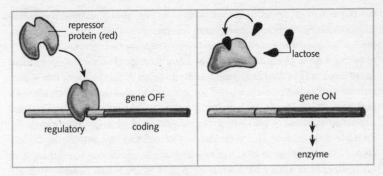

Fig. 17.3 Regulation of an *E. coli* gene in the absence (*left*) or presence (*right*) of lactose.

protein therefore ensures that this gene is kept off. This is the situation when no lactose is around. When lactose is added, however, there is a change in hidden colour. This is brought about by the lactose molecules themselves binding to the red protein. Remember that the shape of a protein can change when another molecule binds to it. In this case, the lactose molecule fits into a small crevice in the red protein, and when it binds there it causes a change in the shape of the protein. In this altered state, the protein can't bind to DNA any more, meaning its colour has effectively changed. The protein is now unable to keep the gene for making the lactose-digesting enzyme switched off, so the cell starts enzyme production (Fig. 17.3, right). An *E. coli* bacterium can regulate itself because lactose from the environment itself enters the cell and brings about a change in the hidden colour, an alteration in the shape of a master protein. Genes in the cell can interpret this change in colour so as to be switched on or off. The bacteria are like tiny chameleons, changing their hidden colours in response to their surroundings.

There is a simple moral to this tale. Changing colour with *time* can be a very effective way of dealing with a variable environment. Every time you eat or drink, the *E. coli* cells living in your gut experience an influx of new conditions. If you drink a pint of milk, the gut bacteria will suddenly find themselves swimming in a medium rich in lactose (lactose is a major sugar in milk, hence the Latin root *lact*, from milk). Individual bacteria that can switch the state of their genes to cope with the new conditions will be better off than those which ignore them. Hidden colours enable *E. coli* cells to change their gene activity in response to circumstances. They provide a one-dimensional pattern, a state that can change in time to reflect the particular surroundings they find themselves in. I have mentioned the availability of lactose to give just one example of how it might be advantageous for a cell to change its colour. But obviously there are many other things that can vary in a cell's situation. A cell can respond to many of these by a change in one or more of its hidden colours. At any one time, there will be an overall combination of colours in a cell, which can be interpreted to ensure that only the most appropriate genes are active.

So far, I have given a rationale for unicellular organisms having hidden colours as a way of changing gene activity in time. But what about scents and sensitivities? Are the mechanisms for producing and detecting scents also present in these organisms? Remember that in multicellular organisms, scents and sensitivities allow cells within the same individual to communicate with each other, providing an important way of refining the pattern of hidden colours. Now it is much less clear how this might be relevant to a unicellular organism. If you are a single-celled individual, there would be little point in giving off a scent just so as to smell yourself. But there is another possibility. A single-celled individual need not live completely alone like a hermit. It might have other individual cells nearby. There could well be a point in it giving off a scent if it were trying to communicate with some of its neighbours. And one of the best reasons for individuals wanting to communicate is sex.

Sexual chemistry in yeast

A good way to illustrate sexual communication between unicellular organisms is with yeast. Yeasts are a type of fungus, the commonest example being the species used by brewers and bakers (*Saccharomyces cerevisiae*). Individuals of this unicellular species come in two opposite sexes or mating types, termed a and α. Although individuals of the different mating types look identical, they can be distinguished by their sexual behaviour. Yeast sex happens when two individuals, each a single cell, come together and fuse to form a hybrid cell. Sex for yeast is an all-consuming event in which the two individual partners become completely united. Most importantly, from our point of view, yeast cells are very discriminating in their choice of partner. An individual will only mate and fuse with a cell of the opposite mating type: an a-type cell will mate with an α-type but not with another a-type. Similarly, α-types will happily conjoin with a-types but will have nothing to do with other α-types.

The ability of a yeast cell to respond to its opposite mating type in such a discriminating manner depends on a set of hidden colours, scents and sensitivities. As with so many of the other examples I have given, this was worked out by analysing mutants. In this case, the mutants were defective in some aspect of their sexual behaviour. To save time, I will not go through all the details of how this was done but go on to give some of the basic conclusions.

The key difference between the two sexual types lies in their hidden colour. Although individuals from the different mating types may look the same from the outside, they express distinct master proteins. We can think of the two mating types **a** and α as corresponding to two colours, say *apricot* and *almond* (Fig. 17.4). We now need to explain how it is that an apricot-coloured cell recognises and mates only with an almond-coloured cell. Remember from Chapter 11 that a hidden colour can lead to the production of particular scents (this is because the genes for making scents can be switched on or off in response to the hidden colour). It is by using such scents that the yeast mating types recognise each other. The apricot and almond cells each produce a characteristic scent. The scents are actually small proteins, called pheromones, but we can imagine them as being like the scents of apricots and almonds. Furthermore, each cell type is only sensitive to the scent of the other type: apricot cells are only sensitive to the almond scent, and almond cells only respond to the whiff of apricots (this is because each cell type expresses a different receptor protein, another consequence of their hidden colour). When an apricot cell encounters another apricot cell nothing happens because neither cell is able to detect the other's scent. However, when an apricot encounters an almond, the cells will respond to each other.

The eventual outcome of this mutual sniffing exercise is that when a pair of apricot and almond cells meet, both cells change their hidden colours. The apricot cell is stimulated by the scent of almonds, setting off a chain of events

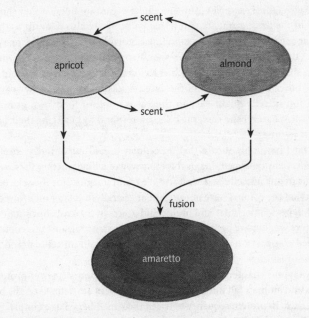

Fig. 17.4 Role of hidden colours, scents and sensitivities in mating in yeast.

within the cell that leads to altered activity of its master proteins, resulting in a change in colour. This new colour in turn leads to the activation of genes needed for cell fusion and mating. In other words, the cell starts to make proteins that encourage it to fuse with its neighbour. A similar response is shown by an almond cell when it detects the scent of apricots. The overall result is that when apricot and almond cells encounter each other, they both respond by becoming locked together, and they fuse to form a hybrid.

In case you are wondering what happens to the hybrid cell, I should briefly complete the life cycle of yeast. The hybrid cell carries a combination of colours from both partners, giving it a new overall colour; we can call it *amaretto* (appropriately enough, legend has it that the Amaretto liqueur originated by a widow flavouring some brandy with a combination of almonds and apricots from her garden). The amaretto hybrid cell can do one of two things, depending on environmental conditions. It may reproduce by cell division (mitosis) to give more amarettos. Alternatively, it can undergo another type of division (meiosis) to produce two types of cellular offspring, the apricot and almond mating types—thus completing the life cycle.

The colours of our ancestors

I have given these examples from unicellular organisms because they show how many of the basic elements of internal painting are also relevant to the unicellular condition. Although a unicellular individual can only have one overall colour, this can change in time in response to its circumstances. Environmental factors may include nutrients, such as lactose for *E. coli*, or specific scents given off by other individuals, as in the case of yeast mating. In other cases, the condition of cells might vary in time as a result of their own growth and division. Whatever the case, the cells can respond by changing their internal colours in time.

So far I have only shown that unicellular organisms of today employ the elements of internal painting. Is there any way of telling whether they were also present in our unicellular ancestors? We could imagine, for the sake of argument, that the colours were not present ancestrally, but later on evolved separately in both unicellular and multicellular life. In this case, our single-celled ancestors would have been colourless. Strong evidence against this comes from detailed comparisons between the hidden colours in unicellular and multicellular organisms, which I shall now describe.

In previous chapters, we saw how the hidden colours (master proteins) of plants and animals fall into families. These families are defined on the basis of similarities in protein sequence between their members. For example, there is a family of related green proteins involved in giving the regions of a fly or human a distinct identity from head to tail. Similarly, there is a red family that provides distinctions between the whorls of a flower.

A very important finding started to emerge in the 1980s: members of these families were not only found in multicellular plants and animals but also in some unicellular organisms. In the case of yeast, for example, members of both the red and green family were found to be involved in the mating response. That is, both multicellular and unicellular organisms were using a *related palette* of basic colours. The simplest explanation is that all of these organisms inherited their basic range of hidden colour genes from a common ancestor. This ancestor was almost certainly unicellular, implying that hidden colours were already being used by ancient unicellular organisms.

More evidence in support of this has come from comparing the hidden colours of modern plants and animals. In spite of their distinct lifestyles and appearance, plants and animals have also been shown to use a related palette of hidden colours. For example, members of the green family of master proteins, involved in giving the regions of a fly or human a distinct identity, have been found in plants such as maize, *Arabidopsis* and *Antirrhinum*. Similarly, members of the red family, involved in distinguishing the identity of whorls of organs in a flower, have been found in animals, including flies and humans.

The simplest explanation for the similarity in hidden colours between plants and animals is that it reflects their shared ancestry. The common ancestor of plants and animals is thought to have been a unicellular organism, living more than a billion years ago. For our present purposes, we can think of this unicellular ancestor as giving rise to two lineages: one that led to plants and the other to animals. Initially, both lineages would have contained only unicellular organisms: there would have been unicellular plants and animals. Complex multicellular forms are thought to have arisen later on along each of these routes. Now the easiest way to account for why modern plants and animals use a related palette of hidden colours is that many of these colours were already present before the two lineages diverged, in their common unicellular ancestor. This implies that hidden colours were being employed more than one billion years ago by single-celled organisms.

The multicellular dimension

Although some of the elements of painting had evolved in our unicellular ancestors, most probably as a way of changing according to circumstance, there was an important limitation in their use of colour. As long as an individual comprises a single cell with one nucleus, it can only interpret one overall hidden colour at any one time. To make this clearer, imagine that a single cell contains two different hidden colours, say blue towards one end and yellow towards the other. This would correspond to two types of master protein in the cell's cytoplasm, each concentrated at a different end of the cell. Now even though there are two colours in the cell, the genes that interpret the colours, located within the nucleus, can only be in one place at any one time. The nucleus could be at the yellow end of the cell or at the blue end, but not at both ends at once. It could of course be in the middle and perhaps experience a combination of some blue and yellow, but this would still correspond to only one overall colour (green). Even if the nucleus physically travelled from one end of the cell to the other, experiencing yellow, green and blue as it went, it could still only interpret one overall colour at a time.

By contrast, the genes of a multicellular organism can interpret different colours in *space* as well as time. Because there are many cells, each carrying its own nucleus, it is possible for different colours to be interpreted at one and the same time. If cells in one part of the organism express a yellow master protein whereas those in another part express blue, it is perfectly possible to interpret these colours separately: the genes in the nuclei in one set of cells can respond to yellow whilst at the same time those in the other region can respond to blue. In other words, it is possible for colours to be used in both space and time. Multicellular life gives a new dimension to the possibilities of internal painting, or, more strictly speaking, three new dimensions. (In principle, the same could also apply to single-celled organisms that have many nuclei: the alga *Caulerpa*, for example, which may comprise a single cell several centimetres long containing many nuclei within a shared cytoplasm. However, this situation is relatively rare in plants and animals, most likely because there are additional advantages to being multicellular, such as having internal cell membranes to limit the movement of molecules within the organism.)

I do not wish to imply that in being able to interpret only one overall colour at a time, unicellular organisms are homogeneous blobs with no spatial pattern to them. Far from it. Single-celled creatures display numerous patterns and types of asymmetry in their structure and shape. They do not seem to achieve this, however, by elaborating spatial patterns of internal colour; they generate their patterns by other mechanisms. In some cases, interactions between particular types of protein in the cell may be involved; in others, proteins may work more indirectly by affecting cell components which then interact to produce a pattern. Whatever the precise mechanisms might be, it is unlikely that the spatial pattern of master proteins has a major role. With only one nucleus to play with, the scope for using spatial patterns of hidden colour are strictly limited. This does not mean that hidden colours are irrelevant. They can influence patterning by affecting *when* particular proteins appear on the scene, but the spatial patterns themselves are, for the most part, not derived from regional differences in colour.

My point is not that our unicellular ancestors lacked patterns, but that their hidden colours were used in a different way from those of their multicellular descendants. The basic elements of painting evolved in unicellular organisms primarily as a way of changing in time. As multicellular organisms arose, these pre-existing mechanisms were put to a new use: to generate internal spatial patterns. It is not known exactly how this happened but we can imagine some possible scenarios.

Suppose that the progeny of a single-celled organism do not separate from each other after division, but remain together temporarily to form a clump or ball of cells. Eventually some or all of the cells in the clump separate and go off to form new clumps, so we end up with an organism that goes through two phases each generation: a clumped phase and a single-cell phase. Now during the clumped phase of life, the cells on the outside of the clump will naturally experience a different environment to those on the inside. For example, if the organism is aquatic, the outer cells will be directly exposed to the surrounding water together with whatever salts, sugars or other molecules are dissolved in it, whereas the inner cells of the clump will be protected and only exposed to other cells. Now because unicellular organisms have the ability to change their hidden colour according to their local environment, we could imagine that the cells on the outside of the clump adopt a different colour from those on the inside (Fig. 17.5). The difference in colour simply reflects the response of these cells to the differing conditions on the outside compared to the inside of the clump. Once there is a difference in colour between inside and outside, this could then be interpreted by genes (through their regulatory regions). We might expect that over many generations of clumps, natural selection would favour interpretations that lead to cells expressing genes that are more suited to their locations in the clump. Outer cells, for example, might come to express genes involved in motility, protection or ingestion; whereas inner cells might specialise in digesting and metabolising food. The detailed specialisations do not matter. What counts is that the hidden colours, which originally evolved

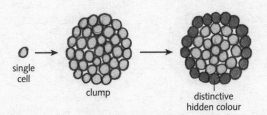

single cell

clump

distinctive hidden colour

Fig. 17.5 Evolution of a simple spatial pattern of hidden colour.

in unicellular organisms as a way of changing in time, can now be used to generate regional differences between cells within the same multicellular individual.

One way of summarising this series of events is to say that a hidden colour, previously serving one function, has been *recruited* to a different use. Instead of providing information about the local environment that changes with time like a chameleon, the hidden colour has been recruited to provide a spatial distinction, between outside and inside. This process of recruitment did not involve some special force telling the hidden colour to change its role. It arose naturally by mutations in genes, changing the way the hidden colour was interpreted. In a population of clumps, each mutation was subject to natural selection so that over many generations you end up with outer and inner cells expressing genes that seem appropriate to their position.

After some initial differences have been established between cells within an individual, more elaborate patterns may evolve. In our example of the cellular clump, there are two cell types, each with a distinct colour: outer cells and inner cells. But now we can also subdivide the cells further. There will be a subset of inner cells that make direct contact with the outer cells, allowing us to classify inner cells into two sorts: those nearer the surface that make contact with outer cells, and those lying further towards the core of the clump. We could imagine further evolutionary steps, leading to these two cell types assuming distinct hidden colours (Fig. 17.6). For example, suppose the hidden colour in the outer cells leads to them producing a short-range scent. The cells just beneath them might respond to this local scent, resulting in a change in hidden colour; whereas the core cells might remain unchanged as they are too far away to detect the scent. This would give three regions, each with its own colour: outer cells, the cells just beneath them and core cells. The point I wish to make here is that once there are different colours within a multicellular organism, they can be built upon during evolution. Cells can begin to experience and respond to an internal environment, the other cells of the organism around them. And they

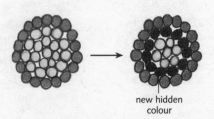

new hidden
colour

Fig. 17.6 Further evolution of an inside–outside axis.

do this not by inventing some wonderful new mechanism, but by using the box of painting tricks that had already evolved in unicellular life.

We could use a similar style of reasoning to account for other asymmetries of the organism. If, for example, our cellular clumps land on a surface, say by sinking onto rocks on the sea bed, a new difference will be experienced between the cells of the clump lying next to the rock surface and those more freely exposed to the water. Again, we could imagine this difference in local environment leading to a difference in hidden colours between one part of the organism and the other, giving it a 'rock end' and an 'exposed end'. These two colours could be just a prelude to the evolution of further elaborations that refine the pattern along this axis. Perhaps this could lead to the cells at one end specialising in gripping the rock surface whereas those at the other end deal with harvesting energy from the surroundings. Or perhaps the organism eventually swims away from the rock, using specialised cells at its 'rock end' to propel it along. As with the inside–outside axis, the organism's environment could initially have provided an important way of producing asymmetries along other axes.

There is, however, a slightly more subtle way of producing further asymmetries. Our clump of cells has to reproduce by generating more clumps. This means that somewhere in our clump there have to be cells that will act as founders for the next generation of clumps. Suppose that the founders come from the outer cells of the clump (the ability to form a founder would be yet another consequence of hidden colour). As a founder cell starts to divide to form a small daughter clump attached to the outside of its mother, the daughter might now experience two asymmetries. As before, its inner cells are in a different environment from its outer cells, giving it an inside–outside axis. But now there is also a difference between the cells attached to the mother and those freely exposed to the outside. The mother is acting as a surrogate rock, providing an initial asymmetry. In other words, asymmetries may be transmitted from mother to daughter, as well as from the external environment.

I have given these examples to show how regional differences in colour might have been initially established in multicellular organisms. We could imagine several other ways in which this might have occurred, but there is little merit in listing them all. What matters is that it is not too difficult to see how the basic elements of painting, which originally provided a way of changing in time, might have been recruited to generate spatial patterns.

This does not mean that the hidden colours of organisms alive today are identical to those of their ancestors. The colours have changed to some extent during evolution, as mutations have accumulated in the genes for master proteins. New types of greens and reds have evolved that have some different features from the original colours. Furthermore, parts of the original palette have been expanded through the process of gene duplication. Recall that this occurs when an extra copy of a gene is incorporated in the DNA by mistake (Chapter 6). This means that if we have an organism with one gene for a red master protein, very occasionally a descendant will be produced with two copies of the gene. This may be of little consequence immediately but, over time, the two gene copies might start to accumulate differences. Perhaps some of these changes allow the proteins produced by these genes to be interpreted in slightly different ways. Having started off with one red colour, the organism is effectively evolving two different versions, say ruby red and garnet red. We could imagine the process repeating itself so that eventually you end up with a whole family of reds, each with a different role in the painting.

The overall outcome is that although we find a common basic palette in plants and animals, the colours differ both in detail and significance for the organism. In animals, some members of the green colour family can provide distinctions from head to tail (Chapter 6). In plants, green colours confer other distinctions, such as between the growing tip and its nearby leaf-buds, or between a hair-forming cell and its neighbours. In other words, the green family is still being used to divide up the organism in space and time, but the detailed significance of the colours, the identities they relate to, are different. The same could be said about the red family of proteins that provide distinctions between the flower whorls. Members of the red family are also found in animals, but the processes they affect, such as muscle development, have no counterpart in plants. Plants and animals use common basic colours but they interpret them in very different ways.

It is tempting to think that this process of expanding and diversifying the palette was something that only happened in multicellular organisms, as they became more complex. Perhaps the common ancestor of plants and animals had only one gene for each colour and the colour families subsequently arose as multicellular organisms evolved. This, however, would be misleading. The ability to duplicate genes would also have been present in our unicellular ancestors and they would almost certainly have had their own families of reds, greens, etc. What seems likely is that some colours from the ancestral palette have been expanded whereas other colours have been lost during evolution. The range of hidden colours that have ended up in the organisms of today is therefore a rather complex assortment, in which some colour families happen to predominate.

I have concentrated on how a range of colours might have evolved but the same could also be said of scents and sensitivities. The genes involved in the various steps in producing and detecting scents also fall into families. As with hidden colours, many of these families are believed to be quite ancient, and probably evolved by duplication of various genes.

The price of multicellularity

In following the transition from unicellular to multicellular organisms, there is a danger of seeing it as an absolute progression. Whereas unicellular organisms can only paint in time, multicellular organisms may appear superior in being able to paint in space as well. It looks like a steady march of progress in which organisms become increasingly more elaborate and better off. As I pointed out earlier in this chapter, however, adaptations bring their own disadvantages with them. Multicellular organisms with their greater size and more elaborate internal patterns may seem to hold all the cards but their lifestyle brings new hazards.

For one thing, their larger size makes them susceptible to a broader range of parasites. In T. H. White's book, *The Sword in the Stone*, the wizard Merlin has a duel of magic with a witch, Madame Mim. In the climax, the witch turns into an enormous monster only to be defeated by Merlin turning into a tiny infectious germ that brings her down in a rash. Internal parasites, many of them unicellular organisms, pose a continual threat to multicellular life. Plants and animals are vulnerable to all sorts of infection and whole batteries of genes, such as those involved in the animal immune system or plant disease resistance, have evolved as a way of trying to combat this problem.

Another price paid by multicellular organisms is that they go through a vulnerable period as they develop. When a unicellular organism reproduces, its progeny can be pretty much ready straight away to survive in their environment. Multicellular organisms, however, typically need to spend extended periods as embryos developing from single cells before they get to the point that they can survive effectively as independent individuals. This has led to the evolution of various protective devices, such as surrounding the developing embryos with a hard shell or seed coat, or keeping them within the mother for as long as possible. In all these cases, new adaptations have evolved to protect the more vulnerable stages brought about by the requirements of development.

Yet another problem for multicellular life is that development itself is prone to go wrong. Development is a highly intricate process that can derail at any time. For example, cancer is caused by particular cells in the body dividing and proliferating in an uncontrolled way. In many cases this is because hidden colours, scents or sensitivities, which have a normal role to play in development, start to be active at inappropriate times or places. The very mechanisms that allow internal patterns to be elaborated become the cause of a disease when they go wrong. Unicellular individuals cannot die from cancer because there are no cells within them that can start to misbehave; the individual and the cell are one and the same.

Perhaps the most important sacrifice of all in adopting a multicellular life has to do with the length of time it takes to reproduce. One year for a dog is often compared to seven human years. If you were to make the same comparison with the bacterium *E. coli*, the figure would be more like a million human years. In one year, it is possible for certain bacteria to go through more generations than is covered by the span of human life on earth. This means that unicellular organisms can evolve much more rapidly in response to environmental conditions.

For all these reasons, unicellular organisms have not died out and been replaced by multicellular life. They are ever present and vastly outnumber everything else. You can find unicellular organisms everywhere, from hot acid springs to the depths of the ocean. These organisms are not inferior creatures, lower down on the scale of being, but highly successful in their own right. Multicellular and unicellular organisms are not different stages on the road to perfection, they simply represent different types of compromise, different ways of surviving and reproducing in an environment.

The history of internal painting is a story of putting old genes to new uses. We have seen how genes for hidden colours, scents and sensitivities allowed unicellular ancestors to change in time according to circumstance. As multicellular organisms evolved, these elements of painting were recruited to elaborate internal spatial patterns. New identities started to be specified as colours were interpreted in space as well as time. As this happened, the range of particular colours and sensitivities also expanded, allowing more elaborate patterns to be built up. Internal painting did not arise in a vacuum but through the modification of previous genes, allowing the same basic paintbox to be used in a whole variety of ways.

CHAPTER 18

The art of Heath Robinson

I have tried to show in this book how organisms are not simply manufactured according to a set of instructions. There is no easy way to separate instructions from the process of carrying them out, to distinguish plan from execution. Development is more like a creative process, in which each step interprets and elaborates what went before, than a process of fabrication. Nevertheless, an objection might still be raised to this viewpoint. Creativity seems somehow more subjective and difficult to pin down than fabrication. Although there may be many respects in which it is inadequate as a metaphor for development, at least fabrication has the merit of being there for everyone to see.

I now want to take a closer look at this issue. To do this, I will begin with the more traditional notion of development as a form of fabrication and try to establish where its logical foundations lie. We shall see that, contrary to what might have been expected, this view is actually on a weaker material footing than the alternative based on comparisons with creativity. Once this is appreciated, we will be able to look in a more coherent way at the fundamental relationships between different forms of making.

The dual analogy

Fabrication is not a self-contained process. To manufacture an object, somebody first has to come up with a plan for how to make it. For example, it took many years of technological innovation to come up with the detailed procedures and instructions for how cars should be made. These ideas and plans, essential for car production today, depend on a long history of human creativity and ingenuity. A similar thing may be said for all cases of fabrication, from the manufacture of washers to computers: they all depend on plans having first been created by humans. Fabrication, then, does not support itself; it is grounded in a form of human creativity.

Now if the development of an organism is a type of fabrication, a process of manufacture according to a plan, where does the plan come from in this case? The traditional answer is that it is a product of evolution. The ability of organisms to develop did not appear out of the blue, it arose gradually by natural selection over many millions of years. As with fabrication, biological

Fig. 18.1 The dual analogy.

development depends on a previous history, but a history of evolution rather than human ingenuity. The analogy between development and fabrication therefore leads to a second analogy: between evolution and human creativity. I shall refer to this double comparison as the *dual analogy*, summarised in Fig. 18.1.

According to this view, the horizontal arrows in Fig. 18.1, which take us from creativity to fabrication (top arrow) or from evolution to development (bottom arrow), should represent parallel pathways. In other words, if there is an analogy between these various processes, then the same should apply to the paths that connect them. To see whether this is indeed the case, I want to take a particular example of human creativity, the act of tinkering, and follow its course as compared to that of evolution.

Tinkering with the past

The anthropologist Claude Lévi-Strauss used the French equivalent of *tinkering* (*bricolage*) to describe how myths evolve. He contrasted the activity of a tinkerer to that of an engineer. Whereas an engineer carefully chooses his tools and raw materials to achieve the desired end, a tinkerer uses whatever is ready to hand. The tinkerer makes do with whatever happens to be in the work-shed to cobble various bits together:

> Consider him at work and excited by his project. His first practical step is retrospective. He has to turn back to an already existent set made up of tools and materials, to consider or reconsider what it contains and, finally and above all, to engage in a sort of dialogue with it . . . A particular cube of oak could be a wedge to make up for the inadequate length of a plank of pine or it could be a pedestal—which would allow the grain and polish of the old wood to show to advantage. In one case it will serve as extension, in the other as material. But the possibilities always remain limited by the particular history of each piece and by those of its features which are already determined by the use for which it was originally intended or the modifications it has undergone for other purposes.

There is a continual interaction, or dialogue, between the tinkerer and his materials, rather than the tinkerer following a clearly defined plan. As a consequence, the tinkerer is constrained by the collection of oddments, each of which may have had a previous use, rather than a set of tools and materials that have been carefully chosen with a particular outcome in mind. Lévi-Strauss used this example to point out how mythical thought is an intellectual form of 'bricolage'. Myths arise by tinkering with previous cultural elements, not by carefully planned design. François Jacob used the same imagery to describe biological evolution:

> In contrast to the engineer, evolution does not produce innovations from scratch. It works on what already exists, either transforming a system to give it a new function or combining several systems to produce a more complex one. Natural selection has no analogy with any aspect of human behaviour. If one wanted to use a comparison, however, one would have to say that this process resembles not engineering but tinkering, *bricolage* as we say in French. While the engineer's work relies on his having the raw materials and the tools that exactly fit his project, the tinkerer manages with odds and ends. Often without even knowing what he is going to produce, he uses whatever he finds around him, old cardboard boxes, pieces of string, fragments of wood or metal, to make some kind of workable object. As pointed out by Claude Lévi-Strauss, none of the materials at the tinkerer's disposal has a precise and definite function. Each can be used in different ways.

Like tinkering, evolution is always constrained by what happens to be available. It builds on the past rather than making plans for the future. The history of internal painting, described in the previous chapter, illustrates this very well. Unicellular organisms did not evolve the elements of internal painting just so that they might be used some time later on to generate complex spatial patterns. They did not anticipate or plan their subsequent use. The later patterns arose by simply tinkering with elements that were previously on the table for other reasons. Hidden colours that were originally an adaptation for one purpose (changing in time) were recruited to an additional use, the production of patterns in space. As multicellular organisms evolved, so the range and significance of hidden colours expanded by tinkering.

The notion of tinkering is perhaps best captured in the art of William Heath Robinson. Heath Robinson became famous for his illustrations of various contraptions; so much so that *Heath Robinson* or *Heath Robinsonian* has become a catch-phrase for complicated and ingenious devices. Look at his drawing of a Pancake Making Machine (Fig. 18.2). The contraption has obviously been cobbled together from various bits and pieces that happened to be around. Some irons and a coal scuttle are used as a weight, and a brick on a piece of string is used to control when the weight is released. It is an obvious case of tinkering, putting all sorts of available objects to a new use. The illustrations work so well because we are familiar with the previous use of many of the objects and enjoy seeing them used in an unexpected way. Irons are normally

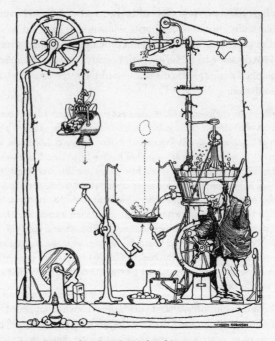

Fig. 18.2 *Pancake Making Machine* (1933), Heath Robinson.

used for pressing clothes and a coal scuttle for moving coal, not for tossing a pancake in the air.

If tinkering, as a creative process, provides a good analogy for evolution, where might fabrication and development fit in? At each stage in the creation of a Heath Robinson machine, the contraption changes in some way, with some part added, rearranged or removed. Every stage therefore looks different from the next, not needing the steps to have been planned in advance. But once you have a particular Heath Robinson contraption and want to make many copies of it *consistently*, the best way would be to write a set of instructions on how to make it. By following the instructions we can manufacture exactly the same contraption every time. We can churn out or fabricate as many copies of the Pancake Making Machine as we want. Perhaps this, then, corresponds to the process of development. According to this view, the transformation of an egg into an adult is like following a set of instructions to manufacture the same contraption every time. In the case of the egg, the instructions are not written down in words but in the DNA text, as genes. Evolution proceeds by creative tinkering but development is a question of consistent manufacture once evolution has produced the contraption to begin with. We have arrived back at the dual analogy: evolution as creativity, development as fabrication.

To my mind, there is a flaw in this reasoning. We have skipped too rapidly along the horizontal arrows of the dual analogy and have therefore missed out some problematic steps. The problems we jumped over trace back to Darwin's own account of evolution, and there is perhaps no better way of revealing them than by looking at how Darwin was himself misled.

A new twist on orchids

Shortly after completing his work *On the Origin of Species* in 1859, Charles Darwin chose to illustrate many of his ideas with orchids. In 1862, he published a volume *On the Various Contrivances by which British and Foreign Orchids are Fertilised by Insects* in which he described how different types of orchid flower could be understood as a series of modifications of previous types. Towards the end of the book he summarised:

> Although an organ may not have been originally formed for some special purpose, if it now serves for this end, we are justified in saying that it is specially adapted for it. On the same principle, if a man were to make a machine for some special purpose, but were to use old wheels, springs and pulleys, only slightly altered, the whole machine, with all its parts, might be said to be specially contrived for its present purpose. Thus throughout nature almost every part of each living being has probably served, in a slightly modified condition, for diverse purposes, and has acted in the living machinery of many ancient and distinct specific forms.

In other words, adaptations were not invented from scratch. They evolved by tinkering, modifying what went before, just as a man might make a new machine by slightly altering some old wheels and springs that were already around. Darwin illustrated the point with many different examples taken from orchids, ranging from the way pollen is released to the overall arrangement and structure of organs in the flower. I want to highlight just one example, to show how Darwin's conception of tinkering compares to our present viewpoint.

The flowers of some orchid species, such as *Malaxis paludosa* (bog orchid, also known as *Hammarbya paludosa*) undergo a change during development that seems to be completely pointless: they twist themselves round in a full circle. To make this clearer, think of an upright main stem, with a flower at the end of a stalk coming out to the side. Now imagine twisting the flower stalk, gradually rotating the flower so that it turns upside-down, and then continuing to twist in the same direction until the flower is upright again, a turn of 360°. You end up by getting the flower back to where it started from, except that it now has a twist in its stalk. This is essentially what happens in some species of

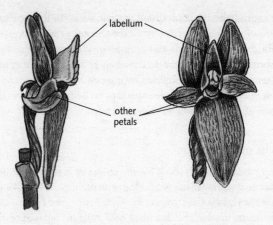

Fig. 18.3 The orchid *Malaxis paludosa* showing a flower with the labellum at the top after it has twisted around by 360°. Note the twist in the flower stalk (modified from Darwin 1862).

orchid to *all* of their flowers; they all do a little pirouette. You can see the evidence in the mature flowers by looking at their twisted stalks (Fig. 18.3).

What appears to be pointless at first sight becomes a bit more comprehensible when you know another fact about orchids: the flowers of most species are upside-down! Each orchid flower has three petals, one of which is distinctively elaborated to form a lip or *labellum*. In the great majority of orchids, the labellum is lowermost and typically acts as a landing platform for pollinators (Fig. 18.4). However, although the labellum is at the bottom of the mature flower, it is uppermost earlier on in the flower-bud. As the flower-bud develops,

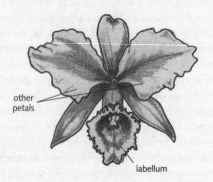

Fig. 18.4 Generalised orchid showing the labellum in the typical lower position.

it goes through a half turn, a 180° twist, so that the labellum eventually comes to lie towards the bottom even though it started at the top (the twist normally occurs in the flower stalk, which also includes the ovary).

This inverted arrangement, called the *resupinate* position, is so common in orchids that it is thought to have been a feature of their ancestors. According to this view, the few exceptional orchid species (e.g. *M. paludosa*) with the labellum uppermost could have evolved later on by counteracting the inversion in some way. There are several ways to turn an upside-down flower the right way up again. One is for the bud not to twist at all but to stay in the same position throughout development. This appears to be the case in a few orchid species. Another way, as noted by Darwin for *M. paludosa*, is by continuing to twist beyond 180° until the flower comes back to having its labellum at the top, a full turn. In Darwin's words:

> in many Orchids the ovarium (but sometimes the foot-stalk) becomes for a period twisted, causing the labellum to assume the position of a lower petal, so that insects can easily visit the flower; but from slow changes in the form or position of the petals, or from new sorts of insects visiting the flowers, it might be advantageous to the plant that the labellum should resume its normal position on the upper side of the flower, as is actually the case with *Malaxis paludosa*, and some species of Catasetum, &c. This change, it is obvious, might be simply effected by the continued selection of varieties which had their ovaria less and less twisted; but if the plant only afforded varieties with the ovarium more twisted, the same end could be attained by the selection of such variations, until the flower was turned completely round on its axis. This seems to have actually occurred with *Malaxis paludosa*, for the labellum has acquired its present upward position by the ovarium being twisted twice as much as is usual.

If orchids were designed according to some plan, the full turn of *M. paludosa* makes no sense. In Darwin's view, though, adaptations were not carefully planned from scratch; they arose by continual tinkering with what was there before. Should it become advantageous for an orchid flower to have its labellum uppermost, natural selection would act on what was already available. It would modify the inverted flower to restore the upright condition in one of two ways: either by reducing the extent of twisting or by increasing the twist until the flower is the right way round again. When seen in the light of evolutionary tinkering, as a modification of what went before, the 360° turn becomes much more understandable.

There is, however, an important way in which Darwin's conception of this sort of tinkering was misleading. To appreciate this, I need to explain Darwin's views on heredity. I will then come back to his notion of how the orchid got its twist.

Darwin's hypothesis of pangenesis

As mentioned in Chapter 13, Darwin's ideas on inheritance were very different from our present day outlook based on genes. For one thing, he believed that characters acquired during the lifetime of an individual could be inherited. If, for example, you had an accident as a child and lost your thumb, then according to this notion, as a result of this accident your own children would have a greater chance of being born without a thumb. At first sight it may come as a surprise that Darwin would have believed in such an idea, as we normally associate it with his predecessor, Jean Baptiste Lamarck. Lamarck, though, not only believed in the inheritance of acquired characters, he also thought that the variation in these characters was directed according to the organism's needs. If a deer-like animal, for example, found its surroundings bare of vegetation, it might need to stretch its neck to reach the leaves higher up. Over many generations, the neck would gradually be extended more and more to satisfy this need, accounting for the evolution of giraffes. Unlike Lamarck, Darwin believed that the acquired characters were essentially *random* with respect to adaptation. He did not think that adaptive or beneficial variations arose preferentially over non-adaptive ones. The vagaries of the environment were simply a source of heritable variation that was then acted upon by natural selection.

But what sort of mechanism would allow characters acquired during the lifetime of an individual to be passed on to its offspring? Darwin reasoned that there had to be some way for the sex cells to monitor the state of the whole body. Otherwise, it is difficult to see how the accidental loss of a thumb, say, would be able to influence the reproductive system. Somehow, hereditary information from the maturing body tissues had to travel to the egg or sperm cells. Darwin's provisional solution to how this might occur was his hypothesis of *pangenesis*.

According to pangenesis, as the cells of the body divide and grow, they continually throw off minute granules or gemmules, which are dispersed throughout the individual. As the organism reaches the reproductive age, gemmules from all parts of the body are collected within the sex cells. By this means they can be passed on to the next generation. The essential property of Darwin's gemmules sounds rather peculiar to us now: they tend to reproduce the structures from which they were derived. That is, the gemmules thrown off by eye cells will be those involved in making eyes in the next generation; while gemmules produced by ear cells will be of a different type, those needed to make ears. That is why gemmules from all over the body need to be collected together in the sex cells if a complete organism is to be produced in the next generation. Failure to collect the eye gemmules, for instance, would lead to offspring without eyes (strictly speaking this would have to happen in both parents for complete eye loss, as both contribute the gemmules). The corollary is that if, for example, an animal were to lose its eyes through injury, then eye-making gemmules would cease to be thrown off. Fewer of these gemmules would then be collected in the sex cells, leading to a tendency for the eyeless feature to be transmitted to its offspring. The loss of eyes or thumbs are rather extreme examples of environmental accidents; most of the acquired variations that Darwin had in mind as being important for evolution were much more subtle than this. As organisms developed, each would experience slight differences in their environment that would lead to minor changes to their gemmules, and so to small heritable changes:

> organisms have often been subjected to changed conditions of life at a certain stage of their development, and in consequence have been slightly modified; and the gemmules cast off from such modified parts will tend to reproduce parts modified in the same manner.

He was aware, however, of some striking exceptions that seemed to argue against his idea of pangenesis. For example, the circumcision of Jewish boys did not result in their sons inheriting the modification; although even here Darwin expressed some doubts:

> With respect to Jews, I have been assured by three medical men of the Jewish faith that circumcision, which has been practised for so many ages, has produced no inherited effect. Blumenbach, however, asserts that Jews are often born in Germany in a condition rendering circumcision difficult, so that a name is given them signifying 'born circumcised' . . . But it is possible that all these cases may be accidental coincidences, for Sir J. Paget has seen five sons of a lady and one son of her sister with adherent prepuces; and one of these boys was affected in a manner 'which might be considered like that commonly produced by circumcision;' yet there was no suspicion of Jewish blood in the family of these two sisters.

How could his theory account for an operation repeated over many generations having at best only sporadic effects on the offspring? It was no good postulating that the mother provided the foreskin gemmules because she would not have had a foreskin to cast them off. Darwin's answer was that the gemmules are not only thrown off by parts of the body, they also have an innate ability to divide and multiply themselves. This means that removal of the foreskin need not completely eliminate foreskin gemmules because there might still be lots of them kicking around from earlier generations, when circumcision had not been practised. In such cases, it might take many generations to significantly alter the course of inheritance.

Back to orchids

The reason that Darwin's conception of heredity is so important to us here is that it had a profound effect on his notion of evolutionary tinkering. If acquired characters can be inherited, then evolutionary change can result from direct tinkering with the organism. The modifications you see over an evolutionary timescale can be built up from physical rearrangements of the adult parts themselves. In the case of orchids, for example, if the flower happened to twist a bit more or a bit less for some environmental reason, the seed from that flower might be expected to give plants that also showed the altered twist. In other words, new degrees of twisting acquired during the lifetime of a plant might be transmitted to its progeny. A final twist of 360° would result from an accumulation of twists brought about by the environment. Each extra little twist in the flower would have been acquired during the lifetime of one or more plants, and then been subject to natural selection. It is as if all the tinkering could happen at the level of the flower itself. I have given the twist of the orchid flower to illustrate this point but the same principle would apply to all other evolutionary changes: if acquired characters are inherited, all features can be thought to have evolved through direct tinkering with the organism.

We now know that this is not the way inheritance works. A change in the developed parts of an organism will not of itself be transmitted to its progeny. This is because it is variation in the genetic material, DNA, rather than in the structure of the final organism, that is passed on from one generation to another. A novel character can be inherited only in so far as it results from a mutation in DNA, because it is only variation in DNA that gets replicated each generation. This means that evolutionary tinkering cannot be understood simply by considering the final organism: we have to look at the relationship between genes (DNA) and the way the organism grows and develops. I shall try to illustrate this by taking another look at the orchid's twist. I should warn you in advance, though, that the genes involved in orchid evolution are very far from being understood (orchids have a long generation time and are avoided by most geneticists like the plague), so I shall only be able to show the *sort* of explanation that might now be given. Before doing this, I need to give a little more background on how orchids twist.

The twisting and turning that leads to most orchid flowers adopting an inverted (resupinate) condition, with the labellum lowermost, is thought to be a response to gravity. If, for example, you turn an orchid plant upside-down so that its main stem points downwards, the flowers that develop will fail to twist: they retain the same orientation as when they were in bud. In other words, whichever way up the plant is, the flowers always develop oriented with the labellum at the bottom. If the plant is upright, the flower has to twist by 180° to achieve this; but if the plant is upside-down the labellum will already be in the lower position to begin with, so no twist is needed. The twisting of the flower therefore seems to reflect a mechanism that orients the flower in a consistent way with respect to gravity.

This sort of *orienting response* is found in many other plant species. It is just that the final orientation in most species is similar to the orientation in the bud. For example, members of the pea family (legumes), such as lupins, beans and peas, have flowers with an enlarged petal called the *standard* (Fig. 18.5). Like the labellum of orchids, the standard starts off uppermost in the bud, but unlike the labellum, this orientation is maintained through to maturity. Nevertheless, if a stem of a pea plant is bent or happens to be tilted, the flowers it produces will still be oriented with the standard upwards; flowers of the pea family still adjust their orientation with respect to gravity. This makes adaptive sense because it ensures that the flower is suitably oriented for visits by pollinators, even when the stem bearing the flowers happens not to be growing straight up. A dramatic illustration of this is given by laburnum, a member of the pea family that bears its flowers on drooping (pendant) stems. Even though the flowering stem now points downwards, the flowers of laburnum are upright—they twist through 180°, restoring the standard to the uppermost position (Fig. 18.5). Laburnums and orchids are almost inverted images of each other: laburnum flowers twist so as to keep the standard uppermost, whereas orchids twist so as to keep the labellum lowermost. For a stem growing upwards, this means that orchid flowers need to twist, but for a stem growing downwards, it is flowers of the pea family that need to twist. If most plant stems grew pointing down, as eventually happens in species with pendant stems, orchids would seem perfectly natural and untwisted, whereas members of the pea family would look peculiar in always twisting to readjust the standard to be uppermost. Incidentally, this

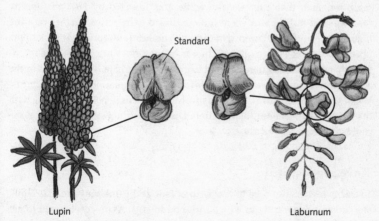

Standard

Lupin

Laburnum

Fig. 18.5 Flowers of lupin (upright stem on left) compared to laburnum (pendant stem on right). Note that in both cases the mature flowers have the same orientation, with the standard uppermost.

suggests one explanation (amongst others) for why most orchids twist by 180°: perhaps the common ancestor of orchids had pendant stems and therefore evolved an orienting response while the flowers were upside-down. Later on, when orchid species evolved with upright stems, the orienting response ensured that the flowers would twist to maintain their original orientation.

The precise mechanism responsible for the orientation response of flowers is not known. Nevertheless, by analogy with previous examples in this book, we can imagine how genes might be involved. Perhaps orientation response genes code for some sort of receptor protein in the cells of the flower that can respond to gravity. When the flower is oriented 'incorrectly', some receptors are stimulated, setting off a chain of events that lead to parts of the flower stalk growing more than others, giving it a twist. When the flower reaches the 'correct' orientation (i.e. labellum lowermost for most orchids), the receptors are no longer stimulated in this way and the flower stops twisting any further. Of course, I am not pretending that this explanation is at all satisfactory: it begs a lot of questions, like how a receptor can detect the direction of gravity. I only present it because if you want to understand how a twist might have evolved, you need to have some idea of how genes could contribute to it.

We can now look again at the evolution of the 360° twist in orchids such as *M. paludosa*. The extra twist cannot be explained simply by tinkering with the final flower; we have to tinker with the *genes* involved in the orientation process. Perhaps the simplest model is that the same mechanism that normally turns the flower through 180° is somehow reactivated once the flower is upside-down, turning it the right way round again. If mutations altered the orienting response, such that after the flower became inverted by 180° the gravity receptors did not become silent but continued to be active, further twisting might occur until the flower went through another 180°, turning it back to the upright position. This might be achieved by mutations in genes for the receptors, changing the way they respond to gravity, or perhaps by mutations in the genes involved in the chain of events following stimulation of the receptors. Whatever the mechanism, it had to involve mutations in genes, tinkering with the DNA, to bring about the variation. Direct tinkering with the flower alone could not have been the cause.

Tinkering revisited

To clarify matters, we need to take a closer look at the tinkering process. Think of a tinkerer playing about with various oddments. As objects happen to get juxtaposed, say an old spring and a piece of wood with a hook in it, the tinkerer thinks of some sort of simple contraption, like a rudimentary weighing machine. Further tinkering with the contraption, combining it with other oddments by trial and error, leads the tinkerer to produce a different device, say a simple musical instrument that makes a sound when the spring is hit. The tinkerer continues in this way, changing bits and pieces, assessing them and moving on to combine more oddments, modifying the contraption in some way each time. Of course, the tinkerer need not always be adding new bits; perhaps as the contraption gets more elaborate some old bits are eventually taken off or simply rearranged. The outcome of the tinkering is a series of contraptions, each with a slightly different structure and function. The final contraption may be a rather complicated device that bears little resemblance to the original simple weighing machine.

To see how this relates to evolutionary change, it will help to distinguish between two types of activity during the tinkering process. One is the physical activity of juxtaposing and rearranging objects, putting new combinations together. The other is assessing how well they work, deciding whether or not they make sense as part of a contraption. You need both activities to tinker successfully. If you combine bits and pieces without any sort of assessment you end up with a hotchpotch of objects put together in any old fashion. If you can assess things but are unable to combine any bits and pieces, there will be no way of moving from one contraption to the next. Clearly the two activities are intimately dependent on each other: the tinkerer's assessment depends on what happens to have been put together, and each assessment in turn sets the scene for the next combination of objects.

Now when it comes to the evolution of organisms, these two types of activity, alterations and assessment, occur in a different way. Almost all of the alterations have to occur at the level of DNA. You can rearrange the parts of an organism any way you want, but these alterations will be of little significance for evolution unless they are passed on from one generation to the next. As I have just described, this effectively means that the alterations have to occur in DNA: they have to be mutations in genes. They could involve substituting one base for another in the DNA sequence, say an A for a G, or the duplication of a gene, or some other change in the DNA. Whatever the details, the alterations during tinkering—the changes in the bits and pieces—occur at the level of genes. DNA provides the storehouse of raw material for tinkering.

The second aspect of tinkering, the assessment of the alterations, occurs for the most part at the level of the organism, or more strictly speaking at the level of populations of organisms. The significance of any change in DNA only becomes apparent as the organism develops and reproduces. If a mutation results in an organism that is less able to survive and reproduce relative to other members of the population, it will tend to be selected against. If the mutation confers some advantage to survival and reproduction, it is likely to be selected for in the population. The continual assessment of organisms, through natural selection, is an essential part of evolutionary tinkering. Without some sort of assessment, evolution would not give rise to organisms with adaptive features.

We can now begin to see how the tinkering metaphor can be misleading. In human tinkering the parts that are physically played with are also the components of the final contraption. Each of the bits and pieces that have been tinkered with to make a Heath Robinson device are visible as the working parts of the machine. In biological evolution, the parts of an organism are not the raw material for tinkering. The parts cannot be directly rearranged; instead, some feature in the genes has to be altered which then influences the way parts form. Evolution is a more indirect form of tinkering. Perhaps this is one of the main reasons that Darwin was attracted to the idea that acquired characters are inherited. It is after all so much easier to think about tinkering if the alterations and the assessment are all carried out at the same level. Tinkering with an orchid by simply rearranging some of its component parts seems a lot easier to conceive than first having to fiddle about with something else that affects the parts.

The indirect nature of evolutionary tinkering helps to explain an otherwise curious observation: as we have acquired a deeper knowledge about genes, so hitherto unsuspected degrees of tinkering have been revealed. The reason is that as long as you look purely at the anatomy of organisms, the arrangement of their various parts, you can appreciate tinkering only indirectly. You are looking at only part of the story, the visible manifestation of tinkering rather than the underlying raw material. This means that two structures may seem to be entirely distinct when in fact they have evolved from tinkering with similar genes. The organisation of vertebrates and insects from head to tail looks entirely different from the point of view of anatomy, yet they depend on a remarkably similar pattern of hidden colours, coded for by related genes. It is just that these colours are interpreted and become manifest in very different ways. Similarly, the sexual chemistry of yeast bears no obvious relationship to the parts of a flower, yet they are both based on tinkering with a common palette of basic colours (Chapter 17). As the relationship between genes and development has become clearer, so our conception of how organisms compare with each other has had to change. This is because we can begin to see both sides of the tinkering process, the genes together with their significance and meaning for the organism. In my view, we are still at a very early stage of chipping away at the iceberg of tinkering, revealing a few pieces here and there that lie beneath the surface. One of the great challenges for the future will be to get a sense of what the iceberg as a whole looks like, the true manner in which the various ways organisms develop are related to their genes.

Development and fabrication

With this view of evolution and tinkering in mind, I now want to return to the basis of fabrication and development. Recall that we compared evolution to the way a tinkerer comes up with a contraption, whereas development corresponded to the consistent manufacture of the contraption once it was devised. Now to get from tinkering to manufacture, we have to devise a set of instructions for making the same contraption again and again. The instructions might be a series of pictures showing how to build the contraption, or they could be written down as a series of commands, like 'attach spring to a piece of wood' or 'join spring to a hook', etc.

Now the key point is that whatever the nature of the instructions, they are always devised *after* the event has been carried out or conceived. A tinkerer can write down 'attach spring to a piece of wood' only after he has already come up with the idea of taking a piece of wood and attaching a spring to it. He may have had the idea while fiddling around with some bits of wood and a spring, or he may have simply had the idea first and then looked around for the pieces. Either way, the instruction comes after the event, not before. Try to imagine, for example, that it was the other way round: that the instruction came first. This would mean that the tinkerer was actually following the instruction 'attach spring to a piece of wood' before he came up with the idea. But then who is giving this instruction? Is there a little instructor inside the tinkerer's head telling him what to do or think? And if so, who is instructing the tiny instructor? We end up with an infinite regress.

To plan something, you need to have in advance a conception of what it is you are trying to achieve. The product has in some sense already to be there before you devise the detailed plan. Without foresight, planning is impossible. That is why, once you have a Heath Robinson contraption before you, it is possible to devise a set of instructions for how to make it again and again; you already know what is being aimed at.

Compare this with the parallel process of going from evolution to development. If the evolution of DNA corresponded to devising a set of instructions, the changes in DNA would have to somehow anticipate their outcome. Information about the final state of the organism would need to come before any change in the hereditary material. This is exactly the way Darwin saw things: variation in the organism came first, and was then transmitted to the offspring. In this view it might be legitimate to talk of development as fabrication because the instructions arise after the event. But as we have seen, this is simply not the way that development has evolved: alterations in the DNA *precede* their manifestation in the organism. It is not the case that parts of the organism arose first and were then planned for by DNA. Any modification to the parts is a later manifestation of tinkering with genes: mutating, rearranging or duplicating bits of DNA.

As soon as we look more closely at how development has evolved, we see that the analogy between development and fabrication starts to come apart at the seams. Rather than arising in comparable ways, they are based on opposite principles: instructions for fabrication are devised after the event, whereas changes in DNA precede developmental modifications. Of course, it might still be argued that by their very nature, such comparisons between biological and human processes are bound to be imperfect. Perhaps the dual analogy is the best of a bad job. In my view, there is a better way of looking at the problem; one that is more coherent and more firmly grounded.

The art of development

The dual analogy assumes that there are two primary processes: human creativity and evolution. What are these processes themselves based upon? In the case of evolution, it was Darwin's great achievement to show how this was grounded in the process of natural selection. Over many millions of years, the action of natural selection could account for the features of organisms today.

But when we turn to the human parallel, creativity, things are less clear. We tend to think of creativity as coming from nowhere—it just happens spontaneously. This is after all what it feels like when we are inspired with a new idea or create something for the first time; it seems to come from out of the blue. If we were to interpret this literally, we would be driven to believe in some form of vitalism; that there really is something out there providing the source of our creativity. Of course, there is a more scientific alternative: although we are not aware of the source of creativity as we experience it, it is nevertheless grounded in the way our brain works. Now because our brain is itself something that has evolved, this means that creativity is grounded in evolution. Darwin pointed this out in *The Descent of Man*, where he argued that human mental powers had evolved gradually:

> If no organic being excepting man had possessed any mental power, or if his powers had been of a wholly different nature from those of the lower animals, then we should never have been able to convince ourselves that our high faculties had been gradually developed. But it can be shewn that there is no fundamental difference of this kind.

We have ended up with a curious double standard. On the one hand, creativity can be seen as *parallel* to evolution (dual analogy); on the other, creativity appears to be an *outcome* of evolution. I believe that the reason we have been caught in this double bind stems from our failure to understand the nature of development, because it is this that provides the key link between evolution and creativity. So long as we think of development as akin to fabrication, we will tend to think of creativity as something that comes earlier, as analogous to evolution. But as we have learned more about the internal mechanisms of development, we have seen that it is quite unlike fabrication: there is no clear separation between plan and execution. Rather, it involves a continual process of interpretation and elaboration, in which genes, organism and environment interact. Once this is appreciated, our brain processes no longer seem divorced from development; rather, they can be seen as an extension of it (Chapter 12). In other words, creativity becomes grounded in development, just as development is grounded in evolution.

We can summarise the relationship between the various processes as follows. At an early stage of evolution, life consisted solely of unicellular organisms. Eventually, some of these organisms adopted a multicellular lifestyle, bringing the possibility of elaborating structures in a new way: through the process of development. Hidden colours, scents and sensitivities, already present in unicellular organisms, were now put to a new use—the elaboration of spatial patterns as a single cell developed into a multicellular individual. In some evolutionary lineages, the developmental process led to the production of individuals with complex nervous systems. The most elaborate of these to have evolved so far is the human brain, with its remarkable creative potential. Human creativity in turn ushered in new possibilities as each generation could learn and build upon the creative discoveries of the previous one. Through this, a novel form of making was arrived at—the process of fabrication according to a plan. This enabled humans to churn out many copies of their creations in a systematic and reproducible way.

The order of events was therefore: *evolution–development–creativity–fabrication*. I should emphasise that this is not a sequence of separate processes; rather, each process is woven into the fabric of what went before. I have tried to illustrate this diagrammatically in Fig. 18.6, where each process is shown as a circle embedded within and permeated by its predecessors.

According to this viewpoint, there is only one process that underlies all others: evolution. In contrast, the dual analogy invokes two primary processes: one of which, creativity, seems to be a free-floating entity. The analogy between fabrication and development is therefore based on a notion of creativity as a mystical force that comes from nowhere. In contrast, the view of development and creativity that I have tried to present here is more firmly grounded.

I do not pretend that the view I have given can lead to a completely self-contained account of creativity. This is because our theories of evolution and development are themselves products of creative minds: scientific understanding is based on a creative process in which we continually come up with hypotheses and test them against observations. Our very conception of

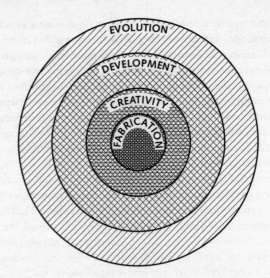

Fig. 18.6 Four processes shown as circles embedded within each other. The shading illustrates how each process is permeated by its predecessors.

evolution therefore depends on the way our mind works and interacts with the world around us. In this sense, creativity precedes our notion of evolution. This does not mean, however, that we cannot investigate the source of our creativity. As highly creative organisms, we have the privilege of being able to question the world in which we find ourselves, including the origin of our own abilities; always accepting that the answers we obtain will be coloured by our creative outlook. The theory of evolution is the best explanation we have come up with for our biological origins. In my view, our recent understanding of development takes us a step further by providing a bridge between evolution and creativity. Rather than thinking of human creativity as a black box that exists in isolation, we can begin to see it as a process that shares features with a more general form of making—development—from which it has itself arisen.

We can also see more clearly how fabrication fits in. As creative organisms, humans have come up with a special way of making things consistently. They create a plan that is then executed. By separating plan from execution, they ensure that anyone can comprehend and follow the instructions and will end up with the same product. But although this is a convenient way for humans to arrange things, it is not the way that evolution works. Evolution is a much more interactive process which does not plan ahead, as is so aptly captured by the tinkering metaphor. This may seem rather complicated to us because we like to order things in terms of plans, but evolution does not care about making itself easy for us to comprehend. The same applies not just to evolution but also to the processes that have arisen from it: development and creativity. The interactions of evolution pervade development and the way our own brain works. It is only in the case of fabrication that human comprehension is essential: otherwise no one could carry it out. That is why we need to separate plan from execution so clearly in this case. But we make a fundamental error when we try to impose the same sort of logic on biological processes such as development. To my mind, trying to see development in this way is reminiscent of another type of Heath Robinson contraption, the *Magnetic Apparatus for Putting Square Pegs into Round Holes* (Fig. 18.7).

In the light of all these considerations, why is it that comparisons between development and creative processes have been so firmly resisted in the past? I

Fig. 18.7 *Magnetic Apparatus for Putting Square Pegs into Round Holes* (*c.*1943), Heath Robinson. Private collection.

believe the answer has to do with our previous ignorance about the mechanisms of development. As long as there was no clear grasp of how development worked, comparisons with human creativity were at best unrevealing; at worst, they raised the spectre of vitalism. Predictable fabrication seemed like a much safer option. But as I have tried to show in this book, the more we have learned about development, the more we have come to see it as a highly interactive process in which there is no clear separation between plan and execution, between software and hardware, or between the maker and the made. Genes do not provide an instruction manual that is interpreted by a separate entity; they are part and parcel of the process of interpretation and elaboration. Genes both generate master proteins (hidden colours) and interpret them through their regulatory regions; they both produce signalling (scents) or receptor (sensitivity) proteins and respond to their effects; they both influence and respond to the way cells divide, grow, move or die. In all these cases, there is a process of continual interaction and elaboration that can eventually lead to a very complex structure.

This is very similar to what happens when humans create something. We do not know the outcome in advance when we are being creative, but this does not mean that creativity comes out of the blue; it is grounded in the biological and cultural heritage of each individual. How else can we explain the fact that artists have a recognisable style? We usually have no difficulty in distinguishing Leonardo's works from those of others. What Leonardo produced was highly constrained; he could no more have produced a Picasso than a frog's egg could develop into a prince. Yet this does not detract from the fact the Leonardo was being fully creative when painting a picture. If we try to standardise the conditions enough, as with the example of Leonardo the amnesiac (Chapter 10), we can even imagine someone continually churning out the same sort of picture whilst at the same time being genuinely creative. Although the outcome may be reproducible, it is not arrived at by the execution of a plan.

In my view, our recent understanding of development provides the clearest example of how such a process can work, how a complex viable outcome can result without its having been manufactured according to a plan. Not only does this liberate us from misconceptions about development, but I believe that it also helps us to see creativity in a clearer light. We no longer need regard creativity as a spontaneous process that comes from nowhere. Nor do we need to go to the other extreme of seeing it as the execution of a separable plan or program in the brain. Rather, like the developmental process from which it has arisen, we can view creativity as a highly interactive process that continually interprets and builds on what went before in a historically informed manner. By coming to a better understanding of development, we have therefore arrived at a new framework for looking at ourselves.

Sources of quotations

Chapter 1

p. 7 Glass, B. (1968). Maupertuis, Pioneer of Genetics and Evolution. In *Forerunners of Darwin 1745–1859* (eds B. Glass, O. Temkin and W. L. Straus), p. 77. Johns Hopkins Press, Baltimore, Maryland.

p. 13 Collingwood, R. G. (1938). *The Principles of Art*, p. 21. Oxford University Press, Oxford.

Chapter 2

p. 16 MacCurdy, E. (1954). *The Notebooks of Leonardo da Vinci*, Vol. 2, p. 269. Reprint Society, London.

p. 20 Hughes, A. (1959). *A History of Cytology*, p. 38. Abelard-Schuman.

p. 29 Watson, J. D. and Crick, F. H. (1953). A structure for deoxyribose nucleic acid. *Nature* 171, 737–8.

p. 37 Haeckel, E. (1874; trans. 1906 by J. McCabe). *The Evolution of Man* (5th edn), p. 11. Watts & Co., London.

Chapter 3

p. 43 Stern, C. (1968). Developmental Genetics of Pattern. In *Genetic Mosaics and other Essays*, p. 140. Harvard University Press, Cambridge, Mass.

p. 47 Richter, I. A. (1977). *Selections from the Notebooks of Leonardo da Vinci*, p. 182. Oxford University Press.

Chapter 4

p. 56 Goethe, J. W. (1817; trans. 1952 by B. Meuller). *Natural Sciences in General; Morphology in Particular*, Vol. 1, No. 1. In *Goethe's Botanical Writings*, p. 161. University of Hawaii Press, Honolulu.

p. 57 Goethe, J. W. (1786–1788; trans. 1970 by W. H. Auden and E. Mayer). *Italian Journey*, p. 366. Penguin, London.

p. 59 Rousseau, J. J. (1782; trans. 1807 by T. Martyn). *Letters on the Elements of Botany*, pp. 27–8. John White, London.

p. 60 Goethe, J. W. (1817; trans. 1952 by B. Mueller). *Natural Sciences in General; Morphology in Particular* (Vol. 1, No. 1). In *Goethe's Botanical Writings*, pp. 178–9. University of Hawaii Press, Honolulu.

p. 70 Goethe, J. W. (1790; trans. 1946 by A. Arber). Goethe's Botany: The Metamorphosis of Plants. *Chronica Botanica* (Waltham, Mass.), **Vol. 10**, p. 92.

p. 71 Bateson, W. M. (1894). *Materials for the Study of Variation*, p. 146. Macmillan, London.

pp. 73, 74 From interview with E. Lewis.

Chapter 7

p. 106 Kafka, F. (1933; repr. 1992). *Metamorphosis*, p. 9. Mandarin, London.

p. 109 Doyle, A. C. (1891; repr. 1987). *A Study in Scarlet*, p. 15. Chancellor Press, London.

p. 110 Coleman, W. (1964). *Georges Cuvier Zoologist*, p. 68. Harvard University Press, Cambridge, Mass.

p. 115 Appel, T. A. (1987). *The Cuvier–Geoffroy Debate*, p. 109. Oxford University Press, New York.

p. 116 Darwin, C. (1859; repr. 1968). *On the Origin of Species*, p. 233. Penguin, Harmondsworth, Middlesex.

p. 116 Darwin, C. (1859; repr. 1968). *On the Origin of Species*, p. 416. Penguin, Harmondsworth, Middlesex.

p. 117 Darwin, C. (1859; repr. 1968). *On the Origin of Species*, p. 233. Penguin, Harmondsworth, Middlesex.

p. 119 Lawrence, P. A. (1992). *The Making of a Fly*, p. 218. Blackwell Scientific, Oxford.

p. 120 From interview with R. Krumlauf.

Chapter 8

p. 133 Coates, T. (1989) *Creating a Self-Portrait*, p. 56. Mitchell Beazley, London.

p. 133 Needham, J. (1934). *A History of Embryology*, p. 174. Cambridge University Press, Cambridge.

Chapter 9

p. 146 From interview with C. Nüsslein-Volhard.

p. 155 Morgan, T. H. (1897). Regeneration in Allolobophora foetida. *Roux Archiv für Entwickelungsmechanic der Organismen* **5**, p. 582.

Chapter 10

p. 174 MacCurdy, E. (1954). *The Notebooks of Leonardo da Vinci*, Vol. 2, p. 226. Reprint Society, London.

Chapter 11

p. 188 Needham, J. (1939). Biochemical aspects of organizer phenomena. *Growth* **Suppl.**, p. 52.

Chapter 12

p. 207 Richter, L. (1909). *Lebenserinnerungen eines deutchen Malers*. Quoted in Gombrich, E. H. (1977). *Art and Illusion*, p. 55. Phaidon, Oxford.

p. 208 Hughes, P. and Brecht, G. (1975). *Vicious Circles and Infinity*, legend to Fig. 7. Penguin, London.

p. 221 Réamur, R. A. F. de (1712). Sur les diverses reproductions qui se font dans les Ecrevisse, les Omars, les Crabes, etc. et entr'autres sur celles de leurs Jambes et de leurs Ecailles. *Mem. Acad. Roy. Sci.* Quoted in Skinner, D. M. and Cook, J. S. (1991). New limbs for old: some highlights in the history of regeneration in Crustacea. In *A History of Regeneration Research* (ed. C. E. Dinsmore) p. 31. Cambridge University Press, Cambridge.

p. 226 Popper, Karl R. (1963). *Conjectures and Refutations*, p. 192. Routledge and Kegan Paul, London.

Chapter 13

p. 230 MacCurdy, E. (1954). *The Notebooks of Leonardo da Vinci*, Vol. 1, p. 104. Reprint Society, London.

p. 243 Gustafsson, Å. (1979). Linnaeus' Peloria: The History of a Monster. *Theoretical and Applied Genetics* **54**, p. 242.

Chapter 15

p. 287 Goethe, J. W. (1790; trans. 1946 by A. Arber). Goethe's Botany: The Metamorphosis of Plants. *Chronica Botanica* (Waltham, Mass.), **Vol. 10**, p. 107.

p. 289 Bower, F. O. (1911). *Plant-Life on Land Considered in some of its Biological Aspects*, p. 67. Cambridge University Press, London.

p. 302 Wittgenstein, L. (1958). *Philosophical Investigations*, p. 35. Blackwell, Oxford.

Chapter 16

p. 306 Conway, W. M. (trans. and ed.) (1958). *The Writings of Albrecht Dürer*, p. 241. Peter Owen, London.

p. 308 Thompson, D'A. (1942). *On Growth and Form*, p. 1085. Cambridge University Press, Cambridge.

p. 309 Woodger, J. H. (1945). On biological transformations. In *Essays on Growth and Form: Presented to D'Arcy Wentworth Thompson* (eds W. E. Le Gros Clark and P. B. Medawar), p. 115. Clarendon Press, Oxford.

p. 309 Medawar, P. B. (1945). Size, shape and age. In *Essays on Growth and Form: Presented to D'Arcy Wentworth Thompson* (eds W. E. Le Gros Clark and P. B. Medawar), p. 168. Clarendon Press, Oxford.

p. 310 Medawar, P. B. (1945). Size, shape and age. In *Essays on Growth and Form: Presented to D'Arcy Wentworth Thompson* (eds W. E. Le Gros Clark and P. B. Medawar), p. 185. Clarendon Press, Oxford.

p. 311 Thompson, D'A. (1942). *On Growth and Form*, p. 1049. Cambridge University Press, Cambridge.

p. 320 Levi-Montalcini, R. (1988). *In Praise of Imperfection*, p. 96. Basic Books, New York.

Chapter 17

pp. 325 Gombrich, E. H. (1950). *The Story of Art*, pp. 392–3. Phaidon, London.

Chapter 18

p. 344 Lévi-Strauss, C. (1966; trans. 1996). *The Savage Mind*, p. 18. Oxford University Press.

p. 345 Jacob, F. (1982). *The Possible and the Actual*, p. 34. Pantheon, New York.

p. 347 Darwin, C. (1862). *On the Various Contrivances by Which British Orchids are Fertilised by Insects*, p. 283. John Murray, London.

p. 349 Darwin, C. (1862). *On the Various Contrivances by Which British Orchids are Fertilised by Insects*, p. 284. John Murray, London.

p. 351 Darwin, C. (1868; repr. 1905). *The Variation of Animals and Plants Under Domestication*, Vol. 2, p. 473. John Murray, London.

p. 351 Darwin, C. (1868; repr. 1905). *The Variation of Animals and Plants Under Domestication*, Vol. 1, p. 558. John Murray, London.

p. 358 Darwin, C. (1871; repr. 1901). *The Descent of Man*, p. 99. John Murray, London.

Bibliography

Akam, M. E. and Martinez-Arias, A. (1985). The distribution of *Ultrabithorax* transcripts in *Drosophila* embryos. *EMBO Journal* 4, 1689–1700.

Alberts, B., Bray, D., Lewis, J., Raff, M., Roberts, K. and Watson, J. D. (1994). *Molecular Biology of the Cell*. Garland, New York.

Allard, H. A. (1945). Clockwise and counterclockwise spirality in the phyllotaxy of tobacco. *Journal of Agricultural Research* 73, 237–42.

Appel, T. A. (1987). *The Cuvier–Geoffroy Debate*. Oxford University Press, New York.

Arber, A. (1950). *The Natural Philosophy of Plant Form*. Cambridge University Press, Cambridge.

Barth, F. G. (1991). *Insects and Flowers*. Princeton University Press, New Jersey.

Bates, H. W. (1862). Contributions to an insect fauna of the Amazon Valley. Lepidoptera: Heliconidae. *Transactions of the Linnean Society of London* 23, 495–566.

Bender, W., Akam, M., Karch, F., Beachy, P. A., Peifer, M., Spierer, P., Lewis, E. B. and Hogness, D. S. (1983). Molecular genetics of the Bithorax Complex in *Drosophila melanogaster*. *Science* 221, 23–9.

Boncinelli, E., Somma, R., Acampora, M. P., D'Esposito, M., Faiella, A. and Simeone, A. (1988). Organization of human homeobox genes. *Human Reproduction* 3, 880–6.

Bowman, J. L., Smyth, D. R. and Meyerowitz, E. M. (1991). Genetic interactions among floral homeotic genes of *Arabidopsis*. Development 112, 1–20.

Brown, N. A. and Wolpert, L. (1990). The development of handedness in left/right asymmetry. *Development* 109, 1–9.

Carpenter, R. and Coen, E. S. (1990). Floral homeotic mutations produced by transposon mutagenesis in *Antirrhinum majus*. *Genes and Development* 4, 1483–93.

Christie, A. H. (1929; repr. 1969). *Pattern Design*. Dover, New York.

Coates, T. (1989). *Creating a Self-Portrait*. Mitchell Beazley, London.

Coen, E. S. and Meyerowitz, E. M. (1991). The war of the whorls: genetic interactions controlling flower development. *Nature* **353**, 31–7.

Collingwood, R. G. (1938). *The Principles of Art*. Clarendon Press, Oxford.

Conway, W. M. (trans. and ed.) (1958). *The Writings of Albrecht Dürer*. Peter Owen, London.

Cook, T. A. (1914). *The Curves of Life*. Constable, London.

Darwin, C. (1859). *On the Origin of Species*. John Murray, London.

Darwin, C. (1862). *On the Various Contrivances by Which British Orchids are Fertilised by Insects*. John Murray, London.

Darwin, C. (1868). *Variation of Animals and Plants Under Domestication*. John Murray, London.

Driever, W. and Nüsslein-Volhard, C. (1988). The *bicoid* protein determines position in the Drosophila embryo in a concentration-dependent manner. *Cell* **54**, 95–104.

Driever, W., Thoma, G. and Nüsslein-Volhard, C. (1989). Determination of spatial domains of zygotic gene expression in the *Drosophila* embryo by the affinity of binding sites for the bicoid morphogen. *Nature* **340**, 363–7.

Driever, W., Siegel, V. and Nüsslein-Volhard, C. (1990). Autonomous determination of anterior structures in the early *Drosophila* embryo by the *bicoid* morphogen. *Development* **109**, 811–20.

Duboule, D. (1994). *Guidebook to the Homeobox Genes*. Oxford University Press, Oxford.

Duboule, D. and Dollé, P. (1989). The structural and functional organization of the murine HOX gene family resembles that of *Drosophila* homeotic genes. *EMBO Journal* **8**, 1497–1505.

Ellis, R. E., Yuan, J. and Horwitz, R. H. (1991). Mechanisms and functions of cell death. *Annual Review of Cell Biology* **7**, 663–98.

Fink, K. J. (1991). *Goethe's History of Science*. Cambridge University Press, Cambridge.

Garber, R. L., Kuroiwa, A. and Gehring, W. J. (1983). Genomic and cDNA clones of the homeotic locus *Antennapedia* in *Drosophila*. *EMBO Journal* **2**, 2027–36.

Gardner, M. (1967). *The Ambidextrous Universe*. Allen Lane, London.

Gilbert, S. F. (1997). *Developmental Biology*. Sinauer, Sunderland, Mass.

Glass, B. (1968). Maupertuis, Pioneer of Genetics and Evolution. In *Forerunners of Darwin 1745–1859* (eds B. Glass, O. Temkin and W. L. Straus), pp. 51–83. Johns Hopkins Press, Baltimore, Maryland.

Goethe, J. W. (1786–1788; trans. 1970 by W. H. Auden and E. Mayer). *Italian Journey*. Penguin, London.

Goethe, J. W. (1790; trans. 1946 by A. Arber). Goethe's Botany: The Metamorphosis of Plants. *Chronica Botanica* (Waltham, Mass.), Vol. 10, pp. 67–115.

Gombrich, E. H. (1977). *Art and Illusion*. Phaidon, Oxford.

Goodwin, B. (1994). *How the Leopard Changed Its Spots*. Weidenfeld and Nicolson, London.

Gould, S. J. (1977). *Ontogeny and Phylogeny*. Harvard University Press, Cambridge, Mass.

Graham, A., Papalopulu, N. and Krumlauf, R. (1989). The murine and Drosophila homeobox gene complexes have common features of organization and expression. *Cell* 57, 367–78.

Hafen, E., Kuroiwa, A. and Gehring, W. J. (1984). Spatial distribution of transcripts from the segmentation gene *fushi tarazu* during Drosophila embryonic development. *Cell* 37, 833–41.

Hafen, E., Basler, K., Edstroem, J.-E. and Rubin, G. M. (1987). *Sevenless*, a cell-specific homeotic gene of Drosophila, encodes a putative transmembrane receptor with a tyrosine kinase domain. *Science* 236, 55–63.

Haldane, J. B. S. (1927). *Possible Worlds and Other Essays*. Chatto & Windus, London.

Hamilton, J. (1992). *William Heath Robinson*. Pavillion, London.

Howard, J. (1982). *Darwin*. Oxford University Press, Oxford.

Ingham, P. W., Baker, N. E . and Maritnez-Arias, A. (1988). Regulation of segment polarity genes in the *Drosophila* blastoderm by *fushi tarazu* and *even skipped*. *Nature* 331, 73–5.

Jacob, F. (1982). *The Possible and the Actual*. Pantheon, New York.

Johnstone, I. L. (1994). The cuticle of the nematode *Caenorhabditis elegans*: a complex collagen structure. *Bioessays* 16, 171–8.

Judson, H. F. (1979). *The Eighth Day of Creation*. Jonathan Cape, London.

Krämer, H., Cagan, R. L. and Zipursky, S. L. (1991). Interaction of *bride of sevenless* membrane-bound ligand and the *sevenless* tyrosine-kinase receptor. *Nature* 352, 207–12.

Lawrence, P. A. (1992). *The Making of a Fly*. Blackwell Scientific, Oxford.

Le Mouellic, H., Lallemand, Y. and Brûlet, P. (1992). Homeosis in the mouse induced by a null mutation in the *Hox-3.1* gene. *Cell* 69, 251–64.

Levin, M., Johnson, R., Stern, C. D., Kuehn, M. and Tabin, C. (1995). A molecular pathway determining left–right asymmetry in chick embryogenesis. *Cell* **82**, 803–14.

Levine, M., Hafen, E., Garber, R. L. and Gehring, W. J. (1983). Spatial distribution of *Antennapedia* transcripts during *Drosophila* development. *EMBO Journal* **2**, 2037–46.

Lewis, E. B. (1978). A gene complex controlling segmentation in *Drosophila*. *Nature* **276**, 565–70.

Luo, D., Carpenter, R., Vincent, C., Copsey, L. and Coen, E. (1996). Origin of floral asymmetry in *Antirrhinum*. *Nature* **383**, 794–9.

Maupertuis, P.-L. M. de (1753; trans. 1966 by S. Boas). *The Earthly Venus*. Johnson Reprint Corporation, New York.

Medawar, P. B. (1945). Size, shape and age. In *Essays on Growth and Form: Presented to D'Arcy Wentworth Thompson* (eds W. E. Le Gros Clark and P. B. Medawar), pp. 157–87. Clarendon Press, Oxford.

Melville, R. and Wrigley, F. A. (1968). Fenestration in the leaves of *Monstera* and its bearing on the morphogenesis and colour patterns of leaves. *Botanical Journal of the Linnean Society* **62**, 1–16.

Monod, J. (1972). *Chance and Necessity*. Collins, Glasgow.

Needham, J. (1934). *A History of Embryology*. Cambridge University Press, Cambridge.

Neville, A. C. (1976). *Animal Asymmetry*. Edward Arnold, London.

Nüsslein-Volhard, C. (1996). Gradients that organize embryo development. *Scientific American*, August, pp. 38–43.

Nüsslein-Volhard, C. and Wieschaus, E. (1980). Mutations affecting segment number and polarity in *Drosophila*. *Nature* **287**, 795–9.

Nüsslein-Volhard, C., Lohs-Schardin, M., Sander, K. and Cremer, C. (1980). A dorso-ventral shift of embryonic primordia in a new maternal-effect mutant of *Drosophila*. *Nature* **283**, 474–6.

Olby, R. C. (1966). *Origins of Mendelism*. Schocken, New York.

Osborne, H. (ed.) (1997). *The Oxford Companion to Art*. Oxford University Press, Oxford.

Popper, K. R. (1963). *Conjectures and Refutations*. Routledge and Kegan Paul, London.

Roe, S. A. (1981). *Matter, Life, and Generation: Eighteenth-century embryology and the Haller–Wolff debate*. Cambridge University Press, Cambridge.

Russell, E. S. (1916). *Form and Function*. John Murray, London.

Scott, M. P., Weiner, A. J., Polisky, B. A., Hazelrigg, T. I., Pirrotta, V., Scalenghe, F. and Kaufman, T. C. (1983). The Molecular organization of the *Antennapedia* locus of Drosophila. *Cell* **35**, 763–6.

Singer, C. (1950). *A History of Biology*. Henry Schuman, New York.

Sinnott, E. W. (1936). A developmental analysis of inherited shape differences in cucurbit fruits. *American Naturalist* **70**, 245–54

Sinnott, E. W. (1960). *Plant Morphogenesis*. McGraw-Hill, New York.

Sommer, H., Beltrán, J.-P., Huijser, P., Pape, H., Lönnig, W.-E., Saedler, H. and Schwarz-Sommer, Z. (1990). *Deficiens*, a homeotic gene involved in the control of flower morphogenesis in *Antirrhinum majus*: the protein shows homology to transcription factors. *EMBO Journal* **9**, 605–13.

Spemann, H. (1938). *Embryonic Development and Induction*. Yale University Press, New Haven, Conn.

Steeves, T. A. and Sussex, I. M. (1989). *Patterns in Plant Development*. Cambridge University Press, Cambirdge.

Stern, C. (1968). Developmental Genetics of Pattern. In *Genetic Mosaics and other Essays*, pp. 130–73. Harvard University Press, Cambridge, Mass.

Stewart, I. and Golubitsky, M. (1993). *Fearful Symmetry*. Penguin, London.

Stubbe, H. (1972). *History of Genetics* (trans. T. R. W. Waters). MIT Press, Cambridge, Mass.

Sulston, J. E. and Horvitz, H. R. (1977). Post-embryonic cell lineages of the nematode *Caenorhabditis elegans*. *Developmental Biology* **56**, 110–56.

Supp, D. M., Witte, D. P., Potter, S. S. and Brueckner, M. (1997). Mutation of an axonemal dynein affects left–right asymmetry in *inversus viscerum* mice. *Nature* **389**, 963–6.

Thompson, D'A. (1942). *On Growth and Form*. Cambridge University Press, Cambridge.

Turner, J. R. G. (1984). Mimicry: The Palatability Spectrum and its Consequences. In *The Biology of Butterflies* (eds R. I. Vane-Wright and P. R. Ackery) pp. 141–61. Academic Press, London.

Vasari, G. (1568; trans. 1965 by G. Bull). *The Lives of the Artists*. Penguin, Harmondsworth, Middlesex.

Weberling, D. F. (1989). *Morphology of Flowers and Inflorescences* (trans. R. J. Pankhurst). Cambridge University Press, Cambridge.

Weyl, H. (1952). *Symmetry*. Princeton University Press, New Jersey.

Wilmut, I., Schnieke, A. E., McWhir, J., Kind, A. J. and Campbell, K. H. S.

(1997). Viable offspring derived from fetal and adult mammalian cells. *Nature* **385**, 810–13.

Wittgenstein, L. (1958). *Philosophical Investigations*. Blackwell, Oxford.

Wölfflin, H. (1952). *Principles of Art History* (trans. M. D. Hottinger). Dover, New York.

Wolpert, L. (1991). *The Triumph of the Embryo*. Oxford University Press, Oxford.

Wolpert, L., Beddington, R., Brockes, J., Jessell, T., Lawrence, P. and Meyerowitz, E. (1998). *Principles of Development*. Current Biology, London/Oxford University Press, Oxford.

Yanofsky, M. F., Ma, H., Bowman, J. L., Drews, G.N., Feldmann, K. A. and Meyerowitz, E. M. (1990). The protein encoded by the *Arabidopsis* gene *AGAMOUS* resembles transcription factors. *Nature* **346**, 35–9.

Glossary

(Words in italics indicate cross-references within the glossary.)

affinity   Strength with which two molecules stick or bind to each other.

amino acids   Fundamental subunits of *proteins*, comprising 20 different types.

apoptosis   Controlled form of *cell* death.

arthropods   Jointed animals with a hard outer skeleton, such as insects or crustaceans.

bases   Fundamental subunits of *DNA* and *RNA*. There are four types of bases in DNA: A (adenine), G (guanine), C (cytosine) and T (thymine).

bilateral symmetry   Having a single plane of *reflection symmetry*.

binding site   Short sequence of *DNA* which is recognised by a *master protein*.

bract   Small leaf-like organ below each flower.

bud   Early stage in the development of a shoot, flower or leaf.

carpels   Female flower organs that when pollinated will grow to form fruit containing seed.

catalyst   Substance which facilitates a chemical reaction, without itself being used up in the process (e.g. an *enzyme*).

cell   Fundamental unit of life. For plants and animals it comprises a *nucleus*, *cytoplasm* and a surrounding membrane. Plant cells also have a *cell wall*.

cell division   Process whereby a single *cell* divides to give two cells.

cell wall   Rigid structure surrounding the *cell* membrane, common in cells of plants, fungi and bacteria.

chromosome   Molecule of *DNA* packaged within a *cell*. Most human cells contain 23 pairs of chromosomes within their *nucleus*.

cleavage   Early stage in the development of many animals in which the fertilised egg undergoes many rounds of *cell division* without an overall increase in size.

coding region   Part of a *gene* that codes for a *protein*.

collagen   Family of fibrous *proteins*, being a major component of skin, bone and the *cuticle* of nematode worms.

concentration   The number of molecules of a substance per unit volume.

cuticle   Hardened outer skin or covering (e.g. of an insect or nematode worm).

cytoplasm   Contents of a *cell* apart from the *nucleus*.

dedifferentiation   Return of a specialised *cell* to a more embryonic condition.

deformation   Method for relating two different shapes by stretching and/or compressing one so that it comes to resemble the other.

denticle   Tooth- or thorn-like outgrowth, as seen on the outer surface of a fruit fly larva.

differentiation   Process whereby *cells* become finally specialised (e.g. becoming mature nerve cells, liver cells, leaf cells, etc.).

DNA   Hereditary material, consisting of a long double-stranded molecule carrying a sequence of *bases*.

dorsal   Back of an animal, or the part of a flower nearer to the growing tip of the stem (usually the upper part of the flower).

ectoderm   Outer *germ layer* of an animal *embryo*, giving rise to the skin and nervous system.

embranchements   The major categories of animal proposed by Cuvier (*vertebrates*, articulates, molluscs and radiates).

endoderm   Innermost *germ layer* of an animal *embryo*; it gives rise to the inner linings of the body, such as the gut and lungs.

enzymes   Diverse group of (mainly) *proteins* that catalyse chemical reactions (see also *catalyst*).

epigenesis   Theory that the adult gradually emerges from the fertilised egg through a process of fresh formation (as opposed to *preformation*).

evolution   Process whereby populations of individuals may change over many generations to give new types of organisms.

expression pattern   Regions of an organism where a *gene* is switched on or off at any given time.

gastrulation   Process whereby the different *germ layers* of an animal *embryo* are established in their respective positions.

gene   Basic unit of inheritance, corresponding to a segment of *DNA* which usually includes a *coding region* and a *regulatory region*.

gene activity   Extent to which a *gene* is on or off.

gene cloning   Isolation of a *gene* through multiplying it in bacteria.

genetic code   Rules which relate *amino acids* in a *protein* to triplets of *bases* in *DNA*.

germ layers   Regions of an animal embryo that will give rise to various tissues. Most animals have three germ layers: *ectoderm*, *endoderm* and *mesoderm*.

germline *Cells* that will give rise to sperm or eggs.

handedness Cases of left–right asymmetry in which there is a consistent bias towards one mirror-image form over the other.

hardware Machinery of a computer, such as printed circuits and wiring.

homeobox Particular type of *DNA* sequence (about 180 *bases* long) that was originally found as a conserved region in various segment *identity genes* of the fruit fly, and has subsequently been found in many other genes.

homeodomain Particular type of *protein* sequence (about 60 *amino acids* long) encoded by the *homeobox*. It is diagnostic of a large family of *master proteins*, including those involved in establishing distinctions along the head–tail axis in *vertebrates* and *arthropods*.

homeosis Type of variation in which one member of a repeating series assumes features that are normally associated with a different member (e.g. *petals* replacing *stamens*).

identity gene *Gene* involved in making regional distinctions within an organism, such as between *whorls* of organs in a flower, or between segments in a fly.

interpreting gene *Gene* that responds to a pattern of *master proteins* (hidden colours) through its *regulatory region*.

labellum Petal of an orchid flower, often highly elaborate, that typically ends up in the lowermost position.

lactose Type of sugar molecule, abundant in milk.

MADS-box Region of *DNA* that is conserved in many flower organ *identity genes*.

master protein *Protein* that can influence the expression of a *gene* by binding to its *regulatory region* (equivalent to a hidden colour).

meiosis Type of *cell division* that results in the daughter *nuclei* ending up with half the number of *chromosomes* of the parent.

meristem Group of *undifferentiated* plant *cells* that divide and replenish themselves while at the same time adding more tissue to the plant.

mesoderm Middle *germ layer* of an animal *embryo*, giving rise to most of the flesh and bone in the body (e.g. muscle, blood, heart, kidneys, skeleton).

mitosis Normal process of *cell division* in which each daughter cell ends up with the same number of *chromosomes* as the parent cell.

multicellular organism Organism in which each mature individual comprises many cells.

mutation Alteration in *DNA* sequence that gets transmitted each time the DNA is replicated (see *replication*).

nucleus Small body located within a *cell* that contains the *genes*.

ommatidium Individual facet of an insect eye.

pangenesis Now discarded theory that hereditary characteristics are transmitted by 'gemmules' cast off by the body.

petals Organs in the second *whorl* of a flower; typically the most attractive and obvious parts of a flower.

preformation Theory prevalent in the seventeenth and eighteenth centuries that the fertilised egg contains a preformed miniature version of the adult (as opposed to *epigenesis*).

prepattern Underlying pattern within an organism that can be interpreted in various ways by *genes*, equivalent to a pattern of hidden colours.

probe Type of molecule used to detect particular sequences of *DNA*, *RNA* or *protein*.

protein Molecule made up of a sequence of *amino acids* strung together, typically forming a complex shape.

radial symmetry Having a single axis of *rotational symmetry* with more than one plane of *reflection symmetry* passing through it (e.g. bottle, buttercup).

receptor protein Protein that sets off a chain of reactions in a *cell* when triggered by a specific stimulus, such as a *signalling protein*, scent, light or movement.

reflection symmetry Invariance following reflection in a plane.

regeneration Renewal of a body part following injury or loss, derived from *cells* of mature tissue that undergo *dedifferentiation* (compare to *restoration*).

regulatory region Region of a *gene* that contains a series of *binding sites* to which *master proteins* may bind and influence the gene's expression.

replication Process whereby a *DNA* molecule is copied to give two DNA molecules; this usually precedes *cell division*.

restoration Renewal following injury or loss, derived from cells that were already in an embryonic or undifferentiated condition (sometimes this is also classified as *regeneration*).

resupinate Upside-down, as in most orchid flowers, which twist through 180°.

RNA Single-stranded molecule carrying a sequence of *bases*, typically derived by *transcription* of a segment of *DNA*.

rotational symmetry Invariance following rotation about an axis.

sepals Organs in the outermost *whorl* of a flower, typically small and leaf-like.

signalling protein *Protein* that can trigger activity of a *receptor protein* by matching its shape.

situs inversus Condition in which all body organs are arranged as a mirror image of the usual situation.

software Programs that can be run on a computer.

spherical symmetry Having an infinite number of axes and degrees of *rotational symmetry*, and an infinite number of planes of *reflection symmetry* (as in a sphere).

stamens The male organs of a flower, which bear pollen.

transcription Process whereby a segment of *DNA* directs the synthesis of a corresponding *RNA* molecule.

transformation A geometrical operation such as reflection, rotation or *translation*.

translation (in geometry) Movement in a straight line.

translation (of RNA) Process whereby an *RNA* molecule directs the synthesis of a corresponding *protein*.

translational symmetry Invariance following movement by a certain amount in a straight line (as in an infinite repeating pattern).

undifferentiated cells *Cells* in a more embryonic condition, before they have become specialised.

unicellular organism Organism in which each individual is a single *cell*.

ventral Belly of an animal, or the part of a flower further from the growing tip of the stem (usually the lower part of the flower).

vertebrates Animals with internal bony or cartilaginous skeletons.

whorl Circular region that includes plant organs of one type (e.g. *petals* in a flower).

Figure acknowledgements

Acknowledgement is due to the following for illustrations: **1.1** © 1998 Cordon Art B V–Baarn–Holland. All rights reserved; **2.1** From Vaughan, J. (1903) *Nelson's New Drawing Course*, Thomas Nelson and Sons, London; **3.1** Photo Scala, Florence; **3.2, 3.3** After Stern (1968); **3.9, 3.10, 3.11** Oxford University Press; **4.3** From P. J. Redouté (1817) *Les Roses*; **4.5** © ADAGP, Paris and DACS, London 1998. Photograph from the Bridgeman Art Library; **7.2** The Royal Collection © Her Majesty Queen Elizabeth II; **4.12** After Lawrence (1992); **5.1** After Barth (1991); **8.1** Francis Bowyer; **8.2** Francis Bowyer; **8.4** © 1998 Cordon Art B V–Baarn–Holland. All rights reserved; **8.6** The Bridgemand Art Library; **9.2** © ADAGP, Paris and DACS, London 1998. Photograph courtesy of Galerie Christine et Isy Brachot, Brussels; **9.7, 9.15** After Nüsslein-Volhard (1996); **10.1** (top) The Bridgeman Art Library; **10.1** (bottom) © Photo RMN; **11.4** © ADAGP, Paris and DACS, London 1998. Photograph from The Tate Gallery; **12.1** © ADAGP, Paris and DACS, London 1998. Photograph from the Bridgeman Art Library; **12.2** After Sinnott (1960); **12.5** After Steeves and Sussex (1989); **13.1** The Bridgeman Art Library; **13.2** © Arnold Roth; **14.1** The Bridgeman Art Library; **14.5** After Wolpert *et al.* (1998); **14.7** The Bridgeman Art Library; **14.9** (left) Photo Scala, Florence; **14.9** (right) Amsterdam, Van Gogh Museum (Vincent van Gogh Foundation); **15.7** (top) © The Bridgeman Art Library; **15.7** (bottom) © Succession Picasso/DACS 1998. Photograph from the Bridgeman Art Library; **15.8** (left) © The Bridgeman Art Library; **15.8** (right) © Succession Picasso/DACS 1998. Photograph from the Bridgeman Art Library; **16.1** Madonna and Child on a Curved Throne, Andrew W. Mellon Collection, © 1998 Board of Trustees, National Gallery of Art, Washington; **16.2** © ADAGP, Paris and DACS, London 1998. Photograph from The Tate Gallery; **16.5** Oxford University Press; **16.7** © 1998 Cordon Art B V–Baarn–Holland. All rights reserved; **16.8** After Sinnott (1936); **18.2** W. Heath Robinson (1933). Professor Branestraum's Pancake Making Machine. In *The Incredible Adventures of Professor Branestraum*, Bodley Head. Reproduced by permission of Laurence Pollinger Limited and the Estate of Mrs J. C. Robinson; **18.7** W. Heath Robinson (*c*.1943). Magnetic Apparatus for Putting Square Pegs into Round Holes. In *Absurdities*, Gerald Duckworth. Reproduced by permission of Laurence Pollinger Limited and the Estate of Mrs J. C. Robinson.

Index